ENVIRONMENTAL LAW *for* ENGINEERS *and* GEOSCIENTISTS

ENVIRONMENTAL LAW *for* ENGINEERS *and* GEOSCIENTISTS

Robert Lee Aston

CRC Press
Taylor & Francis Group
Boca Raton London New York

CRC Press is an imprint of the
Taylor & Francis Group, an **informa** business

First published 2002 by CRC Press

Published 2020 by CRC Press
Taylor & Francis Group
6000 Broken Sound Parkway NW, Suite 300
Boca Raton, FL 33487-2742

First issued in paperback 2020

© 2002 by Taylor & Francis Group, LLC
CRC Press is an imprint of the Taylor & Francis Group, an informa business

No claim to original U.S. Government works

ISBN 13: 978-0-367-57866-4 (pbk)
ISBN 13: 978-1-56670-575-2 (hbk)

Library of Congress Cataloging-in-Publication Data

Aston, Robert Lee.
    Environmental law for engineers and geoscientists / Robert Lee Aston.
        p.   cm.
    Includes bibliographical references and index.
    ISBN 1-56670-575-4 (alk. paper)
    1. Environmental law--United States. 2. Engineers--United States--Handbooks, manuals, etc. 3. Earth scientists--United States--Handbooks, manuals, etc. I. Title.

KF3775.Z9 A85 2001
344.73′046--dc21
                                                   2001038557

Library of Congress Card Number 2001038557

**Visit the Taylor & Francis Web site at
http://www.taylorandfrancis.com**

**and the CRC Press Web site at
http://www.crcpress.com**

Abstract

This work is intended as a textbook of instruction in environmental law for engineers and geoscientists, primarily to be used in the geological, mining, petroleum, civil, and environmental engineering departments, and in the earth sciences curricula of universities. It will also be useful (with the exception of Chapters 1 and 2) in instructing lawyers entering environmental law practice

Because the subject matter is so extensive, it cannot be fully covered in one book. Hence, this work can only treat the more important highlights of the subject. For that reason, the student or fledgling engineer and geoscientist is introduced to a general orientation of American jurisprudence with an overview of the legal system, its courts, terms, phrases, administrative law, and regulation by the regulatory agencies administering environmental law to projects with which engineers and geoscientists will likely be involved in their professional work.

Because of the space limitation here, the highlights of environmental subjects are the main areas treated, viz., the National Environmental Policy Act (NEPA), which is the master plan for all ensuing environmental statutes; the two basic statutes for improvement of the quality of the environment, i.e., the Clean Air Act and the Clean Water Act are treated in detail; additional, important statutes follow in Chapter 6. With these basic law chapters as a foundation for more specific applicability, in Chapter 7 the work passes on to more practical applications of environmental law for geoengineers, on the subject of water pollution by abandoned mine sites (acid mine drainage [AMD]). The Comprehensive Environmental Response, Compensation and Liability Act (CERCLA), a/k/a Superfund) and Superfund Amendments and Reauthorization Act of 1986 (SARA) play a very large part in Chapter 7. The final Chapter 8 treats the rapidly increasing and important practical application of environmental law for geoengineers and geoscientists in expert witnessing and admissible evidence in environmental litigation. The success or failure of environmental litigation increasingly depends on the preparation and testimony of expert scientific evidence.

Preface

The purpose of this work is to create a textbook of instruction for a course in environmental law for engineers and geoscientists. With the exception of Chapters 1 and 2, it will also be useful in instructing lawyers entering environmental law practice. Its prime purpose, however, is to prepare graduate engineers, geoscientists, and students in acquiring a passing knowledge and familiarity with American law, its courts, and the legal process; certainly not to make "curbstone" or "cracker barrel" lawyers of them; and, to prepare them for the numerous environmental regulatory encounters in their work dealing with various statutes, laws, regulations, and agency rules that govern, affect, and apply to environmental engineering and earth disturbing projects. It is also intended to aid practicing geoscientific engineers in their work with clients and employers to comply with the controlling laws, regulations, and rules; to have some knowledge of what must be done for their project work to be in compliance with regulations; and to determine whether the engineering work planned or performed is in compliance.

Even then, engineering workplace situations may arise where environmental compliance methods were followed, but have been challenged by enforcement agencies. Although statutes, rules, and regulations may seem worded clearly and specifically, there are often questions in application, and occasionally, varied interpretations of rules are, or may be, acceptable as compliance. At times, regulatory-stated requirements may be challenged through agency appeal procedures.

Questions of law and ambiguities of terms may also occur in contracts for mining, landfills, hazardous site reclamation, waste site depositories, cleanup sites and programs, land leases, operating agreements, joint ventures, employment, and various other contractual agreements. Questions may arise about whether there has been a breach of the contractual terms by either or any of the parties thereto. A determination may have to be made as to whether the contractual bargain has been kept, or broken; to whether "curing" corrections can be made to continue the performance of the contract; and to what extent damages may have occurred.

With the tremendous increase in environmental laws controlling earth-affecting activities of all types, whether highway and bridge, or public housing construction, mining, energy producing sites, natural resources, dam site and flood control, harbor and lake marinas, et al., the engineer will most assuredly be confronted by environmental laws and regulations governing professional engineering work. The engineer and scientist may have to prepare permitting forms, data and environmental impact statements, assessments, or testimonial and evidentiary information; to work with attorneys in litigation as an expert witness; to testify as an expert witness in court or agency hearings as to certain geological, scientific, and engineering terms and procedures; or to explain an occurrence that caused damage, or how planning in an engineering project may minimize or avoid damage to the environment; or to critique an opposing expert's testimony at trial or his deposition, or to be deposed on his own work.

In addition to upgrading engineering practice in environmental areas, with the recent emphasis on enforcement of increased penalties for environmental violations and crimes, it behooves engineers to have some familiarity with the law, not only for compliance but also for self-protection.

Indeed, the subject area of environmental law is so very broad and extensive, and its coverage is pervasive into all aspects of engineering and human works that a one-semester course severely limits the time and treatment that can be allowed for each environmental subject introduced. Volumes of books have been written about water pollution and the Clean Water Act (CWA) alone, as well as many volumes on clean air and other environmental subjects. Thus, only a cursory treatment of each subject has been allowed for the engineer to gain a general exposure to environmental regulation.

This work is only a beginning point, or an introduction, for the geoengineer and geoscientist to the legal aspects of environmental law and regulation. Hopefully, through this general exposure, the engineer will have a better orientation of the subject, thus enabling a better vantage point to explore and enlarge on the subject area of specialization in environmental regulation.

In addition, it is hoped, with the exception of Chapter 2 on basic law for engineers and geoscientists, that the remaining chapters of the text may be of use to attorneys who are new to the field of environmental law.

R. Lee Aston, Ph.D., D.Eng., J.D., LL.M.

About the Author

Dr. R. (Robert) Lee Aston was born in Richmond, Henrico County, Virginia, in 1924. He received his primary education in the schools of the Deep and Upper South (Georgia and Missouri). He is a veteran of World War II, having served 3 years in the U.S. Army Air Corps and Army Air Force as an aviator and attaining the active duty rank of First Lieutenant. He flew a tour of 36 combat missions with the Eighth Air Force in Europe, receiving the Distinguished Flying Cross and the Air Medal with six Oak Leaf clusters. During the postwar, he was in the Air Force Reserves for 8 years. His higher education has extended over his adult life, with his having earned eight university degrees as follows:

B.S. 1948 (preengineering), College of William & Mary, Williamsburg, Virginia (Dean's List 1 year)

B.S. 1950 Mining Engineering, University of Missouri, School of Mines — Rolla, Missouri (Dean's List 1 year)

J.D. 1984 (Juris Doctor), Woodrow Wilson College of Law, Atlanta, Georgia (Dean's Award)

LL.M. 1985 (Master of Laws in Civil Litigation); Atlanta Law School, Atlanta, Georgia

M.S. 1992 (geological engineering) (thesis), 1992, University of Missouri — Rolla; graduated with 4.0 GPA (nominated to Tau Beta Pi, National Engineering Honor Society)

P.D./E.M. 1993 Professional Mining Engineer Degree; University of Missouri — Rolla; (1956) Engineer of Mines (E.M.) degree thesis, submitted on Shaft Raising; belatedly granted in 1993 as honorary degree for distinguished performance in the mining engineering profession

Ph.D. 1996, University of Aston, Birmingham, England; Civil Engineering (engineering law); research/dissertation area in Surface Mining Law in the United Kingdom, Canada, and the United States

D.E. 2000, Doctor of Engineering (in geological-mining engineering), University of Missouri — Rolla; research and dissertation area in Environmental Law

LL.D. 2002 [candidate], Doctor of Laws, University of Adelaide Law School, South Australia, Australia

Dr. Aston is an active attorney and a member of the state bars of Georgia, Virginia, Indiana, and Montana. He has been admitted to the bars and law practice of the supreme courts and all federal courts of those four states; to four U.S. Courts of

Appeal; and to the U.S. Supreme Court. Attorney Aston has carried cases to the state supreme courts of Georgia, Montana, and Missouri; and on *certiorari* (denied) to the U.S. Supreme Court.

He has briefly taught mining geology and petroleum geology at the University of Chattanooga, Tennessee. He became an Adjunct Professor in the Mining Engineering Department of the University of Missouri — Rolla in 1991 when he created and taught a 3-hour course in mining and environmental law. He is a prolific author of mining and environmental law articles for professional engineering and legal publications. He had his first mining law book published in 1999 on surface mining law and mined land reclamation by landfilling. He is also the Mining Law and Environmental (Enviromining) Law Editor of the *Mineral Resources Engineering Journal*, Imperial College, London, and a contributirng editor for *Law Mine*, Vancouver, British Columbia <http:www.infomine.com/lawmines>. Aston has been active in underground and surface mining since 1950. He formerly owned three stone quarries and has been operating a stone quarry in Virginia since 1964. His plans include completion of a Doctor of Laws (LL.D.) from the Unversity of Adelaide, Australia, in international mining and environmental law.

PROFESSIONAL ASSOCIATIONS

Mining, Geology and Engineering

Nominated to National Engineering Honor Society, Tau Beta Pi at UMR
Member, American Institute of Mining, Metallurgical and Petroleum Engineers/SME (1948–present)
Member, Society of Mining Engineers; East Tennessee Section; Central Appalachian Section; Georgia Section
Charter Member, Georgia Geological Society
Member, Carolina Geological Society (1952)
Member, Association of Engineering Geologists (AEG)
Georgia Professional Registration, No. 362
Missouri Professional Geologist Registration, No. Rg 0911
Virginia Mine Foreman (surface-open pit) Competency Certificate No. 2253
Adjunct Professor, Faculty/Mining Engineering and Geological Engineering Departments, University of Missouri — Rolla (School of Mines)(since December 1991)
Member, Editorial Board, *Mineral Resources Engineering Journal*, Imperial College, London, United Kingdom

Law

State Bar Memberships: Indiana, Georgia, Montana, and Virginia
Member, American Bar Association (ABA); Natural Resources Section
Member, Association of Trial Lawyers of America (ATLA)
International Bar Association (London, United Kingdom)
International Mining Professionals Society (IMPS)

Australian Mining and Petroleum Law Association (AMPLA)
Law Editor, Mining Law and Environmental (Enviromining) Law, *Mineral Resources Engineering Journal*, Imperial College, London, United Kingdom

Admitted to Practice before State and Federal Courts
U.S. Supreme Court
U.S. Court of Appeals, Fourth Circuit
U.S. Court of Appeals, Seventh Circuit
U.S. Court of Appeals, Ninth Circuit
U.S. Court of Appeals, Eleventh Circuit
U.S. Disctrict Court for the Northern District of Indiana
U.S. District Court for the Southern District of Indiana
U.S. District Court for the Northern District of Georgia
U.S. District for the Middle District of Georgia
U.S. District Court for the District of Montana
All state courts of Indiana including Supreme Court
All state courts of Georgia including Georgia Supreme Court
All state courts of Montana including Montana Supreme Court
All state courts of Virginia including Virginia Supreme Court

Acknowledgments

The author wishes to acknowledge and express appreciation to certain colleagues who were of much help in rendering their services and advice during the compilation of this manuscript, thereby making the overall preparation easier.

At the top of the list is Dr. C. Dale Elifrits, Professor Emeritus of Geological Engineering at University of Missouri — Rolla (UMR), who has once more served in the capacity of advisor for another doctoral degree, but mainly for his omnipresent cheerfulness, enthusiasm, encouragement as to the need and purpose of this study and work; and generous giving of time and advice.

The support of my former University of Missouri doctoral degree committee, Drs. Donald E. Modesitt, Professor Emeritus, Civil Engineering; David A. Summers, Director and Curator's Professor/Rock Mechanics; and Donald D. Myers, Esq., Engineering Management, is also much appreciated in venturing with me on this new direction of engineering work involving law that has become so much an integral part of the environmental engineering and geoscientific professions.

As to UMR colleagues, I express appreciation and thanks to Drs. John W. Wilson and John D. Rockaway, Chairmen of the Mining Engineering and Geological Engineering Departments, respectively, for providing the past opportunities for me to lecture as an adjunct professor in mining law and environmental law in their departments, especially in creating, setting up, and instructing the 3-hour course curriculum for Mining 434 in enviromining law. Much of that preparation and ground work served as the foundation for this text.

Finally, much gratitude and heartfelt thanks go to my faithful, supportive, loving wife and secretary, Mary Pierce-Aston, for help over the years at the law libraries and office that made the task lighter; and for her ever-present, cheerful support and encouragement in completing the work.

R. Lee Aston

Acknowledgments

Contents

List of Cases

A

B

C

D

E

F

G

L

M

N

V

W

Y

List of Figures

List of Commonly Used Abbreviations and Acronyms

AMD acid mine drainage
APA Administrative Procedure Act
AQA Air Quality Act (1965)
ARD acid rock drainage
ASCML abandoned surface coal mined land
BAT best available technology
BACT best available control technology
BCT best cost-reasonable technology
BLM Bureau of Land Management
BOD biochemical oxygen demand
BPT best practicable control technology (currently available)
CAA Clean Air Act
CBOD carbonaceous biochemical oxygen demand
CEQ Council of Environmental Quality
CERCLA Comprehensive Environmental Response Compensation and Liability Act
CFR Code of Federal Regulations
Corps U.S. Army Corps of Engineers
CWA Clean Water Act
EA environmental assessment
EIR environmental impact report
EIS environmental impact statement
EPA Environmental Protection Agency/Administration/Act
FONSI finding of no significant impact
FS feasibility study
ICS individual control strategies
LAER lowest achievable emission rate
MEND mine environment neutral drainage
MSW municipal solid waste
NAAQS national ambient air quality standards
NCP National Contingency Plan
NEPA National Environmental Protection Act
NESHAP National Emission Standards for Hazardous Air Pollutants
NIMBY not in my backyard (syndrome)
NPDES National Pollutant Discharge Elimination System
NPL National Priority List
NSPS new source performance standards
OSHA Occupational Safety and Health Act
PCB polychlorinated biphenyl
POTW public owned treatment works
PRP potentially responsible party
PSI Pollution Standards Index
RACT reasonable available control technology
RCRA Resource Conservation and Recovery Act (1976) as amended

RI remedial investigation
ROD record of decision
SARA Superfund Amendments and Reauthorization Act (1986)
SCR selective catalytic reduction
SIP state implement plan
TSCA Toxic Substances Contol Act (1976)
TMDL total maximum daily load
USBR U.S. Bureau of Reclamation
U.S. EPA United States Environmental Protection Agency/Administration/Act
VOC volatile organic compound

CHAPTER 1

Introduction to Environmental Law

1.1 NEED FOR A NATIONAL ENVIRONMENTAL PROTECTION POLICY

After more than two centuries of unbridled, carefree development and free-swinging exploitation of the natural resources over hundreds of millions of acres in the new land of America, one that had lain relatively untouched by an inconsequential number of native inhabitants since God and nature had so richly endowed it, a dawning of concern for the environment settled on Americans. In retrospect, the rich land was being squandered and allowed to waste. A program of control for environmental concern and protection was needed, and for some aspects, overdue.

The United States was born of English parentage in 1776. As an Anglo-offspring, it lived with and under parental rule for nearly two centuries before setting up its own housekeeping and home rules. In claiming its inheritance of Crown lands after the war for separation, it found itself a very wealthy young nation in terms of space, natural beauty, and richly endowed resources with a moderate clime in which to grow.

In formulating its own mode of home rule or government it threw off the regulatory paraphernalia and trappings of the Crown. Instead of holding lands in the name of the United States or Sovereign, whereby reservations were made in title, interest, and use of the land, as is done to a greater extent by our British cousins for Crown lands in British Commonwealth nations, the new government, being one "of the people, by the people and for the people," sought to distribute its lands to the people for their benefit, enjoyment, and development. Self-enterprise was greatly fostered, which hastened to make the new nation prosperous and greatly accelerated its development.

The young government embarked on its new enterprise of nation building with a typical flare of American enthusiasm and ingenuity for developing an enterprise. In its rush to attract settlers and develop its newly acquired wealth of land and resources, it made few reservations in the land it was willing to deed away or grant to private ownership. Immediately after the cessation of the rebellion against its parents, the infant federal government began giving away parcels of "western" lands (still east of the Mississippi River) to its soldiers of the American Revolution in payment and gratitude for services rendered in the war. Public lands just west of the Appalachian Mountains were easily available for pioneering settlers to push back the eastern frontier. As the young nation acquired more lands in the Far West, it became even more eager to develop them by rapid settlement.

After initial development of the United States extending from the Atlantic to the Pacific shores had become a reality, subsequent governmental land programs were devised and enacted to fill in the sparsely settled areas of the Great Plains and the Rocky Mountains that lay between the two coasts. Before 1841, large blocks of public lands were sold to private land companies to split up and sell. In 1841, the General Preemption Law was passed to open up public lands for settlers and secure patents for the land. The Department of the Interior was created in 1849 to manage the vast public domain. The Homestead Act of 1862 hastened development by allowing settlers to take a quarter-section (160 acres/64.75 ha) for a small fee, build a home on it, and cultivate the land as conditions for obtaining title. The railroads were playing an important part in settling the West by transporting settlers from the East. Land grants were made to railroads for building new lines and for helping in western development. Railroad grants varied from right-of-ways to fee ownership of alternating land lots long the lines. Over 94,000,000 acres (38,041,278.8 ha) in the West are estimated to have been granted to the railroads by various acts (e.g., the Northern Pacific Railway Act of 1864, and the Railroad Right of Way Act of 1875).

Thus, during the period from the end of the British Colonial period to the present, the federal government succeeded in giving away nearly half of the original and acquired land area of the United States. Today, the federal government still owns, or holds title to one third of the land area of the nation.

Somewhat concurrently with the rapid giveaway schemes of land, around 1864, the progenitors of environmentalism, termed conservationists and preservationists, sowed their seeds of concern for modern environmentalism. The seeds grew at a slower pace than did the more exciting development of the lands. One of the progenitors was a lawyer and diplomat by name of George Perkins Marsh. In his travels over Europe and the Middle East, he had noted massive deforestation with resultant soil erosion, and man's interference with the balance of nature. Marsh published a book, titled *Man and Nature*, in which he not only described his observations of the results of man's poor stewardship of nature through wasteful practices in forestry and agriculture but also advocated restorative measures.

This early environmental concern and advocacy of measures to conserve our natural resources did not take hold until the 1890s when Congress authorized the President by the General Revision Act of 1891 to establish forest reserves on the public lands. A few of the early federal laws of conservation that were enacted illustrate a beginning of environmental concern, though under the name of conservation and preservation at that time:

1. 1872: Yellowstone National Park was created.
2. 1902: The Reclamation Act for water resources in the West was enacted.
3. 1906: The Congress-authorized mineral lands were withdrawn from entry until valuations could be made by the Department of Interior.
4. 1906: Land-grant colleges were established to promote improved farming practices in the states.

5. 1908: Theodore Roosevelt "fathered" a conservation spirit for planning of natural resources and national parks under a conference producing a "Declaration of Principles on Conservation."
6. 1909: Theodore Roosevelt hosted a North American conference for advocating natural resources conservation.
7. 1916: The National Park Service was created, although 16 national parks and about 18 national monument sites had already been created by that time.

Along with concerns for preservation and conservation in the early twentieth century, the advanced state of development and higher standard of living of the industrial nation brought with it environmental problems. Deterioration of water and air quality over the past century steadily mounted in intensity and volume until the American society would no longer tolerate uncontrolled pollution by industry and public waters contamination by government waste disposal services. High living standards in this deteriorating environment triggered society to cry out for quick relief, which in turn brought stern corrective measures for the allegedly culpable industrial segment of our society. The higher standard of living with its superfluous amenities for an easier lifestyle also brought with it massive volumes of liquid, gaseous, and solid waste resulting in the problem of how to dispose of it without further deterioration of the environment. Advanced technology of the developed nations, in response to the general public's demand and craving to make a utopian dream of luxury a reality for every individual, resulted in increased production to meet the continuous demand of higher living standards for all persons without regard to societal costs. Consequently, the pursuit of such luxury and abundance ideals generated at the same time tremendous volumes of hazardous and toxic wastes. Simultaneously, there was a demand for immense amounts of energy supplied by fossil fuels, or alternatively by radioactive minerals, to meet the power requirements for the increased production. The developed societies, having lived in the lap of luxury for decades since the turn of the twentieth century, without great concern for the deterioration of the environment, had come to realize that the time has come "to pay the piper" and make amends to the environment.

Mining, as a basic industry, produces the essential raw materials necessary for the manufacturing of society's necessities and luxuries for its well-being. This is well stated by the mining industry's maxim; and although it may be defensive, it states a *nudem veritas* (a naked truth), "*If it can't be grown, it has to be mined.*" Those professing a deep concern for the environment would do well to keep this basic tenet in mind. The advanced or developed society is unwilling to revert to a lesser way of life. Unfortunately, for society to continue in the manner of life to which it has become accustomed, it is caught between the demands of cleaning up its deteriorating environment, and having a continuing supply of basic raw materials for its mode of good living, or continuing exploitation without concern for the environment. An advanced society must have "clean living" conditions at the same time with "good living," neither of which is expendable. The price tag for this combination must be paid by society. The price is, indeed, high and becoming higher every year with increasing environmental constraints placed on the raw materials providers and manufacturing industries, and now on the newer waste disposal industry.

As indicated, a part of the problem of restoring the standard of clean living has resulted from the volumes of waste generated in the manufacturing of products from the raw materials taken from the earth. In addition, society generates tremendous volumes of disposable waste from its consumption of products that have a short-term life, such as plastic and paper products. The resulting problem is disposal of the waste without degrading the environment, or at least minimizing the degradation.

Unclassified and unrestricted landfills, dumping in the oceans, rivers, and large lakes have been earlier solutions for disposing of society's waste. In a few decades those methods boomeranged to haunt society. The current trends for waste disposal are toward incineration of bulk waste, controlled and classified landfills, and recycling of reusable materials. Although incineration greatly reduces the bulk of waste, incineration has not been readily accepted. Recycling does not adequately resolve the necessary volume reduction for disposal of society's waste. Thus, large amounts of waste for disposal continue to be a problem. Placing the waste in controlled landfills appears to be the most accepted and logical solution for disposal of the bulk volume. The waste disposal industry has mushroomed as a result. New excavations of pits, either for the placing waste or for excavating clay cover and lining material for surface mounding of waste, must be made. With the enormous volumes of waste, larger numbers of increased size landfills are needed. Along with this need comes the problem of location, siting, and permitting for landfills.

From the early decades of the twentieth century, natural resources concern has gradually grown and a considerable number of federal laws have been enacted in an effort to conserve and to preserve. With the advent of, and pressures from, the environmental "green" movement of the 1960s, Congress responded by passage of the National Environmental Policy Act of 1969 (NEPA), effective January 1, 1970, which was followed shortly by Earth Day, April 22, 1970.

Since the enactment of NEPA, a plethora of environmental acts, laws, and regulations have been created to carry out the national crusade for a cleaner environment. The professions of law and engineering have been deluged, swamped, and inundated by environmental regulations and paperwork, starting with a staggering number of application forms, environmental assessments, impact studies, and comprehensive reporting, all from the start of a project through monitoring years after the project has ceased to function The regulations and environmental controls for all phases of industry, including waste disposal, continue to increase yearly in number and become more pervasive of nearly every facet of American life and livelihood.

Punishment for environmental violations is no longer limited to fines, but violations have become environmental crimes in some cases, subject to imprisonment for the responsible party or corporate officer. Thus, a critical need is established for engineers to have some insight into environmental law and its workings to cope with and properly dispense their engineering duties and problems.

CHAPTER 2

Basic Law for Engineers and Geoscientists

2.1 GENERAL ORIENTATION TO AMERICAN JURISPRUDENCE

To study various laws and regulations affecting the practice of engineering and engineering projects, the case study method as normally employed in American law schools is used in this text. The case study method examines how the law in a litigated action with a particular issue, or issues, is argued, interpreted, and applied by the hearing court or hearing board in the case of a regulatory agency review. Environmental case studies generally review the environmental issue being litigated from its inception at the agency level, to hearing board appeals, and through the appeal to the trial court. From the case study, the purpose and parameters of a regulation at issue are examined, defined, and construed by the court for implementation as intended by the enacting legislative body. Case studies offer the opportunity to learn from the misinterpretations, mistakes, and challenges of others; and to know what is expected in performance to comply with environmental regulations and laws.

It is important to stress to the engineer when reading a court's decision of a legal question in a case, to avoid becoming entangled in the legal procedures involved. For example, in a case where reviewability of an Environmental Protection Agency (EPA) action or decision is the question of law before the court, the written decision may involve civil or legal procedure in law, or procedure for administrative regulations. That procedure should be left to attorneys to study, and glossed over by the engineer. The part of the court's written decision that is of interest and enlightening to the engineer is often given in the recitation of background information and facts that led to the controversy and brought the action into court. The allegations, interpretations, claims, and arguments put forth by the parties to the suit, similarly, are enlightening to the engineer and should be noted. The court will consider and normally treat the parties' arguments in the decision, pointing out the merits and the flaws. These should be read by the engineer. Finally, the court conclusions of law and findings are of greatest interest to both engineer and attorney.

In reading court decisions, environmental agency procedures are often described and the engineering reader should take note of the details given because they often will apply to environmental engineering procedures that may be encountered. For example, agency rules frequently contain time limits for filing information, environmental data, reports, and various other submissions in obtaining extensions and permitting, all of which are important to the engineer. However, the engineer must learn to discern between agency requirements of administrative regulatory procedure and those of environmental curative procedure, that is, to discern information that is of interest as an engineer, not as an attorney. Again, use caution. Do not become entangled in the details of legal procedure while studying a court's decision.

2.2 DIVISIONS OF LAW

Law, as with branches of engineering, is divided into different subject areas, or categories of practice. For example, divisions of law are named: constitutional law, contract law, torts, evidence, civil and criminal law, corporate law, real property law, wills and estates, naval law, administrative law, et al. The branch of law of main concern in this text is environmental law. The course will touch on other divisions of law such as contract law, real property law, torts, constitutional law, administrative law, and natural resources law, and on a new, specialized branch of law that I refer to as enviromining law, for which the name is self-explanatory. It is defined as those federal and state environmental laws and regulations that have particular application to, and affect, mining activities in the United States.

2.2.1   The American Legal System — Legal Terms and Phrases

The definition of **common law***, or English common law is: "that body of law and juristic theory which was originated, developed, formulated, and administered in England and has obtained amongst most of the states and peoples of Anglo-Saxon stock" (Black, 1968).

By virtue of the early predominantly English and British Isles settlers, ancestry, and roots of the United States, English common law became the law of the American colonies and the United States. It is distinguished from Roman law (civil law/code), which is the basic law of countries of Latin origin and several non-Latin European nations. In the modern United States, it designates that portion of the common law of England in force in the 13 American colonies before and at the time of their separation from the mother country, and is still recognized as an organic part of the jurisprudence of the United States. It is also that part of U.S. law that has not been expressly abrogated by enacted statutes.

That part of the common law of environmental concern is certain tortious acts, such as nuisance, negligence, trespass, and liability arising from such acts. Historically, grounds for grievances and complaints for what would now be called environmental damages from industrial pollution and contamination were found in the legal tort doctrines of both private and public nuisance, trespass, nontrespassory invasion of property, negligence, liability, and remedies found through the **equitable**

---

* Legal terms used in this text are generally explained parenthetically at the time of introduction in the text and are emphasized in boldface print.

doctrine of injunctive relief. (**Equity** is a system of jurisprudence or branch of remedial justice administered by certain tribunals called chancery courts, or courts of equity, distinct from the common law courts and empowered to decree equity.)

Except for claims brought by government entities in the name of the public, the burden of seeking damages and relief for private claims fell on the individual. Where existing statutes fell short in providing penalties for private or public environmental damages caused by polluting operations, the English common law system filled the statutory shortfall. Common law liability for damages to the environment is strictly of local and personal concern, i.e., for damaging the environment of one's neighbor, or neighbors, or the neighborhood. Nevertheless, today they still play a lesser part in controlling pollution and contamination of air, water, public health, and property. The still valid application of nuisance, trespass, and negligence tort claims for local environmental pollution in spite of the presence of modern statutory environmental law, such as the Clean Air Act (CAA) and the Clean Water Act (CWA), is illustrated by the following case filed in an Ohio state court in 1999, removed by the defendants to federal court for trial under the CAA and CWA statutes. The federal court remanded the case to be tried in the Ohio state court under common law claims. For another recent water pollution suit where state claims were filed for nuisance, trespass, and negligence, see *Driscoll v. Adams*, 181 F.3d 1285 (11th Cir. 1999), Chapter 5, Section 5.8.10.1. Also, see *Geraghty and Miller, Inc. v. Conoco Inc.*, and *Condea Vista Chemical Co.*, 234 F.3d 917 (5th Cir. 2000), where common law claims were filed in a Texas state court, Chapter 6, Section 6.4.2.7.

Technical Rubber Co., *et al*, Plaintiffs
v.
Buckeye Egg Farm, L.P. *et al*, Defendants

Case No. 2:99-CV-1413
United States District Court For The Southern District Of Ohio, Eastern Division
2000 U.S. Dist. LEXIS 8602; June 16, 2000, Decided

Plaintiffs originally filed this action in the Court of Common Pleas for Licking County, Ohio, and defendants thereafter removed the action to this Court on the basis of federal question jurisdiction pursuant to 28 U.S.C. § 1441. This matter is now before the Court on plaintiffs' motion to remand.

I. Background

Plaintiffs, property owners near the defendant's egg farm, bring this suit alleging that defendants have failed to properly manage and operate the farm, causing air, water, and soil contamination in and around plaintiffs' properties. Plaintiffs name as defendants the Buckeye Egg Farm ["Buckeye Egg"], a Delaware limited partnership with its principal place of business in Licking County, Ohio; Anton Pohlmann, a limited partner of Buckeye Egg and resident of Licking County, Ohio; PM Egg Corporation ["PM Egg"], an Ohio corporation with its principal place of business in Licking County, Ohio; Marcus Pohlmann, a former president and/or general manager of Buckeye Egg and the sole shareholder of PM Egg; HLA Egg Corporation ["HLA Egg"], an Ohio corporation with its principal place of business in Franklin County, Ohio; Andrew L. Hansen, a former president and sole shareholder of HLA Egg; Elliot B. Jones, President of Buckeye Egg; MPR Egg Corporation, an Ohio corporation; CWS Egg Corporation, a canceled Ohio corporation; and Croton Farm, L.L.C., a Delaware limited liability company and general partner of Buckeye Egg.

On November 19, 1999, plaintiffs filed this suit in the Licking County Court of Common Pleas seeking monetary and injunctive relief in connection **with claims of nuisance, trespass and negligence.** On December 30, 1999, defendants removed this action to this Court as asserting claims under the Federal Water Pollution Control Act ["Clean Water Act"], 33 U.S.C. §§ 1251–1387 and under the Air Pollution Prevention and Control Act ["Clean Air Act"], 42 U.S.C. §§ 7401–7671q. Defendants also contend that these federal statutes completely preempt any state law claims regarding air, water and soil pollution. Plaintiffs move to remand on the grounds that the Clean Air Act and the Clean Water Act do not completely preempt state law, and that the complaint merely raises state law causes of action. Plaintiffs also move for costs and attorney's fees under 28 U.S.C. § 1447.

II. Remand

A. Standard

28 U.S.C. § 1441 governs removal of actions from state to federal court. The statute provides in pertinent part:

(a) Except as otherwise expressly provided by Act of Congress, any civil action brought in a State court of which the district courts of the United States have original jurisdiction, may be removed by the defendant ... to the district court of the United States for the district and division embracing the place where such action is pending....
(b) Any civil action of which the district courts have original jurisdiction founded on a claim or right arising under the Constitution, treaties or laws of the United States shall be removable without regard to the citizenship or residence of the parties.... 28 U.S.C. § 1441(a), (b). The removal provisions are strictly construed against removal and in favor of remand. The burden of establishing the right to removal rests on the party seeking removal. Because the removal petition is to be strictly construed, all doubts are to be resolved against removal (see *Her Majesty the Queen*, 874 F.2d at 339; *Wilson v. USDA*, 584 F.2d 137, 142 (6th Cir. 1978).

Removal of plaintiffs' claims to this Court is proper if one or more of the claims fall within this Court's original jurisdiction, i.e., if they "arise under" federal law. A cause of action arises under federal law for purposes of original jurisdiction and removal if the plaintiffs' "well-pleaded complaint" presents a federal issue. The "well-pleaded complaint" rule makes the plaintiff the master of the claim; "he may avoid federal jurisdiction by exclusive reliance on state law." If a complaint, on its face, is devoid of an issue of federal law, then a federal court lacks federal question jurisdiction. If the federal court lacks both diversity and federal question jurisdiction, then the court must remand the suit to the state court. The removing party bears the burden of establishing that removal was proper. (citations omitted).

"A claim arises under the laws of the United States when 'the complaint seeks a remedy expressly granted by a statute or a distinctive policy of a federal statute requires the application of federal legal principles for its disposition.'" In order for an action to "arise under federal law, a right or immunity created by that law must be an essential element of the plaintiff's claim; the federal right or immunity that forms the basis of the claim must be such that the claim will be supported if the federal law is given one construction or effect and defeated if it is given another." **The "mere presence of a federal issue in a state cause of action does not automatically confer federal-question jurisdiction,"** *Merrell Dow Pharm. Inc. v. Thompson*, 478 U.S. 804, 813, 92 L. Ed. 2d 650, 106 S. Ct. 3229 (1986), because "not every question of federal law emerging in a suit is proof that a federal law is the basis of the suit" (see *Gully v. First Nat'l Bank in Meridian*, 299 U.S. 109, 115, 81 L. Ed. 70, 57 S. Ct. 96 (1936).

B. Application

In arguing that this litigation was properly removed, defendants point to specific language used by the plaintiffs in the complaint. Defendants also contend that the "artful pleading" doctrine applies to plaintiffs' complaint because plaintiffs' nuisance, trespass and negligence claims are essentially federal in nature given that they are based upon air, soil and water pollution. In this regard, defendants maintain that the Clean Air Act and the Clean Water Act completely preempt plaintiffs' state law claims. The Court will consider each of these arguments separately.

In the *sub judice* [meaning the 'case below' transferred from the state court], plaintiffs' complaint did indeed allege that the defendants are "in violation of state and/or federal law" due to their mismanagement of the storage and spreading of manure, the storage of ammonia and the improper disposal of chicken carcasses. (Complaint, PP 42, 48-49, 55-56). However, the claims including such language are clearly labeled as nuisance claims and make no further discussion or even mention of federal law. Plaintiffs affirmatively represent in their memorandum in support of the motion to remand that they do not intend, and they will therefore not be heard, to assert federal causes of action in making passing references to "federal law." Furthermore, **given that the citizen-suit provisions of the Clean Air Act and the Clean Water Act simply provide for injunctive remedies, the fact that the complaint seeks monetary damages as well as injunctive relief is inconsistent with an intentional assertion of claims under those federal statutes.** [author's emphasis added] In addition, plaintiffs' complete failure to comply with, or even to attempt to comply with, the notification requirements mandated by the citizen-suit provisions of the Clean Air Act and the Clean Water Act, see, e.g., 42 U.S.C. § 7604(b)(1)(A) and 33 U.S.C. § 1365(b)(1)(A), indicates that plaintiffs, the masters of their complaint, did not intend to assert claims arising under federal law.*

Defendants also argue that the Clean Air Act and the Clean Water Act completely preempt plaintiffs' state law claims, and that the action was properly removed regardless of whether or not plaintiffs intended to rely on federal law to support their claims. Although the "well-pleaded complaint" rule generally makes plaintiffs the master of their claims, plaintiffs cannot defeat removal by masking, or "artfully pleading," a federal claim as a state claim. (See *Her Majesty the Queen in Right of the Province of Ontario v. City of Detroit*, 874 F.2d 332, 339 (6th Cir. 1989.) If the only remedy available to the plaintiffs is a federal remedy, the case is removable regardless of the express language of the complaint. Ordinarily, because federal preemption does not appear on the face of the well-pleaded complaint, removal on that basis is improper. Under the "complete pre-

* In the Answer, defendants assert that "Plaintiffs have failed to take the prerequisite steps necessary for them to maintain 'citizen-suits' under some or all of the federal environmental statutes [allegedly] referenced in the Complaint, and the Court thus lacks jurisdiction over such claims." (Answer, P 45).

emption doctrine," however, the preemptive force of a federal statute may be so extraordinary that it "converts an ordinary state common-law complaint into one stating a federal claim for purposes of the well-pleaded complaint rule." *Id*. at 65. Stated another way, if an area of state law has been completely preempted by federal law, any claim purportedly based on that preempted state law is considered, from its inception, a federal claim arising under federal law. The complete preemption doctrine is narrowly interpreted. In fact, the Court must begin this analysis with the presumption that Congress did not intend to supplant state law.

Defendants rely on *International Paper Co. v. Ouellette*, 479 U.S. 481, 93 L. Ed. 2d 883, 107 S. Ct. 805 (1987), in support of their proposition that the Clean Water Act completely preempts the field of state water pollution law. In *International Paper*, Vermont landowners brought suit against a New York paper mill under the Vermont common law of nuisance for the company's alleged pollution of Lake Champlain. The United States Supreme Court held that, although the savings clause in the Clean Water Act indicates that individuals may bring causes of action under certain state laws, the "CWA precludes a court from applying the law of an affected State against an out-of-state source." *Id*. at 494. If an out-of-state source were to be held liable for violations of the law of the affected state, "that law could effectively override both the permit requirements and the policy choices made by the source State." *Id*. at 495. The Court went on to state, however, that "nothing in the [Clean Water] Act bars aggrieved individuals from bringing a nuisance claim pursuant to the law of the source State." *Id*. at 497. The Court reasoned that an action brought against [a source] under the [source state's] nuisance law would not frustrate the goals of the CWA as would a suit governed by [the affected state's] law. First, application of the source State's law does not disturb the balance among federal, source-state, and affected-state interests.... Second, the restriction of suits to those brought under source-state nuisance law prevents a source from being subject to an indeterminate number of potential regulations. *Id*. at 498–99.

In *Arkansas v. Oklahoma*, 503 U.S. 91, 100, 117 L. Ed. 2d 239, 112 S. Ct. 1046 (1992), the Supreme Court interpreted *International Paper* as standing for the proposition that "the Clean Water Act taken 'as a whole, its purposes and its history' preempted an action based on the law of the affected State and that the only state law applicable to an interstate discharge is 'the law of the State in which the point source is located.'" Thus, it is clear that *International Paper* is limited to situations involving interstate pollution. It established federal preemption only to the extent necessary to resolve potentially conflicting state laws. This litigation, however, involves Ohio plaintiffs complaining about alleged water pollution in Ohio, allegedly caused by the Ohio activities of persons and entities, most of whom are residents of Ohio. *International Paper* is, quite simply, inapposite. This Court concludes that the Clean Water Act does not preempt plaintiffs' state law claims.

Defendants likewise argue that the Clean Air Act preempts plaintiffs' state law claims predicated on defendants' alleged air pollution: The CAA sets out a comprehensive regulatory scheme designed to prevent and control air pollution.... Pursuant to the CAA, each state is required to adopt an implementation plan of its own for that state. If such a state implementation plan ("SIP") is approved by the EPA, its requirements become federal law and are fully enforceable in federal court. As a result, assuming arguendo that agricultural activities are subject to state air emissions standards contained in Ohio's federally-approved SIP, then clearly Plaintiffs' allegations of air pollution cannot be construed without resort to federal law.

(Defendants' Memorandum contra Plaintiffs' Motion to Remand, at 9)(citations omitted). Plaintiffs do not raise claims under the Ohio SIP; rather, they raise causes of action under Ohio common law tort theories of nuisance, trespass, and negligence. Even if plaintiffs did assert causes of action under the Ohio SIP, plaintiffs' claims would not necessarily be governed by federal law. (See *Her Majesty the Queen*, 874 F.2d at 341; even if the Michigan Environmental Protection Act could be said to be part of the Michigan SIP, it is not thereby transformed into federal law because it "creates a state environmental common law that is unaffected by federal law, and creates an independent state action that is unaffected by anything that happens in the federal sphere of government.")

Furthermore, in *Her Majesty the Queen*, the Sixth Circuit expressly disagreed with defendants' contention that state nuisance claims related to air pollution are preempted by the Clean Air Act:

> Air pollution is, of course, one of the most notorious types of public nuisance in modern experience. Congress has not, however, found a uniform, nationwide solution to all aspects of this problem and, indeed, has declared "that the prevention and control of air pollution at its source is the primary responsibility of States and local governments." To be sure, Congress has largely pre-empted the field with regard to "emissions from new motor vehicles," and motor vehicle fuels and fuel additives. It has also pre-empted the field so far as emissions from airplanes are concerned. So far as factories, incinerators, and other stationary devices are implicated, the States have broad control.... (874 F.2d at 342 (quoting *Washington v. General Motors Corp.*, 406 U.S. 109, 114, 31 L. Ed. 2d 727, 92 S. Ct. 1396 (1972)).

The Sixth Circuit also relied on the Supreme Court's decision in *International Paper* for its conclusion that "Congress did not seek to preempt [state law] actions." The Court noted:

> In *International Paper*, the Court held that a savings clause in the Clean Water Act, 33 U.S.C. § 1365(e), which defendants concede is identical to the savings clause at issue in this case, allows state law actions against water pollution notwithstanding the existence of federal law and standards. The Court held that the savings clause "negates the inference that Congress 'left no room' for state causes of action;" it "specifically preserves other state actions, and therefore nothing in the Act bars aggrieved individuals from bringing a nuisance claim pursuant to the law of the source State." (874 F.2d at 343 (quoting *International Paper*, 479 U.S. at 492, 496).

Although the Sixth Circuit recognized that *International Paper* dealt with water pollution while the case before it concerned air pollution, that court nevertheless determined that "there was no reason to think that the result with regard to air pollution should be different." (874 F.2d at 343.)

This Court therefore concludes that the Clean Air Act does not preempt plaintiffs' state common law nuisance claims, that the action was improperly removed to this Court, and that plaintiffs' motion to remand is meritorious. Plaintiffs' motion to remand (to the State Court of Common Pleas, Licking County, Ohio) is, accordingly, GRANTED.

/s/ Algenon L. Marbley, United States District Judge

Nuisances are classified as public, private, and mixed. A **public nuisance** is one that affects an indefinite number of persons, or all the residents of a particular locality, or all people coming within the extent of its range or operation, although the extent of injury may be unequal [*Burnham v. Hotchkiss*, 14 Conn. 317 (1841)]. A **private nuisance,** as distinguished from public nuisance, includes any wrongful act that destroys or deteriorates the property of an individual, or few persons, or interferes with their lawful use and enjoyment thereof, ... or causes them a special injury different from that sustained by the general public [*Baltzeger v. Carolina Midland R.R. Co.*, 32 S.E. 358 (S.C. 1894)] (Black, 1968, p. 1215).

Trespass is the doing of an unlawful act, or of a lawful act in an unlawful manner to the injury of another's person or property [*Waco Cotton Oil Mill of Waco v. Walker*, 103 S.W. 2d 1071, 1072 (1937)], an unlawful act committed with violence, actual or implied, causing injury to the person, property, or relative rights of another; an injury or misfeasance to the person, property, or rights of another, done with force and violence, either actual or implied in law [*Southern Ry. Co. v. Harden*, 28 S.E. 847 (Ga. 1897)]. In its more limited and ordinary sense, it signifies an injury committed with violence, and this violence may be either actual or implied; and the law will imply violence even though none is actually used, when the injury is of a direct and immediate kind, and committed on the person or tangible and corporeal property of the plaintiff ... a peaceable but wrongful entry upon a person's land is implied.

Trespasses are often described as continuing or permanent. A **permanent trespass** is one that is in its nature a permanent invasion of the rights of another, as where a person builds on his own land so that a part of the building overhangs his neighbor's land [*H.H. Hitt Lbr. Co. v. Cullman Property Co.*, 66 So. 720, 721 (Ala. 1914)]. A **continuing trespass** is one that consists of a series of acts, done on successive days, that are of the same nature, and are renewed or continued from day to day, so that in the aggregate, they make up one indivisible wrong (3 Blackstone's Commentaries 212, Black, 1968, p. 1674).

The term **nontrespassory invasion** is frequently applied where there has been an encroachment by gases, odors, dust, objects, as thrown rock from blasting, airborne particles, migration of hazardous contaminants on to property (e.g., PCBs in soil from a nearby contributor) etc., which have emanated from the acts of the wrongdoer on his property and invaded the land, property or rights of another, but have involved no physical entry by the body of the wrongdoer himself.

Negligence is the absence of care. *Black's Law Dictionary* defines negligence as "the omission to do something *(an act)* which a reasonable man, guided by those ordinary considerations which ordinarily regulate human affairs, would do, or the doing of something which a reasonable and prudent man would not do." "The term refers only to that legal delinquency which results whenever a man fails to exhibit that care which he ought to exhibit, whether it be slight, ordinary, or great" (Black, p. 1184).

To establish a **prima facie** (literally, at first sight; on the face of it) claim for trespass, there must be an act of physical invasion of another's property. The physical invasion need not involve the body of a person, but some object entering the property of another caused by another person. Intent to trespass on another's land is not required, but intent to do the act that constitutes or causes the trespass is sufficient. Gross recklessness resulting in the unlawful invasion of another's property may be sufficient, even in the absence of intent. It is not necessary that the wrongdoer come on to another's land; a trespass will exist where the trespasser throws rocks onto the land or floods it.

As may be inferred, the difference by definition between a trespass and a nuisance may be a gray area at times. A trespass is an interference with a landowner's right to exclusive possession, which includes the quiet enjoyment of his land. Still, a nuisance is also an unreasonable interference with the landowner's use and enjoyment of his land, but lacks the challenge to exclusive possession.

It is noted in the *Restatement of Torts, Second* § 826(a) (1977) that in the case of a socially useful activity, such as a factory or commercial establishment, the defendant's conduct must be unreasonable and cause the plaintiff substantial harm. The determination of reasonableness is a balancing process for the court considering the gravity of the harm and the utility of the defendant's conduct.

To consider the gravity of the harm of a nuisance, American courts consider various factors that are summarized by the *Restatement* (Second) § 827 as follows:

1. Extent of the harm involved
2. Character of the harm involved
3. Social value that the law attaches to the type of enjoyment invaded
4. Suitability of the particular use or enjoyment invaded to the character of the locality
5. Burden on the person harmed of avoiding the harm

The factors used to determine the utility of the defendant's conduct in an alleged nuisance are stated in § 828 of the *Restatement*:

1. Social value that the law attaches to the primary purpose of the conduct
2. Suitability of the conduct to the character of the locality
3. Impracticability of preventing or avoiding the invasion

At common law, the rule was *sic utere tuo ut alienum non laedus*, or "use your own property so as not to injure that of another," and when violated, such harm was actionable. Courts retained some discretion to determine what type of injury was actionable. Modern and present environmental laws have obviated many common law nuisance claims.

Engineers should become familiar with only the more common legal terms encountered as laymen and certainly not with all of the less common legal terms and phrases used by attorneys. At some phase in an engineer's practice, he may be required to give a **deposition** for trial (a **deposition** is an out-of-court examination of a witness or party to litigation by question and answer under oath, giving testimony as to pertinent facts which are admissible at trial or in court. It is a discovery device used by lawyers to discover germane issues as well as learn what information their adversaries have).

Discovery methods are useful tools in the evidentiary process in pretrial work. They have become an important part of modern practice and procedure. They are of interest to this course because engineers frequently appear in environmental litigation as expert witnesses and are often subjected to any of the following discovery devices: (1) depositions on oral examination; (2) depositions on written questions; (3) interrogatories to parties; (4) production of documents; (5) examination of persons; and (6) requests for admissions. Depositions, oral and written, and production of documents are the ones most likely for engineers to encounter.

As stated by the court of appeals in *Hanna Creative Enterprises v. Taylor*, 156 Ga. App. 376 (1980), "The rules of discovery ... are designed *to narrow and clarify the issues* and to remove the potential for secrecy and hiding of material that existed under our previous system. In particular, the rules of discovery are designed to provide parties with the opportunity to obtain material knowledge of all relevant facts, thereby reducing the element of surprise at trial." *The main purpose of discovery, however, is to collect evidence.* Also, a deposed witness is bound to his statement so that he may not change it in trial [emphasis added].

A party can obtain discovery concerning any matter not privileged (as between attorney and client, or between husband and wife) and relevant to the subject matter of the pending action. It can include the existence, description, nature, custody, condition, and location of any books, documents, records, reports, field notes, engineering data, samples, operation and accuracy of instruments, or other tangible things. It can include the identity and location of persons having knowledge of any discoverable matter, and conversations between persons. The information sought and asked does not need to be admissible at trial, but it must be reasonably calculated to lead to the discovery of admissible evidence.

2.2.2 Simplified Overview of the American Court System

Acts, laws, and statutes are enacted by the U.S. Congress and state legislatures; also, county, city, or municipal ordinances are enacted (passed) by their governing bodies, such as a county board of supervisors or commissioners and city councils. Often, a governing body will enact laws that are patterned after those of another location or jurisdiction that have been tested in court and upheld as a valid or constitutional law. Where a statute or ordinance is totally new in its purpose, it may have to undergo challenges as to its validity. Until it has been challenged in a test case, it is unknown as to whether it is a valid (i.e., constitutionally acceptable) law, regulation, or rule. Environmental rules made by government agencies are challenged at times to test their validity. An agency rule may have exceeded its authority granted under some legislative act. First challenges are frequently made to determine whether the particular statute or regulation will pass the test of constitutionality; thus, they are known as **test cases**.

Because major environmental law for the nation is primarily supported by federal acts and statutes, the course will also be primarily concerned with the federal court system and cases. However, state environmental regulations will be examined where pertinent, for many state governments not only have followed the suit of the federal environmental initiative, but also have elected to direct and control their own state environmental activities under the option of primacy with only oversight by the federal environmental agencies.

Article III, Section 1, of the U.S. Constitution establishes the federal court system stating, "The judicial power of the United States shall be vested in one supreme court, and in such inferior courts as the Congress may from time to time ordain and establish."

Under this constitutional authorization, Congress has created, as inferior courts, the federal district courts and the federal appellate courts that are Article III courts.

Congress has been granted the plenary power to delineate the jurisdictional limits of these inferior courts, both original and appellate. (**Jurisdiction = legal power/authority to hear and determine a cause of action.**) However, Congress is also bound by the standards of judicial power delineated in Article III as to the subject matter, the parties, and the limitation of case or controversy. The words *case* and *controversy* derive from Section 2 of Article III in which the U.S. Constitution spells out original jurisdiction for the federal courts. Section 2 states, "The judicial power shall extend to all Cases, in Law and Equity, arising under this Constitution (i.e., constitutional law), the Laws of the United States, etc ...; to Controversies to which the United States shall be a party; to Controversies between two or more States; between a State and Citizens of another State; between Citizens of different States; ... etc." A consequence of the limitation is that Congress cannot require Article III courts to render advisory opinions or perform administrative or nonjudicial functions.

Article I of the Constitution outlines the legislative powers and duties of Congress. Under Section 8, Paragraph 9 of Article I, Congress is charged "To constitute Tribunals inferior to the Supreme Court." Congress has thereby created certain other courts, known as Article I courts, which implement other legislative powers (e.g., the United States Tax Court, the Customs Court, the Claims Court, and inferior courts of the District of Columbia). These courts are vested with administrative powers as well as judicial, but they may not hear any case or controversy vested in Article III courts.

This work will deal mainly with Article III courts, state courts, and federal administrative law courts. The federal courts are those of general jurisdiction, and the lowest order courts of concern will generally be the U.S. district courts.

2.2.3 Jurisdiction

Jurisdiction involves whether the court in which the action is filed is the proper one to hear an action, that is, whether the court has the authorization to hear the class or type of case involved; whether the parties have filed in the right court and are in the court's **territorial jurisdiction** and its **sphere of authority** (**venue**). To hear a case, a court must have both requisite territorial jurisdiction and **subject matter jurisdiction**. State courts of **general jurisdiction** will hear all cases except for those assigned to other courts of limited jurisdiction, specialized courts, so to speak (e.g., juvenile courts, domestic relations courts, probate courts, and magistrates' courts).

2.2.4 Federal Court System

Federal district courts are of restricted jurisdiction. At the trial level, they are limited to (1) cases arising under **federal law**, and (2) cases in which there is **diversity of citizenship** among the parties (**diversity** means different state residency). They have overlapping jurisdiction with the U.S. Claims Court. The U.S. Claims Court was established in 1855. Its jurisdiction is to render judgment on any claim against the United States founded on the Constitution, on any act of Congress, on any regulation of an executive department, on any expressed or implied contracts with the United States, and for liquidated or unliquidated damages in cases not sounding in **tort**. (A **tort** is a private or civil wrong or injury done to another person or persons, other than a breach of contract, for which a court will provide a remedy in the form of an action for damages.) There is a minimal monetary limitation for a case to be filed in the U.S. Claims Court (i.e., the claim must have a value of $10,000 or more to be considered/litigated in that court). The monetary minimum value limit in the federal district court is $20,000, except where the question is one of a constitutional right.

The U.S. District Courts, the U.S. Circuit Courts of Appeal, and the U.S. Supreme Court are not limited to cases against the U.S. government as is the claims court. They hear cases between citizens (corporations being citizens, or persons, for purposes of suing and being sued) where the parties are residents of different states or there is a question involving the constitution or federal law. The district courts may hear cases in tort and federal criminal law, while the claims court may not.

The U.S. Courts of Appeal are intermediate appellate courts created by the Congress in 1891. They have jurisdiction over appeals from the federal district courts, but not over state court decisions. They also review and enforce orders of federal administrative bodies. Decisions of the federal courts of appeal may be subject to appeal and review by the U.S. Supreme Court where the subject matter is mandatory (e.g., where a federal statute has been found invalid in a civil action where the United States or any of its officers or employees is a party or if a federal court of appeals has held a state statute unconstitutional or repugnant to a federal statute). Where an appellate review is not mandatory but discretionary by the U.S. Supreme Court, the appellant may ask for review by a **writ of *certiorari*** (Latin: to be informed of) where the Supreme Court may choose which cases it wishes to hear. (The U.S. Supreme Court grants hearing by *certiorari* to only about 100 cases per year.)

The U.S. is divided into 13 federal judicial circuits (Figure 2.1) for which there is a U.S. Court of Appeals within each circuit. The states within each of the federal judicial circuits have at least one federal district and court (e.g., Indiana has two federal districts and two federal district courts, viz., the federal court for the District of Southern Indiana, and another for the Northern District; Georgia has three federal districts, Northern, Middle, and Southern; Montana has one; Colorado has one; and Missouri has two, Eastern and Western). The federal district courts are subdivisions of the 13 federal judicial circuits.

When a decision of a case in a federal court is appealed, the appeal is generally heard by the U.S. Court of Appeals for the judicial circuit in which the district court is located. For example, referring to Figure 2.1, a U.S. District Court case decision appealed from the Middle District of Georgia would go to the U.S. Court of Appeals for the Eleventh Circuit located in Atlanta, Georgia. One from the Southern District Court of Indiana would go to the Seventh Circuit Court of Appeals located in Chicago, Illinois. One from the U.S. District Court of Montana would go to the Ninth Circuit Court of Appeals located in San Francisco, California.

Appeals of decisions in civil cases are not automatic. An appeal must be made by one of the parties to the decision based on specified alleged errors made by the deciding court. This is true for the state court systems as well as the federal

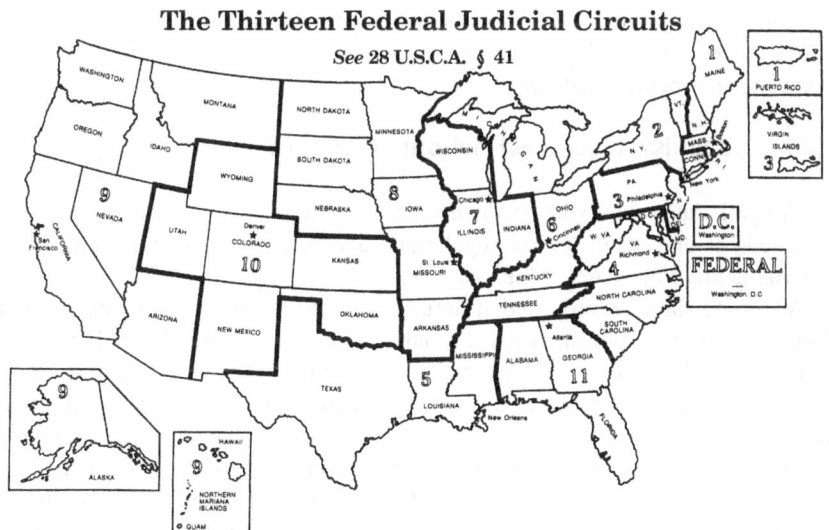

Figure 2.1 Map of 13 U.S. federal judicial districts. (Map Credit — West Publishing Co., St. Paul, Minnesota.)

system. The basis of error must be given and argued by the attorneys representing the appellant.

2.2.5 State Courts

The lowest state trial court of concern for this course is the superior court — in many states called the circuit court, or in Pennsylvania and Ohio, called the court of common pleas — all of which are generally at the county level, that is, one court per county. By comparing the superior court with the federal system, it is the same level as the U.S. District Court. Appeals from the superior and circuit courts, et al. are made to a state appeals court, depending on the subject matter, and generally the final appeal is to a state supreme court, or one of equal standing. The state names vary for appeals courts and supreme courts, for example, in Maine and Massachusetts, they are supreme judicial courts. State supreme court decisions are final unless there is a federal question involved. Then a further appeal may be made to the U.S. Supreme Court. Publication of lower state court decisions is generally not made for superior court level, but only after an appeal decision has been rendered.

2.2.6 Statutory Law vs. Administrative Law

Statutory law governs and limits the powers and procedures of administrative agencies of government. Administrative law is a subdivision of statutory law and is generally nonlegislative and nonjudicial. It is basically procedural in nature and provides the machinery by which an agency implements statutory law. It also places limitations or restraints on the government agencies. **Rule making** by an agency is one of the administrative procedures of great concern in implementing administrative and environmental law. The federal government and most states have an administrative procedure act (APA) that controls administrative law courts, the governmental agencies, and their procedures.

2.2.7 Administrative Law/Environmental Law

Administrative law will be involved to a large extent because we are dealing with federal and state agency regulations and rules, for example, the National Environmental Policy Act (NEPA).

When legislation is proposed, it starts as a bill in Congress or in a state legislature. When passed by the legislature, it becomes an act or a statute. *Act* and *statute* may be used interchangeably. Administrative acts are those that are necessarily carried out by an agency to implement or to effectuate the legislative policies and purposes declared by the legislative body in the creation of the act or statute. The act empowers the particular agency with certain powers to implement and carry out the act as intended by the legislature. Thus, when NEPA was created and enacted by the U.S. Congress in 1969, it also authorized the formation of an agency, the EPA, to carry out the purposes of NEPA envisioned by Congress by giving this agency the power to make regulations and promulgate rules. Hence, administrative law includes that body of law created for administrative agencies in the form of regulations, rules made by rule-making procedure, orders, and decisions.

Because most environmental litigation involves questions of law, and frequently administrative law, the disputes are often between governmental agencies and individuals, corporations, groups of persons in the form of industrial associations, or environmental groups.

The Administrative Procedure Act (APA) (Public Law 89-554; 5 U.S.C. 701 et seq.) serves to outline and guide administrative law procedures for federal agencies to implement federal statutes. Many of the states have patterned their state APAs after the federal model for state agencies. For example, in Montana, state agencies are governed by the Montana Administrative Procedure Act (MAPA). Under MAPA, rules are either **procedural** or **substantive**. (**Procedural rules** merely describe the organization, procedure, and practice requirements of the agency involved. They describe the forms to be used and instructions to be followed in communication with the agency.)

There are a few agencies exempt from the MAPA application, such as the state penal system; the board of regents and the university system; and the financing, construction, and maintenance of public works. Generally, MAPA applies to most agencies.

2.2.8 Agency Rule Making

Procedural rules: (1) do not have the force of law; (2) do not require any authorization other than their original enactment; and (3) are not adopted pursuant to the **rule-making** procedures.

Substantive agency rules may be **legislative** or **interpretive**. The **legislative rules** are equivalent to statutes. They control the rights, duties, and obligations of certain regulated activities (such as mining and landfills) or regulated groups (sellers of alcohol). Violations of legislative rules may result in an assessment of a fine or penalty. Legislative rules: (1) have the force of law; (2) can be adopted only if the specific law administered by the agency expressly authorizes their adoption; and (3) must be adopted according to the rule-making procedure set forth by the APA. Examples of legislative rules are ventilation and sanitary rules for a mine change house and public schools or highway regulations governing gross vehicle weights, axle weights, and load permits.

Interpretive or **adjective rules** merely clarify an agency interpretation of the law it administers. Interpretive rules (1) do not have the force of law; and (2) must be adopted by the rule-making process. Examples are rules defining the terms used in implementing the state's environmental policy act or a rule defining a mineral processing establishment.

Administrative agency rules that are promulgated by an agency by the rule-making process involve (e.g., MAPA):

1. **Notice:** An agency must give prior (30 days) published written notice in some state publication, such as the *Montana Administrative Register* or, for federal rule making, in the *Federal Register.* In Montana, for example, an agency must publish its intent to adopt, amend, or repeal any rule. It must state the intended action, the reasons, and the mode for any persons to present their views. Interested parties must receive 20 days of notice of a public hearing and 28 days of notice to submit data, views, or arguments (orally or written).
2. **Comment:** Comment by the public or interested parties can be in written form or by requested hearing, which requires 10% of, or a minimum of 25 persons, affected. The agency may also request a hearing. Hearings are informal and nonadversarial (supposedly), and recorded, but not under oath; representation is by counsel, if desired, and cross-examination is normally not permitted.
3. **Final report**: On adoption of a rule, the agency must issue a final statement that sets forth the reasons for the adoption and explains why contrary positions were rejected.
4. **Publication:** No newly adopted rule is valid unless it is published within 6 months of publishing of notice of the proposed rule. Rules become effective 10 days after publication in the *Register.* Agency rules must be filed with the Secretary of State. Thus, agencies must follow the procedures for rule making before the rules can be implemented and enforced.

Review of rules, as for example, in Montana, involves MAPA providing four methods: (1) by the agency itself, every 2 years; (2) by legislative review; (3) by petition of an interested party; and (4) by a declaratory judgment of the agency or a court.

2.2.9 Contested Cases

There are procedures for an aggrieved party contesting a decision by an agency's finding of a violation and application of a penalty or fine. The contested case is heard by an agency **hearing officer** or **examiner.** In some states, an administrative law judge (ALJ) is the equivalent of a hearing officer. For a review of a contested case decision to be made, the agency may be petitioned for a review. Most agencies have review boards for that purpose. If the aggrieved party wants an appeal beyond the authority of the agency, it may appeal to a circuit court if a state agency, or to the U.S. District Court if a federal agency. A basic rule in administrative law is that before a court will review a contested agency decision, all administrative remedies must be exhausted; that is, there must be no further administrative hearings available by any review board. Otherwise, the court will dismiss the aggrieved party's claim to be returned to the agency's hearing board. Cases studied in this text will illustrate dismissal by the court because administrative remedies were not exhausted by the party before making an appeal to the courts.

A party who has exhausted all administrative remedies available within the agency and who is aggrieved by a final decision in a contested case is entitled to judicial review. A petition for review may be filed within 30 days after service of the agency's final order.

As professional engineers you may be concerned with the administrative law procedure and the rule-making process because they apply directly to your work. Knowing the process will allow you to participate in the formulation of agency rules that affect and apply to you and your work.

2.3 GENERAL REQUIREMENTS FOR FILING A CIVIL COURT ACTION

A complainant (plaintiff) in filing a civil action must meet certain basic requirements to prevent a dismissal of the action before being heard or tried. Basically, and simplified for the engineer, these requirements are

1. Jurisdiction of the court (both subject matter and parties)
2. Venue of the court
3. Complaint must state a justiciable issue
4. Complaint must have a claim that there is a genuine issue of material fact
5. Plaintiff must have standing to bring the action

For environmental litigation, another requirement must be added, viz., reviewability (i.e., whether the issue to be litigated is subject to judicial resolution).

1. Jurisdiction has been treated earlier in Section 2.2.3.
2. Venue refers to the place where the suit is to be heard. It is distinguished from subject matter jurisdiction of a court, and varies from territorial jurisdiction in that territorial jurisdiction concerns the limits within which a court may have authority and subpoena power over persons.
3. The complaint must be of a justiciable issue (i.e., proper to be examined in a court of justice).
4. The complaint must state the grounds for a valid cause of action showing that the plaintiff has suffered a personal injury, a harm, or a wrong, at the hands of another or by some statute, regulation, or agency rule that is repugnant to existing law. There must be a genuine issue of material fact stated to avoid dismissal of the complaint.
5. Standing to sue doctrine should exist. This is defined in *Black's Law Dictionary* as "the doctrine that an action in a federal constitutional court by a citizen against a government officer, complaining of alleged unlawful conduct where there is no

justiciable controversy unless citizen shows that such conduct invades, or will invade, a private substantive legally protected interest of plaintiff citizen" [Black, 1968, p. 1577].

2.3.1 Bringing an Environmental Administrative Judicial Reviewing Action in Court

Under the APA, § 702 — Right of Review, a party aggrieved by a legal wrong as a result of an agency action is entitled to a judicial review by a federal district court of the agency final order. If the challenged action is for a state agency rule, the judicial review is made in the state circuit or superior court.

Under the APA § 706 — Scope of Review, the reviewing court shall decide all relevant questions of law, interpret constitutional and statutory provisions, and determine the meaning or applicability of the terms of an agency action. The reviewing court shall:

1. Compel agency action unlawfully withheld or unreasonably delayed
2. Hold unlawful and set aside agency action, findings, and conclusions found to be:
   a. Arbitrary, capricious, an abuse of discretion, or otherwise not in accordance with law
   b. Contrary to constitutional right, power, privilege, or immunity
   c. In excess of statutory jurisdiction, authority, or limitations, or short of statutory right
   d. Without observance of procedure required by law
   e. Unsupported by substantial evidence in a case subject to §§ 556 and 557 of this title or otherwise reviewed on the record of an agency hearing provided by statute
   f. Unwarranted by the facts to the extent that the facts are subject to *trial de novo* (**a new trial**) by the reviewing court

As previously stated in Section 2.2.9 on contested cases, a basic rule in administrative law is that before a court will review a contested agency order, all administrative remedies must be exhausted; that is, there must be no further administrative hearings available by any review board. Otherwise, the court will dismiss the aggrieved party's claim to be returned to the agency hearing board. To illustrate this basic administrative law rule, the following case review which appeared as an article in *Pit & Quarry*, bears out the rule:

N.C. "Concerned Citizens" Fail to Stop Quarry Permit*

After Vulcan Materials Company was granted a permit to open a new quarry near Weaverville, North Carolina, a citizens' action group in May 1988 filed a suit to have the permit granting the right to operate a crushed stone quarry declared void. Their failure to void the permit was due to improper procedures.

In *Concerned Citizens v. N.C. DEHNR*, 394 S.E.2d 462 (N.C. App. 1990), the North Buncombe Association of Concerned Citizens (Citizens) in their suit charged: (1) the North Carolina Mining Act violated the U.S. Constitution; (2) the N.C. Department of

* Legal Briefs, *Pit & Quarry*, Jan. 1992, by Legal Editor R. Lee Aston, J.D., LL.M.

Environment, Health, and Natural Resources (DEHNR) had failed to comply with requirements of the Mining Act in issuing a permit to Vulcan; (3) that Buncombe County had
authority to require an Environmental Impact Statement (EIS), and in absence of an EIS,
the permit issued was void; (4) their "due process" rights under the U.S. Constitution's
14th Amendment were violated; and, (5) Vulcan's project was a nuisance.

In August 1987, prior to filing the suit, the Buncombe County Board of Commissioners
held a local hearing. The Citizens alleged and were concerned about Vulcan's failure to
address the effects of dewatering (pumping and drainage, or other removal of impounded
or collected water) in its quarrying operation plans, and its effect on neighboring water
wells. Shortly afterward, Vulcan submitted a private and independent hydrological study
which reported "no adverse impact on neighboring wells."

During 1987 and 1988, the DEHNR had required numerous revisions for Vulcan's
permit application and finally approved it in April 1988, issuing a permit to operate a
crushed stone quarry. Immediately, the Citizens filed their suit against the State and Vulcan
to void the quarrying permit; and a preliminary injunction was issued to halt and restrain
Vulcan from any mining operations on the site.

In December 1988, Vulcan moved for dismissal of Citizens' claims 1, 2, and 3. The
trial court dismissed charges 1 and 2 in favor of Vulcan, but not 3 which required an EIS.
The court declared Vulcan's issued permit "void" and held the claim of a "nuisance" in
abeyance until further order of the court. Vulcan and the State appealed.

The North Carolina Court of Appeals found that, in a dispute between a state agency
and another party which cannot be settled informally, procedures for resolving disputes
are governed by the Administrative Procedure Act (APA). In this case, the Citizens' action
group had failed to exhaust their administrative remedies under the APA before seeking
judicial review. Hence, the trial court lacked subject matter jurisdiction to invalidate the
mining permit. The Appeals Court also found that the DEHNR was not among the state
agencies specifically exempted from the provisions of the APA, and therefore, anyone
challenging decisions made by the director of the DEHNR must follow the guidelines and
provisions of the APA, and may not directly seek court review. The Citizens had an
opportunity to receive formal, evidentiary hearing on the record under the APA in an
appeal from any decision by the Division of Land Resources concerning their rights under
the Mining Act, and thus, the Mining Act was not unconstitutional as it provided the
Citizens with the required "due process."

The Court emphasized that, "As long as statutory procedures provide effective means
for review of an agency action, the courts will require parties to exhaust administrative
remedies before seeking judicial review." It added that, "The claim of failure of the DEHNR
to abide by the requirements of the Mining Act had to be presented by" APA procedures.

The claim of a "nuisance," having been put off by the trial court, was not treated by
the Appeals Court [Aston, 1992].

2.3.2   Standing

Standing requires more attention for environmental matters than the other
requirements for civil actions against government agencies. This federal doctrine
requirement for a plaintiff to bring an action against the government or one of its
agencies has been substantially altered and eased since its earlier origins in federal
constitutional law. Formerly, under the constitutional limitation of Article III, Section

2, where suits in federal courts are confined to cases and controversies, the **injury-in-fact** test was applied to determine a plaintiff's standing to bring the action against a government agency. The plaintiff was required to allege a present, personal, actual injury-in-fact, or threatened injury to his property or a civil right (distinguishable from an injury to the public at large), whereby the governmental action caused a violation or intrusion of his personal interests as protected by the Constitution, statute, or common law. The plaintiff party cannot assert a general complaint or raise an issue that affects another and not himself.

With the advent of NEPA, grievances and, particularly alleged injuries, were more abstract and abstruse, thus, more difficult to show the person's interest or injury required for standing in challenging an agency environmental action. The standing requirement was eased by the APA, which required only that the plaintiff party be adversely affected or aggrieved. In summation, to bring judicial review of a governmental environmental agency order or action by private citizen, corporation, or private citizen environmental group, the plaintiff party must establish standing. Establishment of standing for the plaintiff before the court is critical to maintain the suit and avoid dismissal for lack of standing. Plaintiffs most often are able to establish standing under provisions of federal statutes that allow citizen suits (e.g., the Clean Water Act (CWA), Endangered Species Act, Resource Conservation and Recovery Act, CAA, and the APA).

In a recent citizen suit action brought under the provisions of the CWA, the plaintiff's standing was challenged. In *Friends of the Earth, Inc. v. Laidlaw Environmental Services (TOC), Inc.*, 528 U.S. 167, 120 S. Ct. 693, 145 L. Ed. 2d 610 (2000), Laidlaw was the operator of a hazardous waste incinerator facility. Laidlaw held a National Pollutant Discharge Elimination System (NPDES) permit by a South Carolina state agency. The permit authorized the company to discharge treated water, within limits for several pollutants, into a nearby river. Two environmental organizations alleged the owner's discharges of pollutants had exceeded the limits set by the NPDES permit and filed a citizen suit action against Laidlaw under 505(a) of the CWA in the U.S. District Court for the District of South Carolina. The organizations sought declaratory and injunctive relief and an award of civil penalties. The owner moved for summary judgment on the ground that the organizations had failed to present evidence demonstrating injury in fact and, therefore, lacked standing under the federal Constitution's Article III to bring the suit.

In opposition to this motion, the organizations submitted affidavits and depositional testimony from some members of the organizations who lived near the river and who alleged that the owner's discharges (1) had curtailed those members' recreational use of the river, and (2) would subject those members to other economic and aesthetic harms. In 1993, the district court denied Laidlaw's summary judgment motion on finding that the organizations had standing at the time the action was filed. The district court issued its judgment 4 years later, and the plaintiff's organizations appealed on other grounds. On cross appeal, the defendant Laidlaw argued that the district court erred in its finding that the plaintiffs had standing to bring the suit. The Fourth Circuit Court of Appeals, without deciding the issue of standing, ordered the case be dismissed.

The U.S. Supreme Court granted *certiorari* (review) of the court of appeals decision. On *certiorari*, concerning the issue of standing, the Supreme Court reversed the Court of Appeals judgment and remanded the case for further proceedings. In its opinion, it was held that the organizations had demonstrated sufficient alleged injury in fact, by means of the members' affidavits and testimony, to establish the organizations' standing to bring the suit, even if there was no proof of harm to the environment from the facility owner's alleged violations.

The following case headnotes from the *Laidlaw* opinion are helpful in defining standing:

Headnote: [2A] [2B]
In a citizen suit brought under 505(a) of the Clean Water Act (33 USCS 1365(a)) by three environmental organizations alleging a hazardous waste incinerator facility owner's noncompliance with a National Pollutant Discharge Elimination System permit, the organizations have **standing** to seek civil penalties, where such penalties (1) are for alleged violations that are ongoing at the time of the complaint and that could continue into the future if undeterred; and (2) carry with them a deterrent effect that makes it likely, as opposed to merely speculative, that the penalties will redress the organizations' injuries by abating current violations and preventing future ones. (Scalia and Thomas, J.J., dissented from this holding.)

PARTIES § 22 — citizen suit — **standing** — civil penalties

Headnote: [3A] [3B] [3C] [3D]
Because civil penalties afford redress to citizen plaintiffs who are injured or threatened with injury as a consequence of ongoing unlawful conduct — to the extent that such penalties encourage defendants to discontinue current violations and deter defendants from committing future violations — citizen plaintiffs facing ongoing violations may, under some circumstances, have **standing** to seek civil penalties, although there may be a point at which the deterrent effect of a claim for civil penalties becomes so insubstantial or so remote that such a claim cannot support citizen standing. (Scalia and Thomas, J.J., dissented from this holding.)

COURTS § 762 — mootness — **standing**

Headnote: [13A] [13B] [13C]
A description of the doctrine of mootness as "**the doctrine of standing** set in a time frame" — that is, that the requisite personal interest which must exist at the commencement of the litigation must continue throughout the litigation's existence — is not comprehensive, for (1) there are circumstances in which the prospect that a defendant will engage in or resume harmful conduct may be too speculative to support **standing**, but not too speculative to overcome mootness; and (2) if mootness were simply standing set in a time frame, then the exception to mootness that arises when the defendant's allegedly unlawful activity is "capable of repetition, yet evading review" could not exist. (Scalia and Thomas, J.J., dissented from this holding.)

EVIDENCE § 103 — **standing** — burden

Headnote: [14]
In a lawsuit brought to force compliance with the law, it is the plaintiff's burden to establish **standing** by demonstrating that the defendant's allegedly wrongful behavior, if unchecked by the litigation, will likely occur or continue and that the threatened injury is certainly impending.

PARTIES § 3 — **standing**

Headnote: [17]
Standing doctrine functions to insure, among other things, that the scarce resources of the federal courts are devoted to those disputes in which the parties have a concrete stake.

Under the APA's requirement of **adversely affected** or **aggrieved**, the U.S. Supreme Court established in *Association of Data Processing Service Organizations, Inc. v. Camp*, 397 U.S. 150 (1970), that the plaintiff must show: (1) an injury-in-fact that is (2) within the zone of interest protected by the relevant statute. "This two-part test, having become traditional, must be interpreted in several different areas: first, the nature of the injury sufficient to give the plaintiff a personal interest; second, the required relationship, if any, between the injury alleged and the interest asserted; and, third, the nature of the causal relationship required between the injury alleged and the agency action challenged" (Schoenbaum and Rosenburg, 1991, p. 165).

In Professor Schoenbaum's *Note of Private Enforcement of NEPA*, attention is called to "a series of Supreme Court cases [that] have dealt with the problems [of standing] in environmental cases. In *Sierra Club v. Morton*, 405 U.S. 727 (1972), the court ruled that a plaintiff must allege an actual injury to have standing and that mere interest in a problem, no matter how sincere, is not enough. The court reaffirmed, however, that injuries to aesthetics and environmental well-being, as well as economic well-being, are cognizable under the *Association of Data Processing* test" (Schoenbaum and Rosenburg, 1991, p. 166).

Sierra Club v. Morton became the leading environmental reference case on standing. The brief facts are that in 1965, the U.S. Forest Service (USFS) selected Walt Disney Enterprises, Inc. to prepare a master plan for a recreational resort in the Mineral King Valley adjacent to the Sequoia National Forest. In 1969, USFS approved the Disney plan for a ski resort, which included a $35 million complex of motels, restaurants, and other structures to accommodate an estimated 14,000 visitors per day. A 20-mile long highway and power line easement through the Sequoia National Forest to service motorists and the resort required approval of the USFS. The Sierra Club filed an action seeking an injunction to prevent final approval of the project, and to secure a declaratory judgment that the plan violated federal laws for the preservation of national forests. The USFS argued that Sierra Club lacked standing. The federal district court, however, granted the injunction. On appeal by USFS, the state court of appeals reversed the district court decision and held the injunctive measure in abeyance. On *certiorari*, the U.S. Supreme Court heard the case and affirmed the court of appeals' holding that the Sierra Club lacked standing because neither club nor its members alleged that they were adversely affected by the proposed action.

The deficiency in pleading the complaint to attain standing was a legal technicality. The Sierra Club had merely claimed that the recreational complex would adversely affect the park and impair its enjoyment for future generations. In its decision, the court stated:

Aesthetic and environmental well-being, like economic well-being are important ingredients of the quality of life in our society, and the fact that particular environmental interested are shared by the many rather than the few does not make them less deserving of legal protection through the judicial process (p. 734).

In spite of that pro-standing statement made in the Supreme Court dicta (in the course of its words or speech or opinion), it denied the Sierra Club standing on other grounds. Standing could not be found for Sierra Club merely as a representative for the general public.

Schoenbaum continues,

A short time later in *U.S. v. Students Challenging Regulatory Agency Procedures (SCRAP)*, 412 U.S. 669 (1973), the court held that a group of law students from the Washington, D.C. area who were challenging an ICC order permitting an increase in rail rates applicable to recyclable materials had standing by alleging that its members would be harmed by the ICC action in that it would interfere with its members' use and enjoyment of the forest streams, and other natural resources of the Washington area. The court made it clear that standing is not to be denied because many people suffer the same injury.

In *Duke Power Co. v. Carolina Environmental Study Group, Inc.*, 438 U.S. 59 (1978), a suit which involved a challenge to the constitutionality of a statute that allegedly enabled Duke to construct two nuclear power plants, the court held that the individuals and groups had standing based on some of the injuries alleged, but not on others. The injuries found sufficient to support standing were the thermal pollution of two lakes and the emission of small quantities of non-natural radiation. Other possible injuries the court regarded as questionable included the possibility of a nuclear accident and present apprehension created by that possibility.... Although agency violations of procedural rights granted to plaintiffs by statute have been held to constitute injury-in-fact, e.g., *Defenders of Wildlife, Inc. v. Hodel*, 851 F.2d 1035 (8th Cir. 1988), environmental plaintiffs should be prepared to prove actual use of the affected environment to prevail. See, *e.g.*, *Japan Whaling Association v. American Cetacean Society*, 478 U.S. 221 (1966) where whale watching and studying by members of wildlife conservation groups was held to be adversely affected by the Secretary of Commerce's failure to cite Japan for over-harvesting whales.

In some cases environmental plaintiffs have failed to demonstrate injury with sufficient specificity. In *Wilderness Society v. Griles*, 824 F.2d 4 (D.C. Cir. 1987), the Sierra Club was held to lack standing to challenge the Bureau of Land Management's decision to exclude submerged lands under navigable waters from the total acreage charged to Alaska's land allocations under the Alaska Statehood Act. The plaintiff alleged and demonstrated that its members used various specified federal lands throughout Alaska for recreational and aesthetic purposes, but did not sufficiently specify the land that it intended to use that will be affected by the challenged action. *Id.* at 12. In *Donham v. U.S. Department of Agriculture*, 725 F. Supp. 985 (S.D. Ill. 1989), the plaintiff, who was challenging a proposed timber harvest in a national forest, was held not to have standing because he only alleged a single visit to the area in question (Schoenbaum and Rosenburg, 1991, pp. 166–167).

The federal test for standing has been adopted by the states for suits brought against the states and their agencies.

2.3.3 Reviewability/Judicial Review

The question of whether a court may hear and try an issue brought before it by a party with standing is called **reviewability**. Reviewability of administrative law must be determined before a complainant can be heard in court, even after establishing the required relationship between himself and the environmental issue, or standing. Under the Administrative Procedure Act, 5 U.S.C.A., § 701 provides that the action of every administrative agency is subject to judicial review, except in two instances, viz., (1) where there is a statutory prohibition of the agency actions, and (2) where "agency action is committed to agency discretion by law."

Generally, in environmental law most issues are subject to judicial review because statutory prohibitions for review are rare. Thus, the discretionary actions of agencies are subject to the scrutiny of the courts as to whether their discretionary actions were abused. The standard of review adopted by the reviewing court must be determined next by the court. This involves the extent to which the court will delve into the questions raised by the plaintiff in the complaint. The extent of review is in turn determined by whether the questions raised are those of law, thus making the reviewability easier for the court, or whether the questions are of fact to be determined by the court. In the latter case, a court may find itself in a highly technical or scientific area requiring an expertise that is beyond its knowledge. The determination of facts may be better left to the discretion of the particular agency having the expertise, for example, to establish effluent discharge standards under the CWA. However, the manner, or agency rules, by which those scientific standards are applied to the public are perhaps better determined by the court as a matter of law. Thus, the standard of review is essentially a degree of deference by the court given to an agency in reviewing the action that has been brought before the scrutiny of the court. However, if the case being heard requires adjudication, the review standard for the agency's action to be upheld will be the substantial evidence test. Where the agency's action is a nonadjudicative administrative question, the arbitrary and capricious standard will be used. Although this test's name may seem a rather mild or weak standard, in environmental cases the test has acquired other names, the hard look or the hard scrutiny test. Essentially, to qualify for a finding of arbitrary and capricious, the agency must have strayed afar from its authorized discretionary area.

An early, outstanding and enlightening case involving judicial review of an agency action is that of *Citizens to Preserve Overton Park, Inc. v. Volpe*, 401 U.S. 402 (U.S. 1971) which follows.

Overton Park arose over a petition to have a review of an agency decision by the U.S. Secretary of Transportation (Volpe) where a federal highway construction was approved to pass through Overton Park, a public park near the center of Memphis, Tennessee. The park contained a zoo, a golf course, and numerous recreational facilities. The lower courts found that formal findings of fact were not necessary and that Volpe had not exceeded his authority by the department's informal environmental investigation and findings. The decision was appealed to the U.S. Supreme Court.

The citizens' complaint alleged that the secretary's decision was invalid because it did not reveal any formal findings (e.g., whether there was consideration of alternative routes in its design, the impact and reduction of potential harm to the park, and that Volpe had exceeded his authority). The citizens obtained an injunction to stay construction.

The issues became whether the action was within the discretional scope of a government agency, and whether the action was subject to judicial review.

The Supreme Court held that the agency's action was subject to review; the secretary's action was not one subject to agency discretion. The case was remanded to the lower court to determine whether the secretary had properly investigated alternative routes and relevant environmental impact factors.

This preceding analysis is, in essence, an informal, very concise case brief. If not an acceptable case brief, it at least states what the case stands for. However, reading of a fuller version of the court's decision should give the reader a better understanding of the intended workings of NEPA.

Citizens to Preserve Overton Park, Inc. v. Volpe

Supreme Court of the United States, 1971
401 U.S. 402, 91 S.Ct. 814, 28 L. Ed.2d 136

Opinion of the Court announced by Mr. Justice STEWART.

The growing public concern about the quality of our natural environment has prompted Congress in recent years to enact legislation designed to curb the accelerating destruction of our country's natural beauty. We are concerned in this case with § 4(f) of the Department of Transportation Act of 1966, as amended, and 18(a) of the Federal-Aid Highway Act of 1968, 82 Stat. 823, 23 U.S.C. § 138 (1964 ed. Supp. V) (hereafter, § 138). These statutes prohibit the Secretary of Transportation from authorizing the use of federal funds to finance the construction of highways through public parks if a "feasible and prudent" alternative route exists. If no such route is available, the statutes allow him to approve construction through parks only if there has been "all possible planning to minimize harm" to the park.

Petitioners, private citizens as well as local and national conservation organizations, contend that the Secretary has violated these statutes by authorizing the expenditure of federal funds for the construction of a six-lane interstate highway through a public park in Memphis, Tennessee. Their claim was rejected by the District Court, which granted the Secretary's motion for summary judgement, and the Court of Appeals for the Sixth Circuit affirmed. After oral argument, this Court granted a stay that halted construction and, treating the application for the stay as a petition for certiorari, granted review. (400 U.S. 939). We now reverse the judgement below and remand for further proceedings in the District Court.

Overton Park is a 342-acre city park located near the center of Memphis. The park contains a zoo, a nine-hole municipal golf course, an outdoor theater, nature trails, a bridle path, an art academy, picnic areas, and 170 acres of forest. The proposed highway, which is to be a six-lane, high-speed expressway, will sever the zoo from the rest of the park. Although the roadway will be depressed below ground level except where it crosses a small creek, 26 acres of the park will be destroyed. The highway is to be a segment of Interstate Highway I-40, part of the National System of Interstate and Defense Highways. I-40 will provide Memphis with a major east–west expressway which will allow easier access to downtown Memphis from the residential areas on the eastern edge of the city.

Although the route through the park was approved by the Bureau of Public Roads in 1956 and by the Federal Highway Administrator in 1966, the enactment of § 4(f)of the Department of Transportation Act prevented distribution of federal funds for the section of the highway designated to go through Overton Park until the Secretary of Transportation determined whether the requirements of § 4(f) had been met. Federal funding for the rest of the project was, however, available; and the state acquired a right-of-way on both sides of the park. In April 1968, the Secretary announced that he concurred in the judgement of local officials that I-40 should be built through the park. And in September 1969 the state acquired the right-of-way inside Overton Park from the city. Final approval for the project — the route as well as the design — was not announced until November 1969, after Congress had reiterated in § 138 of the Federal-Aid Highway Act that highway construction through public parks was to be restricted. Neither announcement approving the route and design of I-40 was accompanied by a statement of the Secretary's factual findings. He did not indicate why he believed there were no feasible and prudent alternative routes or why design changes could not be made to reduce the harm to the park.

Petitioners contend that the Secretary's action is invalid without such formal findings and that the Secretary did not make an independent determination but merely relied on the judgement of the Memphis City Council. They also contend that it would be "feasible and prudent" to route I-40 around Overton Park either to the north or to the south. And they argue that if these alternative routes are not "feasible and prudent," the present plan does not include "all possible" methods for reducing harm to the park. Petitioners claim that I-40 could be built under the park by using either of two possible tunneling methods, and they claim that, at a minimum, by using advanced drainage techniques the expressway could be depressed below ground level along the entire route through the park including the section that crosses the small creek.

Respondents argue that it was unnecessary for the Secretary to make formal findings, and that he did, in fact, exercise his own independent judgement which was supported by the facts. In the District Court, respondents introduced affidavits, prepared specifically for this litigation, which indicated that the Secretary had made the decision and that the decision was supportable. These affidavits were contradicted by affidavits introduced by petitioners, who also sought to take the deposition of a former Federal Highway Administrator who had participated in the decision to route I-40 through Overton Park.

The District Court and the Court of Appeals found that formal findings by the Secretary were not necessary and refused to order the deposition of the former Federal Highway Administrator because those courts believed that probing of the mental processes of an administrative decision maker was prohibited. And, believing that the Secretary's authority was wide and reviewing courts' authority narrow in the approval of highway routes, the lower courts held that the affidavits contained no basis for a determination that the Secretary had exceeded his authority.

We agree that formal findings were not required. But we do not believe that in this case judicial review based solely on litigation affidavits was adequate.

A threshold question — whether petitioners are entitled to any judicial review — is easily answered. Section 701 of the Administrative Procedure Act, 5 U.S.C. 701 (1964 ed. Supp. V), provides that the action of "each authority of the government of the United States," which includes the Department of Transportation, is subject to judicial review except where there is a statutory prohibition on review or where "agency action is committed to agency discretion by law." In this case, there is no indication that Congress sought to prohibit judicial review and there is most certainly no "showing of 'clear and convincing evidence' of a [...] legislative intent" to restrict access to judicial review. [*Abbott Laboratories v. Gardner*, 387 U.S. 136, 141 (1967)].

Similarly, the Secretary's decision here does not fall within the exception for action "committed to agency discretion." This is a very narrow exception. Berger, Administrative Arbitrariness and Judicial Review, 65 Col. L. Rev. 55 (1965). The legislative history of the Administrative Procedure Act indicates that it is applicable in those rare instances where "statutes are drawn in such broad terms that in a given case there is no law to apply." S. Rep. No. 752, 79th Cong., 1st Sess., 26 (1945).

Section 4(f) of the Department of Transportation Act and § 138 of the Federal-Aid Highway Act are clear and specific directives. Both the Department of Transportation Act and the Federal-Aid to Highway Act provide that the Secretary "shall not approve any program or project" that requires the use of any public park land "unless (1) there is no feasible and prudent alternative to the use of such land, and (2) such program includes all possible planning to minimize harm to such park...." This language is a plain and explicit bar to the use of federal funds for construction of highways through parks — only the most unusual situations are exempted.

Despite the clarity of the statutory language, respondents argue that the Secretary has wide discretion. They recognize that the requirement that there be no "feasible" alternative route admits of little administrative discretion. For this exemption to apply the Secretary must find that as a matter of sound engineering it would not be feasible to build the highway along any other route. Respondents argue, however, that the requirement that there be no other "prudent" route requires the Secretary to engage in a wide-ranging balancing of competing interests. They contend that the Secretary should weigh the detriment resulting from the destruction of park land against the cost of other routes, safety considerations, and other factors, and determine on the basis of the importance that he attaches to these other factors whether, on balance, alternative feasible routes would be "prudent."

But no such wide-ranging endeavor was intended. It is obvious that in most cases considerations of cost, directness of route, and community disruption will indicate that park land should be used for highway construction whenever possible. Although it may be necessary to transfer funds from one jurisdiction to another, there will always be a smaller outlay required from the public purse when park land is used since the public already owns the land and there will be no need to pay for right-of-way. And since people do not live or work in parks, if a highway is built on park land no one will have to leave his home or give up his business. Such factors are common to substantially all highway construction. Thus, if Congress intended these factors to be on an equal footing with preservation of park land there would have been no need for the statutes.

Congress clearly did not intend that cost and disruption of the community were to be ignored by the Secretary. But the very existence of the statutes indicates that protection of park land was to be given paramount importance. The few green havens that are public parks were not to be lost unless there were truly unusual factors present in a particular case or the cost or community disruption resulting from alternative routes reached extraordinary magnitudes. If the statutes are to have any meaning, the Secretary cannot approve the destruction of park land unless he finds that alternative routes present unique problems.

Plainly, there is "law to apply" and thus the exemption for action "committed to agency discretion" is inapplicable. But the existence of judicial review is only the start: the standard for review must also be determined. For that we must look to § 706 of the Administrative Procedure Act, 5 U.S.C. § 706.... [A] "reviewing court shall ... hold unlawful and set aside agency action, findings, and conclusions found" not to meet six separate standards. In all cases agency action must be set aside if the action was "arbitrary, capricious, an abuse of discretion, or otherwise not in accordance with law" or if the action failed to meet statutory, procedural, or constitutional requirements.... In certain narrow, specifically limited situations, the agency action is to be set aside if the action was not supported by "substantial evidence." And in other equally narrow circumstances the reviewing court is to engage in a de novo review of the action and set it aside if it was "unwarranted by the facts" ...

Petitioners argue that the Secretary's approval of the construction of I-40 through Overton Park is subject to one or the other of these latter two standards of limited applicability. First, they contend that the "substantial evidence" standard of § 706(2)(E) must be applied. In the alternative, they claim that § 706(2)(F) applies and that there must be a de novo review to determine if the Secretary's action was "unwarranted by the facts." Neither of these standards is, however, applicable.

Review under the substantial-evidence test is authorized only when the agency action is taken pursuant to a rule making provision of the Administrative Procedure Act itself, 5 U.S.C.§ 553 ... or when the agency action is based on a public adjudicatory hearing. See 5 U.S.C.§§ 556, 557.... The Secretary's decision to allow the expenditure of federal funds to build I-40 through Overton Park was plainly not an exercise of a rule making function.... And the only hearing that is required by either the Administrative Procedure Act or the statutes regulating the distribution of federal funds for highway construction is a public hearing conducted by local officials for the purpose of informing the community about the proposed project and eliciting community views on the design and route. 23 U.S.C.§ 128.... The hearing is non-adjudicatory, quasi-legislative in nature. It is not designed to produce a record that is to be the basis of agency action — the basic requirement for substantial-evidence review....

Petitioner's alternative argument also fails. De novo review of whether the Secretary's decision was "unwarranted by the facts" is authorized by § 706(2)(F) in only two circumstances. First, such de novo review is authorized when the action is adjudicatory in nature and the agency fact-finding procedures are inadequate. And, there may be independent judicial fact-finding when issues that were not before the agency are raised in a proceeding to enforce non-adjudicatory agency action. H.R. Rep. No. 1980, 79th Cong., 2d Sess. Neither situation exists here.

Even though there is no de novo review in this case and the Secretary's approval of the route of I-40 does not have ultimately to meet the substantial-evidence test, the generally applicable standards of § 706 require the reviewing court to engage in a substantial inquiry. Certainly, the Secretary's decision is entitled to a presumption of regularity.... But that presumption is not to shield his action from a thorough, probing, in-depth review.

The court is first required to decide whether the Secretary acted within the scope of his authority.... This determination naturally begins with a delineation of the scope of the Secretary's authority and discretion. L. Jaffe, Judicial Control of Administrative Action 359 (1965). As has been shown, Congress has specified only a small range of choices that the Secretary can make. Also involved in this initial inquiry is a determination of whether on the facts the Secretary's decision can reasonably be said to be within that range. The reviewing court must consider whether the Secretary properly construed his authority to approve the use of park land as limited to situations where there are no feasible alternative routes or where feasible alternative routes involve uniquely difficult problems. And the reviewing court must be able to find that the Secretary could have reasonably believed that in this case there are no feasible alternatives or that alternatives do involve unique problems.

Scrutiny of the facts does not end, however, with the determination that the Secretary has acted within the scope of his statutory authority. Section 706(2)(A) requires a finding that the actual choice made was not — "arbitrary, capricious, an abuse of discretion, or otherwise not in accordance with law." ... To make this finding the court must consider whether the decision was based on a consideration of the relevant factors and whether there has been a clear error of judgement.... Although this inquiry into the facts is to be searching and careful, the ultimate standard of review is a narrow one. The court is not empowered to substitute its judgement for that of the agency.

The final inquiry is whether the Secretary's action followed the necessary procedural requirements. Here the only procedural error alleged is the failure of the Secretary to make formal findings and state his reason for allowing the highway to be built through the park.

Undoubtedly, review of the Secretary's action is hampered by his failure to make such findings, but the absence of formal findings does not necessarily require that the case be remanded to the Secretary. Neither the Department of Transportation Act nor the Federal-Aid Highway Act requires such formal findings. Moreover, the Administrative Procedure Act requirements that there be formal findings in certain rule making and adjudicatory proceedings do not apply to the Secretary's action here. See 5 U.S.C.§§ 553(a)(2),554(a). And, although formal findings may be required in some cases in the absence of statutory directives when the nature of the agency action is ambiguous, those situations are rare. Plainly, there is no ambiguity here; the Secretary has approved the construction of I-40 through Overton Park and has approved a specific design for the project.

Thus it is necessary to remand this case to the District Court for plenary review of the Secretary's decision. That review is to be based on the full administrative record that was before the Secretary at the time he made his decision. But since the bare record may not disclose the factors that were considered or the Secretary's construction of the evidence, it may be necessary for the District Court to require some explanation in order to determine if the Secretary acted within the scope of his authority and if the Secretary's action was justifiable under the applicable standard.

The court may require the administrative officials who participated in the decision to give testimony explaining their action. Of course, such inquiry into the mental processes of administrative decision makers is usually to be avoided. And where there are administrative findings that were made at the same time as the decision, there must be a strong showing of bad faith or improper behavior before such inquiry may be made. But here there are no such formal findings and it may be that the only way there can be effective judicial review is by examining the decision makers themselves.

The District Court is not, however, required to make such an inquiry. It may be that the Secretary can prepare formal findings that will provide an adequate explanation for his action. Such an explanation will, to some extent, be a "*post hoc* rationalization" and thus must be viewed critically. If the District Court decides that additional explanation is necessary, that court should consider which method will prove the most expeditious so that full review may be had as soon as possible.

Reversed and remanded.

2.4 BRIEFS

There are two types of **briefs**, viz., a **legal brief** and a **case brief**. The **legal brief** is the legal argument that attorneys give and submit to the court in their pleadings at trial. It is based on points of law and precedent case decisions that are relevant to the case at issue.

The legal brief is in the way of a summary of a decided case, prepared for trial judges, principally to summarize what the case and its decision stands for. An attorney may use the legal brief to cite to the court case law in support of his or her argument. In reviewing cases in this text for various environmental topics, a court decision will be given, which will cite the full text of a case (i.e., the facts; the issues and questions for the court to decide; the parties' legal arguments; and the court analysis, opinions, conclusions, and its decision). From an engineer's viewpoint, the main concern is with the background facts that created the environmental situation issues, the questions to be decided, the principal arguments concerning points of law in question, and the decision based on conclusions of law; these will matter and be studied. The learning process should include case handouts where the student will be required to write a case brief, or summary. In reviewing the case, the student should weed out the legal jargon and legal procedures, and reduce the case information to the practical results of the decision that are applicable for an understanding as to how the decision applies to the environmental work from an engineering and scientific viewpoint. The engineer, as well as the attorney, should know and be able to express in a few sentences, or at least a short paragraph, what the case and decision stand for.

The **case brief** is in the way of a total summary of a case. In reviewing cases for the various environmental topics, a court's decision will be given that will give the highlights of a case (i.e., the background history or facts; the issues and questions for the court to decide; the parties' legal arguments; and the court's analysis, opinions, conclusions, and its decision).

The first illustrative example of a case brief for engineers using the suggested case brief form is given later in Chapter 4, Section 4.2.4 about New Source Performance Standards (NSPS). The brief form of the case is stated under the suggested headings of: (1) nature of litigation, (2) background facts, (3) issues, (4) arguments, (5) court's decision, and, (6) conclusions.

Cases for study generally are taken from the West Publishing Company's various case reporters. Case reporters may be found in any university law school library. County courthouses usually have a law library with state cases and occasionally West's federal reporters. Federal agency offices frequently have law libraries containing case reporters.

2.4.1   Case Reporters

For the federal system of published cases, the West Group publishes weekly case decisions in paperback advanced copies for the U.S. Supreme Court in the *Supreme Court Reports*; for the U.S. Courts of Appeal in the *Federal Reporter*; and for the U.S. District Court decisions in the *Federal Supplement*. The paperback advanced weekly copies are later bound into hardback books.

In West's regional system of state case reporting system, cases from the lower trial courts, the circuit and superior courts, are not published. West publishes appealed decisions of the lower, trial courts made to state courts of appeal and to the state supreme courts in a series of reporters grouped according to the geographic location of the states. West publishes seven state case reporters, viz., Atlantic,

Northeastern, Southeastern, Southern, Southwestern, Northwestern, and Pacific reporters. For example, Nevada and Oklahoma State decisions are found in the *Pacific Reporter*; the *Pacific Reporter* covers most of the Rocky Mountain states; Arkansas, Missouri, Kentucky, Tennessee, and Texas decisions are in the *Southwest Reporter*; the New England states and southward through Pennsylvania are in the *Atlantic Reporter*; below the Mason and Dixon Line, Maryland, West Virginia, Virginia, southward to South Carolina are in the *Southeastern Reporter*; the *Southern Reporter* covers the Deep South, Georgia, Florida, and westward to Louisiana. Michigan is included in the *Northwest Reporter*; Illinois in the *Northeastern Reporter* along with Indiana. West publishes state cases for California and New York in separate reporters. All states also have their own decisions reported and published in bound volumes, but not by West.

West Publishing Company has a **keyword** system that helps locate subject matter quickly. For example, if you are interested in finding a case on mining, look up the keywords *mines* and *minerals*. Cases dealing with minerals will be listed. If looking for cases involving the environment, look up the keywords *health* and *environment*; for litigated cases on zoning, see *zoning*; etc. Each reporter volume has its keyword section. Current cases in weekly issues are softbound, called supplements, until the end of the court annual session, and then they are hardbound into volumes.

2.4.2   Finding a Case — Style and Citation of Cases

Examples of case style and citations are *Shiny Rock Mining Co. v. U.S.*, 906 F.2d 1362 (9th Cir. 1990); *Parker v. Reynolds Metals Co.*, 747 F. Supp. 71 (M.D. Ga. 1990); *McDonald v. Snyder Construction Co.*, 744 S.W. 2d 550 (Mo. App. 1988); *Georgia Marble Co. v. Whitlock*, 392 S.E. 2d 881 (Ga. 1990), 260 Ga. 350 (the latter is the state citation); *File v. D. & L. Landfill, Inc.*, 579 N.E.2d 1228 (Ill. App. 5 Dist. 1991); *Simpson Tacoma Kraft v. Department of Ecology*, 835 P.2d 1030 (Wash. 1992); *New York Coastal Fishermen's Assn. v. New York City Department of Sanitation*, 772 F. Supp. 162 (S.D. N.Y. 1991).

2.4.3   Explanation of Citations

The citation (**cite**) gives the **style** of the case in the names of the parties, with the **plaintiff's** name being given first (**plaintiff** is the party bringing the complaint and action), versus (*v.*) the **defendants** of the suit. It is customary to give the style of the case (the parties' names) in italics, or to underline the parties' names.

It should be noted that there are variations of terms in different court systems whereby plaintiff and defendant may be called by different terms (e.g., the plaintiff may also be called the **complainant**, and the defendant also called the **respondent**). Variations of terms for the parties also occur at the appeals level. In federal courts, the parties are termed appellants and appellees. In state court appeals, the appealing parties may be termed (e.g., as **movant** [in Kentucky] and **respondent,** or **appellant** and **appellee**; or in Indiana as the **appellant-respondent, appellee-petitioner**, depending on which party makes the appeal; and in Pennsylvania, simply as **petitioner** and **respondent**).

After the style, the cite is given indicating where it may be found. Explanations for the cases cited earlier follow:

1. *Shiny Rock Mining Co. v. U.S.,* 906 F.2d 1362 (9th Cir. 1990): Shiny Rock Mining Company is the plaintiff suing the U.S. government in federal court. The federal court citation shows that this case report is one that was appealed from the federal district court to the Ninth Circuit U.S. Court of Appeals. Without looking up the case, the state it originated in is unknown, but it is from one of the Rocky Mountain states in the Ninth Judicial Circuit. The appeal was decided in 1990. The case is found in West's *Federal Supplement* (reports for U.S. Courts of Appeal), Second Series (2d), in volume 906 beginning on page 1362.

2. *Parker v. Reynolds Metals Co.,* 747 F. Supp.71 (M.D. Ga. 1990): Parker, plaintiff brought the action against Reynolds Metals Co. in the federal district court for the Middle District of Georgia; the case was decided in 1990. The text of the case and decision is to be found in West's *Federal Supplement* (Federal District Court Reporter), Second Series (2d), in volume 747, starting on page 71.

3. *McDonald v. Snyder Construction Co.,* 744 S.W. 2d 550 (Mo. App. 1988): McDonald was the plaintiff instituting the suit against the defendant, Snyder Construction Co. The text of the case is reported in volume 774 of the *Southwestern Reporter* (S.W.), Second Series (2d), beginning on page 550. It is a state court case before the Missouri Court of Appeals (Mo. App.) and was decided in 1988.

4. *Georgia Marble Co. v. Whitlock,* 392 S.E. 2d 881 (Ga. 1990), 260 Ga. 350: Georgia Marble Co. is the plaintiff and Whitlock, the defendant party. The suit is a state court action (Ga. 1990) having been decided by the Supreme Court of Georgia (as indicated by Ga. Had the case been decided by the Georgia Court of Appeals, the citation would have read Ga. App.). The decision was handed down in 1990. The text of the case is found in the West Publishing Co. *Southeastern Reporter* (S.E.), Second Series, in volume 392, beginning on page 881. A second citation is given for this case for a state reporter series by another publisher of Georgia cases, titled *Georgia Supreme Court Reports,* and found in volume 260, beginning on page 350.

5. *File v. D. & L. Landfill, Inc.,* 579 N.E. 2d 1228 (Ill. App. 5 Dist. 1991): The file is the plaintiff in this action against D & L. Landfill, the defendant in an Illinois state court of appeals suit. The text of the case is found in the *Northeastern Reporter* (N.E.), Second Series (2d), in volume 579, beginning on page 1228. The appeal was heard by the Fifth District Illinois Court of Appeals in 1991.

6. *New York Coastal Fishermen's Association v. New York City Department of Sanitation,* 772 F. Supp. 162 (S.D. N.Y. 1991). The plaintiff, New York Coastal Fishermen's Association, brought the action against the New York City Department of Sanitation as defendant. This action was brought in a federal district court as indicated by being published in the *Federal Supplement* (F. Supp.). The federal district court hearing the action was in the Southern District of New York (S.D. N.Y.) in 1991. The case text is found in volume 772 of the *Federal Supplement,* beginning on page 162.

Figure 2.2 illustrates the cover of a soft-bound advance copy of one of the regional reporters, the *Southwestern Reporter.* Figure 2.3 is an example of a title page of a *Southwestern Reporter* listing the states covered in that regional reporter. Figures 2.4 and 2.5, respectively, give examples of keywords used to locate cases under the keywords, *mines* and *minerals* and *health* and *environment.*

West's

SOUTH WESTERN REPORTER

Third Series

A Unit of the National Reporter System

6 S.W.3d No. 2   January 18, 2000   Pages 74-371

Missouri–Court Rules

—

See Table of Contents

Figure 2.2   Cover of one of the West Publishing Company's advanced copy regional law reporters — the *South Western Reporter*. (Photo Credit — West Publishing Co., St. Paul, Minnesota.)

West's
SOUTH WESTERN
REPORTER

Third Series

A Unit of the National Reporter System

Volume 5 S.W.3d

Cases Argued and Determined
in the Courts of

<table>
<tr><td>**ARKANSAS**</td><td>**MISSOURI**</td></tr>
<tr><td>**KENTUCKY**</td><td>**TENNESSEE**</td></tr>
<tr><td colspan="2" align="center">**TEXAS**</td></tr>
</table>

ST. PAUL, MINN.

WEST GROUP

2000

Figure 2.3 Title page of the *South Western Reporter*. (Photo Credit — West Publishing Co., St. Paul, Minnesota.)

☞441.1 MENTAL HEALTH (Photo Credit - West Publishing Co., St. Paul, Minn.)

commitment had long been considered a civil solution to mental health problems. A.R.S. § 36-3701 et seq.—Id.

Mere uncertainty as to term of commitment under Sexually Violent Persons (SVP) Act did not render commitment punitive and criminal rather than civil or regulatory. A.R.S. §§ 36-3701 et seq., 36-3707 to 36-3709.—Id.

Sexually Violent Persons (SVP) Act might not serve goal of general deterrence and it was not retributive because criminal conviction was not prerequisite for commitment but instead was used for evidentiary purposes, as factors supporting finding Act was civil or regulatory rather than punitive and criminal. A.R.S. §§ 36-3701 et seq., 36-3701, subd. 7(a), 36-3704, subd. B.—Id.

Sexually Violent Persons (SVP) Act's purposes of protecting the public and treating sexually violent persons so they may return to society were non-punitive, as element supporting finding Act was civil or regulatory rather than punitive and criminal. A.R.S. § 36-3701 et seq.—Id.

Restraint resulting from application of Sexually Violent Persons (SVP) Act was not excessive in relation to Act's non-punitive purposes of protecting the public and treating sexually violent persons so they may return to society, as element supporting finding Act was civil or regulatory rather than punitive and criminal. A.R.S. §§ 36-3701 et seq., 36-3707, subd. B, 36-3712.—Id.

Commitment pursuant to Sexually Violent Persons (SVP) Act was for treatment and protection of the public rather than punishment, and commitment was not a prosecution but instead a civil proceeding, and thus, Act did not violate Double Jeopardy protections. U.S.C.A. Const.Amend. 5; A.R.S. Const. Art. 2, § 10; A.R.S. § 36-3701 et seq. —Id.

Requiring sexually violent persons (SVP), but not those subject to general civil commitment, to wear distinguishing clothing in state hospital and to be held in separate section of the hospital satisfied rational basis equal protection test; sexually violent persons suffered from particular mental condition that rendered them likely to commit violent sexual acts, and the procedures helped ensure safety and might prevent escape. U.S.C.A. Const.Amend. 14; A.R.S. Const. Art. 2, § 13; A.R.S. § 36-3701 et seq. —Id.

Wash.App. Div. 2 1999. Convicted sex offender does not have statutory duty to give written notice of a change of address at least fourteen days before moving, where the offender does not know fourteen days in advance that he will be moving and he has no known residence into which to move. West's RCWA 9A.44.130(4)(a) (1997).—State v. Bassett, 987 P.2d 119.

For purposes of convicted sex offender registration statute's requirement that convicted offenders file written notice of a change in residence address, living on the streets, homelessness, does not constitute a "residence." West's RCWA 9A.44.130(4)(a) (1997).—Id.

Evidence was insufficient to support conviction for failing to register as a convicted sex offender, where defendant did not know 14 days in advance that he would be "evicted" from his sister's apartment, defendant was homeless for 11-day-period following eviction, and defendant had been in hotel for only hours before his arrest. West's RCWA 9A.44.130(4)(a) (1997). —Id.

☞442. —— Persons and offenses included.

Colo.App. 1999. Defendant, who pled guilty to contributing to the delinquency of a minor, was a "sex offender," within meaning of sex offender

statute, even though offense was not an enumerated sex offense, where defendant had a "history" of enumerated sex offenses, requiring defendant to register as a sex offender. West's C.R.S.A. § 16-11.7-102(3). People v. Meidinger, 987 P.2d 937.

Defendant was required to undergo mandatory sex offender evaluation before trial court could impose sex offender conditions as part of her probation following her conviction for contributing to the delinquency of a minor. West's C.R.S.A. § 16-11.7-104.—Id.

☞446. —— Hearing and examination.

Ariz.App. Div. 1 1999. Repealed provision of Sexually Violent Persons (SVP) Act stating that "all constitutional rights available to criminal defendants except the right not to be tried while incompetent .. apply to the [commitment] hearing" did not provide a right to prehearing release on bond, but instead applied only to rights at the hearing itself. A.R.S. Const. Art. 2, § 22; A.R.S. § 13-4606(D) (Repealed).—Martin v. Reinstein, 987 P.2d 779.

Colo.App. 1999. Defendant was required to undergo mandatory sex offender evaluation before trial court could impose sex offender conditions as part of her probation following her conviction for contributing to the delinquency of a minor. West's C.R.S.A. § 16-11.7-104.—People v. Meidinger, 987 P.2d 937.

MINES AND MINERALS

I. PUBLIC MINERAL LANDS.

(B) LOCATION AND ACQUISITION OF CLAIMS.

☞38(6). Jurisdiction.

Or.App. 1999. Federal regulation promulgated under the General Mining Law of 1872 does not deprive Oregon courts of concurrent jurisdiction to determine possessory rights to unpatented mining claims on federal land. 30 U.S.C.A. § 22 et seq.; 43 C.F.R. § 4.450-1.—Gibson v. Pacific Summa Capital, Inc., 987 P.2d 1240, 163 Or.App. 321.

III. OPERATION OF MINES, QUARRIES, AND WELLS.

(A) STATUTORY AND OFFICIAL REGULATIONS.

☞87. Licenses; severance and production taxes.

Wyo. 1999. County could present to the State Board of Equalization allegations that natural gas extractor's production valuation from past tax years caused the production to be improperly or unequally assessed for purposes of severance and ad valorem taxes, even though the county did not appeal the valuation certifications for those years to the Board within 30 days. Wyo.Stat.Ann. § 39-11-102.1(c)(x).—Exxon Corp. v. Board of County Com'rs, Sublette County, 987 P.2d 158.

MONOPOLIES

II. TRUSTS AND OTHER COMBINATIONS IN RESTRAINT OF TRADE.

☞12(16.5). —— Efforts to influence governmental action; litigation.

Okla.Civ.App. Div. 4 1999. Under the "Noerr-Pennington doctrine," a party's attempt to influence enforcement or passage of legislation is not a

Figure 2.4   Example of keywords from West Publishing Company's reporters for locating cases listed under keywords *mines* and *minerals*. (Photo Credit — West Publishing Co., St. Paul, Minnesota.)

(Photo Credit - West Publishing Co., St. Paul, Minn.)   **HOMICIDE** ⊜⊐**34**

HEALTH AND ENVIRONMENT

II. REGULATIONS AND OFFENSES.

⊜⊐**25.5(2). Constitutional and statutory provisions.**

Ill.App. 1 Dist. 1993. Amendment to Environmental Protection Act, changing eligibility requirements for reimbursement from Underground Storage Tank Fund, did not apply retroactively; tank owner was eligible for reimbursement from Underground Storage Tank Fund by notifying state agencies of underground storage tank leaks, and embarking upon remediation required by statute and rules then in effect. S.H.A. 415 ILCS 5/1 et seq.; 415 ILCS 5/22.13(a), (a)(5) (1992 Bar Ed.).—Chemrex, Inc. v. Pollution Control Bd., 195 Ill.Dec. 499, 628 N.E.2d 963.

⊜⊐**25.5(7). —— Nuclear projects; radioactivity.**

Ill.App. 1 Dist. 1993. Specific public policy behind Radiation Protection Act is to prohibit and prevent exposure to ionizing radiation in amounts which are or may be detrimental to health. S.H.A. 420 ILCS 40/1 et seq.—Adco Services, Inc. v. Bullard, 195 Ill.Dec. 308, 628 N.E.2d 772.

Purpose of the Central Midwest Radioactive Waste Compact Act is to provide facilities for proper management of low-level radioactive waste generated in the region and to protect the health and safety of citizens of the region. S.H.A. 45 ILCS 140/1.—Id.

⊜⊐**25.5(9). Administrative boards and proceedings.**

Ill.App. 1 Dist. 1993. Among the duties of the Central Midwest Interstate Low-Level Radioactive Waste Commission (CMILRWC) is the obligation to adopt regional management plan designed to ensure a safe and efficient management of waste generated within the region. S.H.A. 45 ILCS 140/1.—Adco Services, Inc. v. Bullard, 195 Ill.Dec. 308, 628 N.E.2d 772.

⊜⊐**25.15(3). —— Preliminary injunction.**

Ind.App. 5 Dist. 1994. County solid waste district and board of commissioners had adequate remedy at law in form of money judgment in their action against landfill owners to recover unpaid solid waste planning fees, so that they were not entitled to preliminary injunction prohibiting owners from accepting additional solid waste, notwithstanding contention that remedy was illusory in that, if owners continued accepting waste, landfill would be full and worthless as asset; there was no evidence of value of landfill once it reached capacity, evidence as to whether owners had additional assets to satisfy any judgment, or evidence that owners ever displayed any intention of liquidating their assets to prevent district and board from recovering.—T.H. Landfill Co., Inc. v. Miami County Solid Waste Dist., 628 N.E.2d 1237.

HOMICIDE

I. THE HOMICIDE.

⊜⊐**5. Cause of death.**

Ill.App. 1 Dist. 1993. To establish second-degree murder, state need not prove that defendant's acts were sole and immediate cause of victim's death, but rather that they contributed to victim's death. Ill.Rev.Stat.1989, ch. 38, ¶ 9-2.—People v. Conde, 194 Ill.Dec. 899, 628 N.E.2d 396.

II. MURDER.

⊜⊐**7. Nature and elements in general.**

Ill.App. 1 Dist. 1993. State of mind for murder is knowledge. S.H.A. 720 ILCS 5/9-1(a).—People v. Lee, 194 Ill.Dec. 939, 628 N.E.2d 436.

⊜⊐**8. Statutory provisions.**

N.Y. 1993. Legislature's failure to include extreme emotional disturbance defense in depraved mind murder statute did not raise due process concerns, as People remained responsible for proving fundamental elements of mens rea and actus reus beyond reasonable doubt. McKinney's Penal Law § 125.25, subd. 2; U.S.C.A. Const.Amends. 5, 14.—People v. Fardan, 607 N.Y.S.2d 220, 82 N.Y.2d 638, 628 N.E.2d 41.

⊜⊐**9. Intent or design to effect death.**

Ill.App. 1 Dist. 1993. To prove accused possessed requisite mental state for murder, state may show that accused performed voluntary and willful act which has natural tendency to cause death or great bodily harm.—People v. Conde, 194 Ill.Dec. 899, 628 N.E.2d 396.

Ind.App. 1 Dist. 1994. Intent element distinguishes involuntary manslaughter, battery, and criminal recklessness from murder; hence, lesser included offense instruction will be warranted with respect to charged crime of murder only if there is serious evidentiary dispute about what defendant intended to do, kill, batter, or simply act in such a way as to create a substantial risk of injury. West's A.I.C. 35–42–1–1(1), 35–42–1–4, 35–42–2–1, 35–42–2–2(b).—Simpson v. State, 628 N.E.2d 1215.

⊜⊐**23(1). In general.**

Ill.App. 1 Dist. 1993. Category of "serious provocation" involving mutual quarrel or combat, for purpose of reducing first-degree murder to second-degree murder, did not apply where defendant and victim did not both enter altercation willingly; rather, victim initially accosted defendant by brandishing weapon and slapping defendant, and defendant later continued confrontation by forcibly removing victim from restaurant and pushing him up against wall. Ill.Rev.Stat.1989, ch. 38, ¶ 9-2(b).—People v. Jefferson, 195 Ill.Dec. 461, 628 N.E.2d 925.

⊜⊐**27. Insanity.**

N.Y. 1993. Extreme emotional disturbance is not affirmative defense to charge of depraved mind murder; rather, is only affirmative defense to charge of intentional murder. McKinney's Penal Law § 125.25, subds. 1, 2.—People v. Fardan, 607 N.Y.S.2d 220, 82 N.Y.2d 638, 628 N.E.2d 41.

"Heat of passion," a common-law concept, and "extreme emotional disturbance," an affirmative defense to intentional murder charge, are not identical concepts. McKinney's Penal Law § 125.25, subd. 1.—Id.

⊜⊐**29. Persons liable.**

Ill.App. 1 Dist. 1993. If defendant intended to promote or facilitate escape of prisoner, he was responsible for any act, including murder of two police officers, that was committed in furtherance of intended crime.—People v. Wilson, 195 Ill.Dec. 8, 628 N.E.2d 472.

III. MANSLAUGHTER.

⊜⊐**34. Elements of involuntary manslaughter.**

Ill.App. 1 Dist. 1993. State of mind for involuntary manslaughter is recklessness. S.H.A. 720

Figure 2.5   Example of keywords from West Publishing Company's reporters for locating cases listed under keywords *health* and *environment*. (Photo Credit — West Publishing Co., St. Paul, Minnesota.)

CHAPTER 3

National Environmental Policy Act

3.1 NATIONAL ENVIRONMENTAL POLICY ACT — AN ENVIRONMENTAL PROTECTION BEGINNING

Modern and present environmental law began with the National Environmental Policy Act (NEPA). Familiarization with its rudiments is beneficial to give an understanding of the purposes of subsequent laws that stemmed from it.

Congress responded to the nation's mounting hue and cry for environmental protection by its passage of the NEPA of 1969, effective January 1, 1970. A 1972 explanation for the creation of NEPA by A. Dan Tarlock, Esquire, a distinguished and prominent water law attorney and authority, is noteworthy:

> Federal legislation was necessary because the creation of program, mission-oriented agencies has insured that these environmental considerations have been systematically under-represented in most short- and long-range decision making. Existing agencies were established to supervise the development of our natural resources consistent with the ethic which had prevailed throughout this country's history and, thus, they tended to overstress the benefits of development and to explore insufficiently the less environmentally damaging alternatives to current methods of meeting their programmed objectives (Tarlock, 1994).

The NEPA preamble describes it as a national policy to "encourage productive and enjoyable harmony between man and his environment; to promote efforts which will prevent or eliminate damage to the environment and biosphere and stimulate the health and welfare of man; to enrich the understanding of the ecological systems and natural resources important to the Nation...."

The NEPA Title I text, is reproduced and follows.

The National Environmental Policy Act of 1969, as Amended*

An Act to establish a national policy for the environment, to provide for the establishment of a Council on Environmental Quality, and for other purposes.

Be it enacted by the Senate and House of Representatives of the United States of America in Congress assembled, that this Act may be cited as the "National Environmental Policy Act of 1969."

Purpose

The purposes of this Act are: To declare a national policy which will encourage productive and enjoyable harmony between man and his environment; to promote efforts which will prevent or eliminate damage to the environment and biosphere and stimulate the health and welfare of man; to enrich the understanding of the ecological systems and natural resources important to the Nation and to establish a Council on Environmental Quality.

Title I — Declaration of National Environmental Policy

SEC. 101. (a) The Congress, recognizing the profound impact of man's activity on the interrelations of all components of the natural environment, particularly the profound influences of population growth, high-density urbanization, industrial expansion, resource exploitation, and new and expanding technological advances and recognizing further the critical importance of restoring and maintaining environmental quality to the overall welfare and development of man, declares that it is the continuing policy of the Federal Government, in cooperation with State and local governments, and other concerned public and private organizations, to use all practicable means and measures, including financial and technical assistance, in a manner calculated to foster and promote the general welfare, to create and maintain conditions under which man and nature can exist in productive harmony, and fulfill the social, economic, and other requirements of present and future generations of Americans.

(b) In order to carry out the policy set forth in this Act, it is the continuing responsibility of the Federal Government to use all practical means, consistent with other essential considerations of national policy, to improve and coordinate Federal plans, functions, programs, and resources to the end that the Nation may —

(1) fulfill the responsibilities of each generation as trustee of the environment for succeeding generations;
(2) assure for all Americans safe, healthful, productive, and aesthetically and culturally pleasing surroundings;
(3) attain the widest range of beneficial uses of the environment without degradation, risk to health or safety, or other undesirable and unintended consequences;
(4) preserve important historic, cultural, and natural aspects of our national heritage, and maintain, wherever possible, an environment which supports diversity, and variety of individual choice;
(5) achieve a balance between population and resource use which will permit high standards of living and a wide sharing of life's amenities; and
(6) enhance the quality of renewable resources and approach the maximum attainable recycling of depletable resources.

* Public Law 91-190, 42 U.S.C. 4321-4342, January 1, 1970, as amended by Public Law 94-52, July 3, 1975, and Public Law 94-83, August 9, 1975.

(c) The Congress recognizes that each person should enjoy a healthful environment and that each person has a responsibility to contribute to the preservation and enhancement of the environment.

SEC. 102. The Congress authorizes and directs that, to the fullest extent possible: (1) the policies, regulations, and public laws of the United States shall be interpreted and administered in accordance with the policies set forth in this Act, and (2) all agencies of the Federal Government shall —

(A) Utilize a systematic, interdisciplinary approach which will insure the integrated use of the natural and social sciences and the environmental design arts in planning and in decision making which may have an impact on man's environment;

(B) Identify and develop methods and procedures in consultation with the Council on Environmental Quality established by Title II of this Act, which will insure that presently unquantified environmental amenities and values may be given appropriate consideration in decision making along with economic and technical considerations;

(C) Include in every recommendation or report on proposals for legislation and other major Federal actions significantly affecting the quality of the human environment, a detailed statement by the responsible official on —

(i) The environmental impact of the proposed action,
(ii) Any adverse environmental effects which cannot be avoided should the proposal be implemented,
(iii) Alternatives to the proposed action,
(iv) The relationship between local short-term uses of man's environment and the maintenance and enhancement of long-term productivity, and
(v) Any irreversible and irretrievable commitments of resources which would be involved in the proposed action should it be implemented.

Prior to making any detailed statement, the responsible Federal official shall consult with and obtain the comments of any Federal agency which has jurisdiction by law or special expertise with respect to any environmental impact involved. Copies of such statement and the comments and views of the appropriate Federal, State, and local agencies, which are authorized to develop and enforce environmental standards, shall be made available to the President, the Council on Environmental Quality and to the public as provided by section 552 of title 5, United States Code, and shall accompany the proposal through the existing agency review processes;

(d) Any detailed statement required under subparagraph (c) after January 1, 1970, for any major Federal action funded under a program of grants to States shall not be deemed to be legally insufficient solely by reason of having been prepared by a State agency or official, if:

(i) The State agency or official has statewide jurisdiction and has the responsibility for such action,
(ii) The responsible Federal official furnishes guidance and participates in such preparation,
(iii) The responsible Federal official independently evaluates such statement prior to its approval and adoption, and
(iv) After January 1, 1976, the responsible Federal official provides early notification to, and solicits the views of, any other State or any Federal land management entity of any action or any alternative thereto which may have significant impacts upon such

State or affected Federal land management entity and, if there is any disagreement on such impacts, prepares a written assessment of such impacts and views for incorporation into such detailed statement.

The procedures in this subparagraph shall not relieve the Federal official of his responsibilities for the scope, objectivity, and content of the entire statement or of any other responsibilities under this Act; and further, this subparagraph does not affect the legal sufficiency of statements prepared by State agencies with less than statewide jurisdiction.

(e) Study, develop, and describe appropriate alternatives to recommended courses of action in any proposal which involves unresolved conflicts concerning alternative uses of available resources;

(f) Recognize the worldwide and long-range character of environmental problems and, where consistent with the foreign policy of the United States, lend appropriate support to initiatives, resolutions, and programs designed to maximize international cooperation in anticipating and preventing a decline in the quality of mankind's world environment;

(g) Make available to States, counties, municipalities, institutions, and individuals, advice and information useful in restoring, maintaining, and enhancing the quality of the environment:

(h) Initiate and utilize ecological information in the planning and development of resource-oriented projects; and

(i) Assist the Council on Environmental Quality established by Title II of this Act.

SEC.103. All agencies of the Federal Government shall review their present statutory authority, administrative regulations, and current policies and procedures for the purpose of determining whether there are any deficiencies or inconsistencies therein which prohibit full compliance with the purposes and provisions of this Act and shall propose to the President not later than July l, 1971, such measures as may be necessary to bring their authority and policies into conformity with the intent, purposes, and procedures set forth in this Act.

SEC. 104. Nothing in section 102 or 103 shall in any way affect the specific statutory obligations of any Federal agency (1) to comply with criteria or standards of environmental quality, (2) to coordinate or consult with any other Federal or State agency, or (3) to act, or refrain from acting contingent upon the recommendations or certification of any other Federal or State agency.

SEC. 105. The policies and goals set forth in this Act are supplementary to those set forth in existing authorizations of Federal agencies.

NEPA's purpose was to reform the decision-making process of federal agencies. Under NEPA, all federal agencies are required to take into account the adverse environmental effects of their programs. Although originally intended to apply to the actions of federal agencies, NEPA has been interpreted to include not only federal projects but also state and local programs when funded by federal assistance, thereby becoming federalized and providing the connection, or **nexus**, to be included under federal projects. Under *Ely v. Velde*, 451 F.2d 1130 (4th Cir. 1971), it was found to extend to block grants (i.e., no-strings-attached programs).

It also extends to all private development authorized by federal permits. The term *federal action* as found within NEPA has been expanded to include nonfederal projects as determined in *Scientists' Institute for Public Information, Inc. v. Atomic Energy Commission*, 481 F.2d 1079 (D.C. Cir. 1973). As stated by the U.S. Court of Appeals in *Scientists' Institute* at 1088–1089:

> There is "federal action" within the meaning of the statute not only when an agency proposes to build a facility itself, but also when an agency makes a decision which permits action by other parties which will affect the quality of the environment. NEPA's impact statement procedure has been held to apply where a federal agency approves a lease of land to private parties, grants licenses and permits to private parties, or approves and funds state highway projects. In each of these instances, the federal agency took action affecting the environment in the sense that the agency made a decision which permitted some other party — private or governmental — to take action affecting the environment.

The heart of NEPA is §§ 101 and 102, called the *procedural duties*, which outline and govern its intended application. It is noted that the objective is "...to create and maintain conditions under which man and nature can exist in productive harmony."

Section 101 states the environmental policies; charges the federal government with the responsibility of the environmental concerns; and states the purposes, *viz.*:

1. Act as trustee of the environment and resources for future generations;
2. Assurance of safe, healthful, productive, aesthetic surroundings;
3. Attain the greatest beneficial use of nature without degradation or risk to health or safety;
4. Preserve cultural and natural aspects;
5. Achieve a balance between people and resource use permitting a high standard of living; and
6. Maintain renewable resources and maximize recycling of depletable resources.

Section 102 authorizes the policy making and promulgation of rules and regulations by federal agencies to carry out the purposes of the act; and charges the executive branch agencies with certain specified duties to carry out the purposes and objectives of the environmental act. It has been described as *action enforcing*. Some of the specifics are:

1. Coordination or interdisciplinary action of agencies for all federal projects, plans, etc.
2. Authorization of the NEPA to coordinate and approve all federal development projects (the Council of Environmental Quality [CEQ] composed of three members, appointed by the President, with the advice and consent of the Senate)
3. Establishing the procedure for all federal planning by considering and reporting
   a. An environmental impact statement (EIS)
   b. Any unavoidable adverse environmental effects, if project is implemented
   c. Alternative proposals
   d. Short-term use vs. long-term impact on environment
   e. Any irreversible and irretrievable detriment to the environment in case of implementation

To carry out the purposes for which NEPA was designed by Congress, an agency had to be authorized and enabled with the necessary powers. For this work, NEPA's Title II established the CEQ. To augment the CEQ, the Environmental Quality Improvement Act of 1970 was enacted. This act specified (1) To assure that each Federal department and agency conducting or supporting public works activities which affect the environment shall implement the policies under existing law; and (2) to authorize an Office of Environmental Quality, which, notwithstanding any other provision of law, shall provide the professional and administrative staff for the Council on Environmental Quality established by Public Law 91-190.

3.2 EARLY TESTING OF NEPA BY TRIAL

As with all new legislative acts, the NEPA had to undergo several years of court testing and trying of its terminology and semantics through litigation to arrive at the court's interpretations of just what Congress intended.

During the first 8 years of NEPA's existence, the act was tested in 938 cases in federal courts. In most of those cases, the basis of complaint concerned the failure of federal agencies to prepare an EIS in violation of § 102(C) or that the EIS prepared had been inadequate. Complainants were usually citizens and environmental groups to stop federal agency actions.

According to the CEQ Ninth Annual Report (1978), of the 938 cases, 62% were completed; of that 62% (584 cases) 221 were decided in favor of the agencies and 75 in favor of the plaintiffs; 170 were dismissed; the balance were dealt with in various other ways.

Injunctions were the usual remedy for the plaintiffs, thus, holding up the federal project. In 202 cases, the courts issued NEPA-related injunctions and 92 of those caused a delay of over a year. The Department of Transportation had the most actions delayed, 50, with 27 delayed over a year. The Corps of Engineers had 47 actions delayed, with 19 delayed longer than a year. For nearly the entire decade of the 1970s, the federal courts were besieged with environmental cases testing NEPA. As with all new laws, they must undergo a period of testing that results in a molding process for constitutional procedures.

Early litigation involved testing of terms of NEPA, e.g., *other major federal actions* as found in NEPA § 102(C), whether an EIS was required, and what constitutes a *major* federal action. The phrase is broadly defined to include "projects and programs entirely or partly financed, assisted, conducted, regulated, or approved by federal agencies; new or revised agency rules, regulations, plans, policies, or procedures; and legislative proposals" [40 Code of Federal Regulations (CFR) § 1508.18(a)].

Another semantics, or interpretation of terminology, question that was litigated involved the words *major* and *significantly* from § 102(2)(C). The regulations state that *major* reinforces, but has no meaning independent of *significantly* [40 CFR § 1508.18(b)]. Another example of a NEPA phrase that caused litigated issues was *the quality of the human environment* [§ 102(2) (C)].

One of the earliest and more significant cases in which NEPA was tested was *Calvert Cliffs Coordinating Committee, Inc. v. U.S. Atomic Energy Commission,* 449 F.2d 1109 (D.C.C. 1971). *Calvert Cliffs* is now of historical legal interest. The case involved a challenge by the petitioners, the Calvert Cliffs Coordinating Committee, to some of the Atomic Energy Commission (AEC) rules concerning licensing for construction and for operating nuclear facilities. Calvert Cliffs alleged that AEC had granted construction permits for several nuclear energy plants without full consideration of environmental issues. The nuclear plants were in various stages of construction, and operating licenses had not yet been granted. The case arguments dealt with AEC rules as applied, in general, alleging conflict with NEPA, and, in particular, as applied to the licensing of the Calvert Cliffs nuclear power plant on Chesapeake Bay in Maryland.

The case involved judicial review of several rules promulgated by the AEC in response to the directives of NEPA and CEQ to determine their compliance. The Calvert Cliffs Committee challenged four AEC rules, in particular, as being in noncompliance with § 102 of NEPA, viz.:

1. In any proposed project, environmental factors were to be considered by both the AEC staff and the applicant. However, the AEC rule continued that "such factors need not be considered by the hearing board conducting an independent review of staff recommendations unless affirmatively raised by outside parties or staff members."

2. Another part of the AEC rules prohibited any party from raising non-radiological environmental issues at any hearing if the notice for that hearing appeared in the *Federal Register* before March 4, 1971.

3. The AEC hearing board was prohibited from conducting an independent environmental evaluation and balancing of certain environmental factors if other responsible agencies have already certified that their own environmental standards are satisfied by the proposed federal action.

4. Environmental factors would not be considered in the case of construction permits already issued, or approved but not issued, before NEPA compliance was required.

In examining the AEC's rule cited in 3, the issue of local concern for the Calvert Cliffs nuclear plant was the AEC's lack of proper environmental concern for a nuclear plant discharge of wastewaters. With regard to the quality of the discharged waters under the AEC license, AEC had deferred to other agency water quality specifications, thereby shirking, or *abdicating* in the word of the court, its duties of "full environmental consideration" under NEPA.

AEC argued for support of this rule on the basis "that an independent environmental evaluation and balancing of certain environmental factors" were unnecessary and superfluous "if other responsible agencies have already certified that their own environmental standards are satisfied by the proposed federal action." AEC based this argument on its belief that § 104 of NEPA authorized an agency to "act ... upon the certification of" any other agency standards, which in this case AEC referred to the water quality standards set under the Water Quality Improvement Act (WQIA) of 1970.

The court rejected this argument finding that, although § 104 and the WQIA clearly required obedience to the water quality standards set by other agencies, the standards set were minimal. AEC could neither abdicate its duties under NEPA by derogation of its duties allowed under § 104 nor defer to another agency water quality standards for its own permitting requirements. It must give full environmental consideration to the water quality standards of its licensing procedure for a nuclear plant where wastewater discharges are made into the navigable waters of the United States. Concern for higher water quality was of the highest order at the waste discharge point. This was indicated by the court calling it "perhaps the most significant impact of nuclear power plants" and should be given fullest consideration for environmental concerns. AEC was not prevented from demanding water pollution controls from its licensees which are more strict than those demanded by another agency simply because they meet the standards of the WQIA.

As to AEC's challenged rules 2 and 4 cited earlier, retroactive environmental concerns became the issue for nuclear plants already licensed or approved before the passage of NEPA. Although NEPA was found not to be made specifically retroactive, its application did not preclude AEC from establishing requirements for retrofitting pre-NEPA nuclear plants when giving full environmental consideration to licenses already issued under its control.

In summation, the court in *Calvert Cliffs* held that the procedural rules required a strict standard of compliance; the NEPA directive was not discretionary, but mandatory. Environmental factors must be considered throughout all steps of the procedural rule-making process by an agency. The challenged rules of the AEC were found deficient and not in compliance with NEPA. The court remanded its findings back to the AEC for compliance with the requirements of NEPA in accordance with the decision.

3.2.1   NEPA Terminology

Section 102 further outlines utopian ideals for international and worldwide concern, coordination, and cooperation; advisements and cooperation with the states for environmental quality; collection and use of ecological information in planning and development of natural resources; and assistance and coordination of actions with the CEQ.

Through NEPA, all federal agencies are required to take into consideration the adverse environmental effects of their policies, plans, programs and projects, sometimes referred to as *the four p's. Any of those p's may require an EIS*. The NEPA process requires environmental considerations be integrated into agency planning.

Subsection (C) of § 102 outlines the requirements for the filing of an EIS, stating:

(C) Include in every recommendation or report on proposals for legislation and other major federal actions significantly affecting the quality of the human environment, a detailed statement by the responsible official on —

(i) The environmental impact of the proposed action,
(ii) Any adverse environmental effects which cannot be avoided should the proposal be implemented,
(iii) Alternative to the proposed action,

(iv) The relationship between the local short-term uses of man's environment and the maintenance and enhancement of long-term productivity, and

(v) Any irreversible and irretrievable commitments of resources which would be involved in the proposed action should it be implemented.

Agency actions can be divided into three categories, viz., (1) categorical exclusions, which are those that clearly do not have any significant effects on the environment, and do not require an EIS; (2) planning that clearly will have a significant effect on the environment and will require an EIS; (3) planning that is unclear about whether there will be any significant impact on the environment. For this category, the proper procedure is to first undertake an environmental assessment (EA).

The result of the EA study will be whether an EIS is required for further agency action, or whether the agency may issue a **finding of no significant impact (FONSI)**. Once an agency determines through an EA that a proposal will significantly affect the environment, it must prepare an EIS. The EIS is not discretionary; it is mandatory, and it is reviewable by federal courts.

The next step is called **scoping**, which determines the scope of issues to be addressed. Scoping includes (1) determination of issues; (2) notification of involved agencies et al.; (3) in-depth or significant issues; and (4) elimination of insignificant issues. The EIS preparation is designed (1) to be a tool for decision making; (2) to be analytical and concise; (3) to have 150 pages for ordinary, and 200 pages for complex studies. The lead agency, or most involved one, prepares the EIS.

The Regulations for Implementing the Procedural Provisions of NEPA, 40 CFR §§ 1500–1508, provide definitions of the act's terminology. A few of the more pertinent are reproduced from the Superintendent of Documents' reprint of 43 Fed. Reg. 55978–56007:

Cumulative impact is the impact on the environment which results from the incremental impact of the action when added to other past, present and reasonably foreseeable future action regardless of what agency (Federal or non-Federal) or person undertakes such other actions. Cumulative impacts can result from individually minor but collectively significant actions taking place over a period of time. (§ 1508.7)

Effects include:
(a) Direct effects, which are caused by the action and occur at the same time and place.
(b) Indirect effects, which are caused by the action and are late in time or farther removed in distance, but are still reasonably foreseeable. Indirect effects may include growth-inducing effects and other effects related to induced changes in the pattern of land use, population density or growth rate, and related effects on air and water and other natural systems, including ecosystems.

"Effects" and **"impacts"** as used in these regulations are synonymous. Effects include ecological (such as the effects on natural resources and on the components, structures, and functioning of affected ecosystems), aesthetic, historic, cultural, economic, social, or health, whether direct, indirect, or cumulative. Effects may also include those resulting from actions which may have both beneficial and detrimental effects, even if on balance the agency believes that the effect will be beneficial. (§ 1508.8)

Human Environment shall be interpreted comprehensively to include the natural and physical environment and the relationship of people with that environment. [See the definition of "effects" (§ 1508.8).] This means that economic or social effects are not intended by themselves to require preparation of an environmental impact statement. When an environmental impact statement is prepared and economic or social and natural or physical environmental effects are interrelated, then the environmental impact statement will discuss all of these effects on the human environment. (§ 1508.14)

Federal agency means all agencies of the Federal Government. It does not mean the Congress, the Judiciary, or the President, including the performance of staff functions for the President in his Executive Office. It also includes for purposes of these regulations States and units of general local government and Indian tribes assuming NEPA responsibilities under § 104(h) of the Housing and Community Development Act of 1974. (§ 1508.12)

Environmental Assessment: An environmental impact assessment is similar to an environmental impact statement but not necessarily as comprehensive or in the same depth. It can determine if a negative declaration is indicated or if an environmental impact statement is required. An environmental assessment is not required if it has been decided to prepare an environmental impact statement (§ 1501.3)

(a) Means a concise public document for which a Federal agency is responsible that serves to:
  (1) Briefly provide sufficient evidence and analysis for determining whether to prepare an environmental impact statement or a finding of no significant impact.
  (2) Aid an agency's compliance with the Act when no environmental impact statement is necessary.
  (3) Facilitate preparation of a statement when one is necessary.
(b) Shall include brief discussion of the need for the proposal or alternatives as required by § 102(2)(E) of the environmental impacts of the proposed action and alternatives, and a listing of agencies and persons consulted. (§ 1508.9)

Environmental document includes the documents specified in § 1508.9 (environmental assessment), 1508.11 (environmental impact statement), 1508.13 (finding of no significant impact), and 1508.22 (notice of intent). (§ 1508.10)

Finding of No Significant Impact (FONSI) means a document by a Federal agency briefly presenting the reasons why an action, not otherwise excluded (§ 1508.4), will not have a significant effect on the human environment and for which an environmental impact statement therefore will not be prepared. It shall include the environmental assessment or a summary of it and shall note any other environmental documents related to it. If the assessment is included, the finding need not repeat any of the discussion in the assessment but may incorporate it by reference. (§ 1508.13)

3.2.2 Environmental Impact Statement

The Environmental Impact Statement (EIS) is a detailed written statement as required by § 102(2)(C) of the act (§ 1508.11); 40 CFR § 1502.10 et seq. are the CEQ regulations for the EIS and include, purpose, format, and alternatives to the proposed action. (§ 1502.14 has been called the heart of the EIS.) § 1502.10 recommends the format for an EIS.

Agencies shall use a format for environmental impact statements which will encourage good analysis and clear presentation of the alternatives, including the proposed action. The following standard format for environmental impact statements should be followed unless the agency determines that there is a compelling reason to do otherwise:

(a) Cover Sheet
(b) Summary (See § 1500.8, NEPA)
(c) Table of Contents
(d) Purpose of and Need for Action
(e) Alternatives Including Proposed Action [§§ 102(2)(C)(iii) and 102(2)(E) of the Act]
(f) Affected Environment
(g) Environmental Consequences [especially §§ 102(2)(C)(i),(ii), (iv), and (v) of the Act]
(h) List of Preparers
(i) List of Agencies, Organizations and Persons to Whom Copies of the Statement Are Sent
(j) Index
(k) Appendices (if any)

A suggested outline and content for an EIS is made by § 1500.8. The outline for the content of a CEQ-prescribed EIS follows:

1. PROJECT DESCRIPTION
   a. Purpose of action
   b. Description of action
      (1) Name
      (2) Summary of activities
   c. Environmental setting
      (1) Environment prior to proposed action
      (2) Other related federal activities.
2. LAND-USE RELATIONSHIPS
   a. Conformity or conflict with other land-use plans, policies and controls
      (1) Federal, state, and local
      (2) Clean Air Act and Federal Water Pollution Control Act Amendment of 1972
   b. Conflicts and/or inconsistent land-use plans
      (1) Extent of reconciliation
      (2) Reasons for proceeding with action

3. PROBABLE IMPACT OF THE PROPOSED ACTION ON THE ENVIRON-
   MENT
   a. Positive and negative effects
      (1) National and international environment
      (2) Environmental factors
      (3) Impact of proposed action
   b. Direct and indirect consequences
      (1) Primary effects
      (2) Secondary effects
4. ALTERNATIVES TO THE PROPOSED ACTION
   a. Reasonable alternative actions
      (1) Those that might enhance environmental quality
      (2) Those that might avoid some or all adverse effects
   b. Analysis of alternatives
      (1) Benefits
      (2) Costs
      (3) Risks
5. PROBABLE ADVERSE ENVIRONMENTAL EFFECTS WHICH CANNOT BE
   AVOIDED
   a. Adverse and unavoidable impacts
   b. How avoidable adverse impacts will be mitigated
6. RELATIONSHIP BETWEEN LOCAL SHORT-TERM USES OF MAN'S
   ENVIRONMENT AND THE MAINTENANCE AND ENHANCEMENT OF
   LONG-TERM PRODUCTIVITY
   a. Trade-off between short-term environmental gains at expense of long-term
      losses
   b. Trade-off between long-term environmental gains at expense of short-term
      losses
   c. Extent to which proposed action forecloses future options
7. IRREVERSIBLE AND IRRETRIEVABLE COMMITMENTS OF RESOURCES
   a. Unavoidable impacts irreversibly curtailing the range of potential uses of the
      environment
      (1) Labor
      (2) Materials
      (3) Natural
      (4) Cultural
8. OTHER INTERESTS AND CONSIDERATIONS OF FEDERAL POLICY
   THAT OFFSET THE ADVERSE ENVIRONMENTAL EFFECTS OF THE PRO-
   POSED PLAN
   a. Countervailing benefits of the proposed plan
   b. Countervailing benefits of alternatives

Guidelines for the preparation of environmental impact statements published by
the CEQ are reported in § 1500.8. Many states have adopted similar guidelines. A
draft environmental impact statement is generally first prepared and circulated for
comment. It must fulfill and satisfy the requirements established in § 102(2)(C). The
comments received are evaluated and considered in the decision process. The EIS
may be for a proposed concept or program, for a planning study, and/or a proposed
construction project.

3.2.2.1 Content of Environmental Impact Statements

The eighth annual report of the CEQ, December 1978, published the prescribed content of the EIS (§ 1500.8), as follows:

(a) The following points are to be covered:
  (1) A description of the proposed action, a statement of its purposes, and a description of the environment affected, including information, summary technical data, and maps and diagrams where relevant, adequate to permit an assessment of potential environmental impact by commenting agencies and the public. Highly technical and specialized analyses and data should be avoided in the body of the draft impact statement. Such materials should be attached as appendices or footnoted with adequate bibliographic references. The statement should also succinctly describe the environment of the area affected as it exists prior to a proposed action, including other Federal activities in the area affected by the proposed action which are related to the proposed action. The interrelationships and cumulative environmental impacts of the proposed action and other related Federal projects shall be presented in the statement. The amount of detail provided in such descriptions should be commensurate with the extent and expected impact of the action, and with the amount of information required at the particular level of decision making (planning, feasibility, design, etc.). In order to ensure accurate descriptions and environmental assessments, site visits should be made where feasible. Agencies should also take care to identify, as appropriate, population and growth characteristics of the affected area and any population and growth assumptions used to justify the project or program or to determine secondary population and growth impacts resulting from the proposed action and its alternatives [see paragraph (3)(ii), of this section]. In discussing these population aspects, agencies should give consideration to using the rates of growth in the region of the project contained in the projection compiled for the Water Resources Council by the Bureau of Economic Analysis of the Department of Commerce and the Economic Research Service of the Department of Agriculture. In any event, it is essential that the sources of data used to identify, quantify, or evaluate any and all environmental consequences be expressly noted.
  (2) The relationship of the proposed action to land use plans, policies, and controls for the affected area. This requires a discussion of how the proposed action may conform or conflict with the objectives and specific terms of approved or proposed Federal, State, and local land use plans, policies, and controls, if any, for the area affected, including those developed in response to the Clean Air Act or the Federal Water Pollution Control Act Amendments of 1972. Where a conflict or inconsistency exists, the statement should describe the extent to which the agency has reconciled its proposed action with the plan, policy, or control and the reasons why the agency has decided to proceed notwithstanding the absence of full reconciliation.
  (3) The probable impact of the proposed action on the environment.
    (i) This requires agencies to assess the positive and negative effects of the proposed action as it affects both the national and international environment. The attention given to different environmental factors will vary

according to the nature, scale, and location of proposed actions. Among factors to consider should be the potential effect of the action on such aspects of the environment as those listed in Appendix II* of these guidelines. Primary attention should be given in the statement to discussing those factors most evidently impacted by the proposed action.

(ii) Secondary or indirect, as well as primary or direct, consequences for the environment should be included in the analysis. Many major Federal actions, in particular those that involve the construction or licensing of infrastructure investments (*e.g.*, highways, airports, sewer systems, water resource projects, *etc.*), stimulate or induce secondary effects in the form of associated investments and changed patterns of social and economic activities. Such secondary effects, through their impacts on existing community facilities and activities, through inducing new facilities and activities, or through changes in natural conditions, may often be even more substantial than the primary effects of the original action itself. For example, the effects of the proposed action on population and growth may be among the more significant secondary effects. Such population and growth impacts should be estimated if expected to be significant [using data identified as indicated in § 1500.8(a)(1)] and an assessment made of the effect of any possible change in population patterns or growth upon the resource base, including land use, water, and public services, of the area in question.

(4) Alternatives to the proposed action, including, where relevant, those not within the existing authority of the responsible agency. [Section 102(2)(D) of the Act requires the responsible agency to "study, develop, and describe appropriate alternatives to recommended courses of action in any proposal which involves unresolved conflicts concerning alternative uses of available resources."] A rigorous exploration and objective evaluation of the environmental impacts of all reasonable alternative actions, particularly those that might enhance environmental quality to avoid some or all of the adverse environmental effects, is essential. Sufficient analysis of such alternatives and their environmental benefits, costs and risks should accompany the proposed action through the agency review process in order not to foreclose prematurely options which might enhance environmental quality or have less detrimental effects. Examples of such alternatives include: the alternative of taking no action or of postponing action pending further study; alternatives requiring actions of a significantly different nature which would provide similar benefits with different environmental impacts (*e.g.*, nonstructural alternatives to flood control programs, or mass transit alternatives to highway construction); alter-

* Air quality, water quality, marine pollution, commercial fishers conservation, and shellfish sanitation, water regulation and stream modification, fish and wildlife, solid waste, noise, radiation, toxic materials, food additives and contamination of foodstuffs, pesticides, transportation and handling of hazardous materials, energy supply and natural resources development, petroleum development, extraction, refining, transport, and use, natural gas development, production, transmission, and use, coal and minerals development, mining, conversion, processing, transport and use, renewable resource development, production, management, harvest, transport and use, energy and natural resource conservation, land use and management, public land management, protection of environmentally critical areas flood plains, wetlands, beaches and dunes, unstable soils, steep slopes, aquifer recharge areas, *etc.*, land use in coastal areas, redevelopment and construction in built-up areas, density and congestion mitigation, neighborhood character and continuity, impact on low income populations, historic, architectural, and archeological preservation, soil and plant conservation and hydrology, outdoor recreation

natives related to different designs or details of the proposed action which would present different environmental impacts (*e.g.*, cooling ponds vs. cooling towers for a power plant or alternatives that will significantly conserve energy); alternative measures to provide for compensation of fish and wildlife losses, including the acquisition of land, waters, and interests therein. In each case, the analysis should be sufficiently detailed to reveal the agency's comparative evaluation of the environmental benefits, costs, and risks of the proposed action and each reasonable alternative. Where an existing impact statement already contains such an analysis, its treatment of alternatives may be incorporated provided that such treatment is current and relevant to the precise purpose of the proposed action.

(5) Any probable adverse environmental effects which cannot be avoided [such as water or air pollution, undesirable land use patterns, damage to life systems, urban congestion, threats to health or other consequences adverse to the environmental goals set out in section 101(b) of the Act]. This should be a brief section summarizing in one place those effects discussed in paragraph (a)(3) of this section that are adverse and unavoidable under the proposed action. Included for purposes of contrast should be a clear statement of how other avoidable adverse effects discussed in paragraph (a)(2) of this section will be mitigated.

(6) The relationship between local short-term uses of man's environment and the maintenance and enhancement of long-term productivity. This section should contain a brief discussion of the extent to which the proposed action involves tradeoffs between short-term environmental gains at the expense of long-term losses, or vice versa, and a discussion of the extent to which the proposed action forecloses future options. In this context short-term and long-term do not refer to any fixed time periods, but should be viewed in terms of the environmentally significant consequences of the proposed action.

(7) Any irreversible and irretrievable commitments of resources that would be involved in the proposed action should it be implemented. This requires the agency to identify from its survey unavoidable impacts in paragraph (a) (5) of this section and the extent to which the action irreversibly curtails the range of potential uses of the environment. Agencies should avoid construing the term "resources" to mean only the labor and materials devoted to an action. "Resources" also means the natural and cultural resources committed to loss or destruction by the action.

(8) An indication of what other interests and considerations of Federal policy are thought to offset the adverse environmental effects of the proposed action identified pursuant to paragraphs (a) and (5) of this section. The statement should also indicate the extent to which these countervailing benefits could be realized by following reasonable alternatives to the proposed action (as identified in paragraph (a)(4) of this section) that would avoid some or all of the adverse environmental effects. In this connection, agencies that prepare cost-benefit analyses of proposed actions should attach such analysis, or summaries thereof, to the environmental impact statement, and should carefully indicate the extent to which environmental costs have not been reflected in such analyses.

(b) In developing the above points, agencies should make every effort to convey the required information succinctly in a form easily understood, both by members of the public and by public decision makers, giving attention to the substance of the information conveyed rather than to the particular form, or length, or detail of the statement. Each of the above points, for example, need not always occupy a distinct section of the statement if it is otherwise adequately covered in discussing the impact or the proposed action and its alternatives — which items should normally be the focus of the statement. Draft statements should indicate at appropriate points in the text any underlying studies, reports, and other information obtained and considered by the agency in preparing the statement including any cost-benefit analyses prepared by the agency, and reports of consulting agencies under the Fish and Wildlife Coordination Act, 16 U.S.C. 661 et seq., and the National Historic Preservation Act of 1966, 16 U.S.C.470 et seq., where such consultation has taken place. In the case of documents not likely to be easily accessible (such as internal studies or reports), the agency should indicate how such information may be obtained. If such information is attached to the statement, care should be taken to ensure that the statement remains an essentially self-contained instrument, capable of being understood by the reader without the need for undue cross reference.

(c) Each environmental statement should be prepared in accordance with the precept in section 102(2)(A) of the Act that all agencies of the Federal Government "utilize a systematic, interdisciplinary approach which will insure the integrated use of the natural and social sciences and the environmental design arts in planning and decision making which may have an impact on man's environment." Agencies should attempt to have relevant disciplines represented on their own staff; where this is not feasible, they should make appropriate use of relevant Federal, State, and local agencies or the professional services of universities and outside consultants. The interdisciplinary approach should not be limited to the preparation of the environmental impact statement, but should also be used in the early planning stages of the proposed action. Early application of such an approach should help assure a systematic evaluation of reasonable alternative courses of action and their potential social, economic, and environmental consequences.

(d) Appendix I prescribes the form of the summary sheet which should accompany each draft and final environmental statement.

SUMMARY TO ACCOMPANY DRAFT AND FINAL STATEMENTS

(Check one) ( ) Draft. ( ) Final Environmental Statement.
Name of responsible Federal agency (with name of operating division where appropriate). Name address, and telephone number of individual at the agency who can be contacted for additional information about the proposed action or the statement.

1. Name of action (Check one) ( ) Administrative Action. ( ) Legislative Action.
2. Brief description of action and its purpose. Indicate what States (and countries) particularly affected, and what other proposed Federal actions in the area, if any, are discussed in the statement.
3. Summary of environmental impacts and adverse environmental effects.
4. Summary of major alternatives considered.

5. (For draft statements) List all Federal, State, and local agencies and other parties from which comments have been requested. (For final statements) List all Federal, State, and local agencies and other parties from which written comments have been received.

6. Draft statement (and final environmental statement, if one has been issued) made available to the Council and the public.

Statement for environmental impacts, effects, and alternatives:

1. Environmental considerations
   a. Status of environmental conditions
2. Environmental analysis and assessment
   a. Matrix
   b. Check list
   c. Weighting
   d. Other
3. Alternative solutions
   a. Advantages
   b. Disadvantages
4. Recommended solution

3.2.3 NEPA EIS Requirement and Technological Development Programs

An EIS is required for "every recommendation or report on proposals for legislation and other major federal actions significantly affecting the quality of the human environment." The history of the NEPA evidences that the term *actions* not only applies to the physical construction of planned facilities but also includes "project proposals, proposals for new legislation, regulations, policy statements, or expansion, or revision of on going programs...."

An early example of technological planning that was litigated with its decision requiring an EIS is the case of *Scientists' Institute for Public Information, Inc. v. Atomic Energy Commission*, 481 F.2d 1079 (D.C. Cir., 1973). The AEC had been developing a technology known as the liquid metal fast breeder reactor (LMFBR) since the early 1960s. By 1970, the AEC was ready to proceed with the construction of an experimental generating plant utilizing the LMFBR. Acting under President Nixon's 1971 National Energy Program directives, along with the Congressional Atomic Energy Joint Committee's approval, AEC entered into negotiations with the Tennessee Valley Authority (TVA) and Commonwealth Edison leading to construction contracts for the first experimental nuclear LMFBR power-producing plant. Here, a government agency, AEC, was developing a technology, which if successful, would permit private utilities to take an action that would affect the environment.

Until challenged in *Scientists' Institute*, AEC had taken the position that "no statement was presently required since the program was still in the research and development stage and no specific implementing action which would significantly affect the environment had yet been taken" (*Scientists' Institute*, p.1079). Judge Skelly Wright responded:

Taking into account the magnitude of the on-going federal investment in this program, the controversial environmental effects attendant upon future widespread deployment of breeder reactors should the program fulfill present expectations, the accelerated pace under which this program has moved beyond pure scientific research toward creation of a viable, competitive breeder reactor electrical energy industry, and the manner in which investment in this new technology is likely to restrict future alternatives, we hold that the Commission's program comes within both the letter and spirit of Section 102(C) and that a detailed statement about the program, its environmental impact, and alternatives thereto is presently required. (*id.*, p. 1079) ... Development of the technology serves as much to affect the environment as does a Commission decision granting a construction permit for a specific plant. (*id.*, p. 1082) ... To wait until a technology attains the stage of complete commercial feasibility before considering the possible adverse environmental effects attendant upon ultimate application of the technology will undoubtedly frustrate meaningful consideration and balancing of environmental costs against economic and other benefits (*Scientists' Institute*, p. 1083).

The court continued that although NEPA does not require forecasting by an agency: "The agency need not foresee the unforeseeable, but by the same token, neither can it avoid drafting an impact statement simply because describing the environmental effects of and alternatives to a particular agency action involves some degree of forecasting. And one of the functions of a NEPA statement is to indicate the extent to which environmental effects are essentially unknown. Reasonable forecasting and speculation is implicit in NEPA, and any attempt to shirk ... discussion of future environmental effects as 'crystal ball inquiry' is rejected" (*Scientists' Institute*, p. 1088).

3.2.4   Impact Statements for Whole Projects vs. Segments

In Professor Frank Grad's *Treatise on Environmental Law* (Grad, 1978, pp. 8–218), the requirement of impact statements for long-range projects is discussed as becoming problematic because the preparation of an EIS for the project may be divided into and initiated in segments, and because decades may pass before completion of the entire project is attained. Environmental impact studies made at the beginning of a long-range project may be radically changed or even outmoded by the time late stages of the project are reached. Interstate and large intrastate highway projects are obvious examples, and environmental issues have been frequently litigated. Each segment of a highway project will have its own environmental problems and impacts on local communities varying from overall project impacts. Nevertheless, to allow major, long-range projects to proceed without an EIS would be to circumvent the NEPA requirement. The problem has been resolved by requiring an EIS for the overall project, supplemented by an EIS for each of the segments as it is executed.

One of the outstanding cases concerning the question of impact statement for a regional vs. local project was *Kleppe v. Sierra Club*, 427 U.S. 390 (1976). In *Kleppe*, the Department of the Interior was charged with the alleged violation of granting mineral (coal) leases on government-owned lands in the Northern Great

Plains region without filing an EIS as required by § 102(C). The government argued (1) that the leases were local in nature, and (2) that there was no regional plan to develop the resource. The district court agreed, holding that there was no regional plan or proposal for development of the Northern Great Plains regional coal fields, and that the four coal leases granted were local in nature. Hence, no EIS was required.

The court of appeals in *Kleppe* reversed the court below and the four approved coal leases were enjoined, preventing operation. The appeals court devised its own four-part "balancing" test for a determination of when the agency must begin forming an EIS. On review by the U.S. Supreme Court, the decision of *Kleppe* by the court of appeals was reversed, criticizing its four-part balancing test and saying that there was no support for it to be found in the language or legislative history of NEPA. It said that NEPA clearly stated when an EIS was required and did not call for a balancing test: "Even contemplation of regional development is insufficient to require an impact statement" (*Kleppe*, p. 398). The Supreme Court felt that such a balancing test would result in the preparation of many unnecessary impact statements.

3.2.5 Procedural Duty of Alternatives

Subsection iii of § 102(C) requires a detailed study and statement of alternatives to the proposed action by the leading and responsible agency. Inadequacy of this NEPA procedural duty has been alleged against agencies at various times.

During the oil crisis of 1972, and in response to a Presidential directive to attempt to alleviate the critical national oil shortage, Secretary of Interior Morton announced a general oil and gas lease sale for tracts on the outer shelf in the Gulf of Mexico, offshore from Louisiana and Mississippi. In *Natural Resources Defense Council, Inc. v. Morton*, 458 F.2d 827 (D.C. Cir. 1972), the EIS filed by the Department of Interior (Interior) recognized possible environmental threats to coastal areas, wildlife, and marine life in the event of an oil spill or well seepage, but stated that there was no feasible alternative available to them. This failure of Interior to discuss and even study other reasonable alternatives became an alleged inadequacy that was successfully upheld as such by the court of appeals.

Cited alternatives that Interior had neglected to pursue in the EIS included elimination of the oil import quotas and development of onshore reserves and other fuel sources such as oil shales, uranium, and coal. As a result, a temporary injunction was issued against the sales bid date for the drilling tracts until the EIS inadequacy was rectified.

The court recognized that the removal of the oil import quota was outside the powers of Interior to effect as a viable alternative, as was the technological development and legislative implementation for other potential fuel sources as uranium and oil shales; the *Morton* court pointed out that the agency is not relieved of the responsibility for reviewing those alternatives in the EIS as required and mandated by NEPA to the fullest extent possible.

3.2.6 NEPA Becomes Established Law

The growing deference to prior decisions of the federal courts in NEPA litigation was expressed in the 1982 case of *Cabinet Mountains Wilderness v. Peterson*, 510 F. Supp. 1186 (1981), affirmed by the D.C. Circuit Court of Appeals in 1982. In this case, the plaintiff environmental group was concerned with the lack of an environmental impact statement to be required by the U.S. Forest Service in a wilderness area for proposed exploratory drilling. The federal district court decided that an EIS was not required, but showed deference as it went on to comment:

> This case is but one of an increasing number of challenges in which sincere groups seek to protect the environment and understandably object to any intrusion of industry into the national forests. Equally sincere groups oppose these challenges, recognizing the need to manage the forests and make use of resources available in them when it is possible to so with a minimum of intrusion. The disputes between these groups are partly political, partly philosophical, and go to the core of decisions of how this nation can best achieve a continuing vitality. But as this debate rages, the courts are confined to a very minor role (*Cabinet Mountains Wilderness*, p. 1190).

NEPA was a very ambitious law at its time, approaching idealistic, yet with basic intent and structure to carry it to fruition. As evidenced, the course of NEPA in its first decade was fraught with controversy, both in courts and in Congress. Nevertheless, it plodded on because of its built-in power structure under the law; because of its idealistic goals for an improved environment; and because it is a noble cause (*cause noblesse*) that had caught the nation's spirit.

By the late 1970s, over the course of the decade from 1970 to 1980, NEPA's powers and procedures had been intensely litigated. The right for citizens and citizen groups to bring an action in the cause of environmentalism and for alleged violations of NEPA, as in *Calvert Cliffs* cited earlier, was not granted by NEPA, but by the authority of the Clean Air Act (CAA) of 1970 that followed shortly after passage of NEPA. Section 304 of the CAA authorized citizen action to enforce environmental law. Subsequent federal pollution control legislation incorporated the right of citizens to bring environmental actions (e.g., § 505 of the Federal Water Pollution Control Act; § 12 of the Noise Control Act of 1972; and § 7002 of the Resource Recovery and Conservation Act of 1976).

With many of the questions and issues settled as to how NEPA would operate and proceed, case law was becoming better established and NEPA became largely self-enforcing. It has been suggested that NEPA case law has become inert with federal courts failing to reexamine their earlier decisions, thus, resulting in no new growth or determinations. Perhaps, the adage should be applied, "if it's working, don't try to fix it." However, later NEPA decisions have not totally failed to show growth. The case law requirement of supplemental impact statements has been a later addition.

Although judiciary deference to established NEPA case law has increased over the years, continued judicial review should be maintained to prevent federal agency rule-making and interpretations to go unchallenged. Nevertheless, with NEPA having become entrenched in enforcement procedures, the eminent environmental attorney and professor, Thomas J. Schoenbaum of the University of Georgia's School of Law, has called NEPA a fading star. He states,

> By the late 1970s, the great era of NEPA litigation had crested and the (U.S.) Supreme Court began to have second thoughts about the monster it had created when it gave environmentalist plaintiffs the right to go into court.... Every time the Supreme Court decided a new NEPA case it whittled down the use environmentalists could make of the statutes. As a result, NEPA is a shadow of its former self....

He points out that with the demise of NEPA, the course of environmental litigation did not end but only changed direction because of a new generation of environmental statutes that were enacted. New enactments, such as the CAA, the Clean Water Act (CWA), the Resource Conservation and Recovery Act (RCRA) of 1976, the Comprehensive Environmental Response Compensation and Liability Act (CERCLA) of 1980 (known as Superfund), the Toxic Substances Control Act (TSCA), the Solid Waste Disposal Act (SWDA), the Endangered Species Act, and various others, supplemented and augmented NEPA in a variety of environmental areas (Schoenbaum, 1991, p. 17). NEPA had been created as a general environmental protection law. The various acts of the new generation of environmental laws were specific in that they each treated a special area or resource.

In addition to NEPA, the remaining acts of the so-called "Big Five" environmental acts, that is, CAA, CWA, CERCLA, and RCRA, are discussed in subsequent chapters. Lesser, but important, environmental acts are treated lightly in a catchall chapter. NEPA litigation will be encountered in the various environmental subject areas.

CHAPTER 4

Air Pollution and the Clean Air Act

4.1 NEED FOR AIR POLLUTION CONTROL

Man's individual daily consumption of air is about 35 lb. Decades ago, man was made aware by scientists that, if life on the earth is to continue, the envelope of air surrounding the earth containing a delicate mix of the gases that sustain us cannot be materially upset or radically changed. For decades, large cities have fought against heavy smoke and smog-filled skies (e.g., London, Los Angeles, Chicago, and New York), but without success under local and state controls. After decades of leaving air pollution control to local and state authorities without substantial improvement in air quality, it eventually became apparent that a central governmental control was needed.

4.1.1 Air Pollution Sources

Sources of air pollutants may be divided between natural and man-made. Natural sources either are beyond control of man or have limited significance. Examples of natural sources immediately bring to mind the eruption of volcanoes spewing volcanic dust, ash, and noxious gases. Other, less spectacular natural events are bacteria and spores; emissions from trees (arboreal); gases from forest and grass fires; lightning emitting nitrogen dioxide and ozone; naturally occurring radioactivity; spores, fungi, molds, and bacteria from plant decay; and a host of miscellaneous natural sources.

Man-made sources of air pollutants are more easily defined and identified, with the main contributors including transportation, stationary fuel combustion, industrial processes, and solid waste. The primary pollutants emitted to the air are particulate matter, sulfur oxides, nitrogen oxides, hydrocarbons (volatile organic compounds), carbon monoxide, and lead. (The first five of these were initially listed by the Environmental Protection Agency (EPA) in setting national ambient air quality standards (NAAQS) that will be cited later. Lead was added later to the list. The man-made source pollutant emissions made by various segments are given in Table 4.1 for the years 1970, 1975, 1985, and 1986.)

Air pollution, technically, includes sound emissions into the atmosphere as well as noxious gaseous and particulate matter. Noise pollution is regulated under the Noise Control Act of 1972 and also more directly by the Occupational Safety and Health Act (OSHA). Admittedly, noise pollution is of lesser concern because quietness is not as essential to life as clean air.

4.1.2 Common Law Control

Prior to mounting national concern and federal government intervention for abating heavy pollution of air, the common law dealt with complaints against noxious air by means of actions in nuisance. A common law action in trespass could also be brought against the bad actor (polluter) if (1) it interfered with the right of exclusive possession of the plaintiff's land, and (2) if the pollution was a direct result of an act committed by the defendant. A major limitation in the approach of a common law trespass action was the absence of a physical corpus in the trespass, as with smoke, gases, or even noise. Complaining that the culprit was a public nuisance was more effective in common law complaints. However, the common law did not really furnish an effective tool in law to combat air pollution. Complaints and common law litigation of public nuisances contaminating the air were frequent, but not very successful in air quality improvement.

As air contamination grew worse in the United States, it often became an interstate, or a regional problem, thereby making enforcement against a polluter under local laws more difficult. As a result, a federal common law of nuisance developed from interstate cases of pollution of air and water.

An early example is the case of *Georgia v. Tennessee Copper Co.*, 206 U.S. 230 (1907), in which the State of Georgia obtained injunctive relief against the Tennessee Copper Company situated on the state line to abate release of sulfur dioxide and noxious gases from its open-air roasting of cupric–ferrous sulfide ores. The noxious gases released from the process were combining with moisture from the air to form sulfurous acid and raining destruction on vegetation across the river on croplands and forests in Georgia. As a result of the injunction, Tennessee Copper Company was forced into collecting the destructive sulfur dioxide. The copper company subsequently put it into the production of sulfuric acid. This forced product outlet eventually made the copper company one of the foremost producers of sulfuric acid in the United States. What had formerly been a destructive waste product became its leading sales product.

4.1.3 Federal Control

The first federal entry into the air pollution field was in 1955 when Congress enacted the Air Pollution — Research and Technical Assistance Act. Under that act, Congress recognized that the prevention and control of air pollution at the sources was the primary responsibility of the state and local governments, and that federal financial assistance and leadership were essential for the development of cooperative air pollution prevention programs.

Accordingly, no police power by the federal government was exercised and the act offered only assistance through the publication of air pollution information and grants-in-aid for state and local government research. Minimum federal involvement in national air pollution drifted along for eight more years until the Clean Air Act (CAA) of 1963 was passed. Ostensibly, it continued the primacy of state control and responsibility, but this act gave a greater role to a federal agency, the U.S. Public Health Service; and the Secretary of Health, Education and Welfare (HEW) was given authority to publish nonmandatory air criteria. Additionally, further federal intrusive power was granted by authorizing the HEW secretary to intervene when pollution was alleged to endanger the health or welfare of the public and the state was unable to cope with the problem. In 1965, the 1963 act was amended along with a new Air Quality Act (AQA) of 1965. The AQA was the final step in placing the federal government in complete charge of national air pollution control. Since then, regulation of man-made air-contaminating sources has been intensified under a command and control doctrine.

4.2 CLEAN AIR ACT — 1970

The CAA of 1970 (42 U.S.C. § 7401–7842) was enacted soon after passage of the National Environmental Policy Act (NEPA). In treating this first and older version of the CAA, only the highlights will be reviewed to gain familiarity with it because it furnished the foundation and basic structure for implementing clean air regulation and control. The later, amended CAA of 1990 was so sweeping and so largely revamped that it should, and will, receive the main attention in this chapter. The basic structure and purposes of the 1970 act were carried forward into the 1990 amended act.

Under the act of 1970, each state has the primary responsibility for assuring air quality for its geographic area and for submitting an implementation plan. By it, the EPA was empowered to review federal actions that require an environmental impact statement (EIS). EPA was the federal agency charged with the administration of the CAA and the Clean Water Act (CWA). Section 109 of the CAA directed the EPA administrator to establish NAAQS for air pollutants endangering public health and welfare.

An **ambient air standard** is a legal expression of the permissible amount of pollutants generally allowed in the atmosphere. It is stated in terms of time and the permissible quantity. Time is expressed as hourly, a 30-day period, or annually; the permissible quantity is expressed in parts per million (ppm) or micrograms per cubic meter ($\mu g/m^3$) of air, or both. Most air quality standards are expressed in short terms, 3 or 24 hours. Long-term standards may not be exceeded on an annual basis.

Section 109(b) of the act directed the EPA to establish a national primary (for public health) ambient air quality standard, and a secondary (for public welfare, e.g., crops, animals, and structures) NAAQS for each air pollutant for which air quality criteria have been established by law.

A **primary pollutant** is one that is found in the atmosphere in the same form as it existed when emitted from an exhaust source. A **secondary pollutant** is one that is formed in the atmosphere as a result of reactions of hydrolysis, oxidation, and photochemistry (e.g., photochemical smog).

In 1970, the EPA's NAAQS designated six primary air pollutants as carbon monoxide, sulfur dioxide, nitrogen dioxide, photochemical oxidants, hydrocarbons, and particulates. Carbon monoxide is a colorless, odorless, poisonous, lighter-than-air gas produced from incomplete burning of carbon contained in fuels. Internal combustion engines from motor vehicles are the chief source of these emissions. Sulfur dioxide is a corrosive, poisonous gas produced from fuel or substances burned that contain sulfur. Coal-burning industrial and utility plants are the chief producers. Nitrogen dioxide is generated by the burning of fuels at very high temperature. Stationary combustion plants are the main sources. Photochemical oxidants, including ozone, formaldehydes, and peroxides, are produced in the lower atmosphere (troposphere) as a result of the reaction of oxides of nitrogen and volatile organics in the presence of solar radiation. Hydrocarbons are a product of incomplete combustion of hydrocarbon fuels. Particulates are air-suspended particles of solid or liquid substances produced from combustion and industrial processes. Precipitation is nature's primary cleansing mechanism for airborne gases and fine particles, but as is the case with SO_2, its solubility leads to faster dispersion, contamination, and degradation on the earth's surface.

The NAAQS standard was set to allow an adequate margin of safety requisite to protecting the public health. By § 110(a)(1), within 9 months after EPA had promulgated NAAQS, each state was required to submit a state implement plan (SIP) for approval. The federal agency also has oversight to enforce an approved SIP. Where the federal agency finds a violation and that state enforcement is not carried out, the federal agency may directly intercede with enforcement. The state has 30 days to take corrective action. The alleged violator also has 30 days to become in compliance. At the end of 30 days, the federal administrator may seek an injunction for cessation of the violator's operation, either temporary or permanent.

Clean air is now thought of as a right to be enjoyed by the nation and general public as a whole. In spite of the passage of the CAA of 1970, it took a period for the force of the new law to be established. However, as a holdover or remnant from previous thinking and court decisions at that time, the idea of pure, clean air had come to mean clean air consistent with the character of the locality and the attending circumstances. A post-NEPA public nuisance case of air pollution in 1974 occurred in *Chicago v. Commonwealth Edison*, 321 N.E.2d 412 (Ill. App. 1974). The city of Chicago attempted to enjoin the coal-burning electric utility located in Gary, Indiana, on the grounds that its pollution of air was a public nuisance. Whether smoke, odors, gaseous fumes, and dust constituted a nuisance depended on the facts in each (ad hoc) case. Several factors were weighed in *Commonwealth Edison* in determining whether there was an unreasonable interference with the right to clean air, for example:

1. What was the extent of harm or injury to the safety, health, peace, or comfort of the public?
2. What is the measure of the company's operational methods as compared with local, state, and federal governmental outlines?

3. What is the suitability of the industrial location?
4. What is the net result in balancing the harm done to the public as measured against the utility of the defendant's business to the community as a whole?

In *Commonwealth Edison*, the court found the plant to be located in a highly industrialized area and unpleasant odors were expected to exist. The utility of the power plant was found to outweigh the nuisance of its emissions. The air pollution continued, unabated, because it was justified under standards of the day.

In contrast, in 1975 in *Reserve Mining Co. v. EPA*, 514 F.2d 492 (8th Cir. 1975), the court upheld a claim of public nuisance where asbestos fibers were being released into the air in violation of Minnesota air quality standards, and there was a severe health danger to the people in the area.

The federal government's environmental strategy had been to make action-forcing rules, impossible to attain in the time frame allowed, believing that progress would be greater in alleviation of air pollutants. For example, the EPA's improbable plan set standards that all emissions were to be eliminated by 1975 to 1977, which did not happen. It also approved an unlikely-to-occur California SIP that was scheduled to reduce automobile traffic in Los Angeles by 70 to 80% in 1975. Nevertheless, action-forcing impossible goals were set as most likely to improve air quality in the shortest time.

Whether a state may, in developing its SIP, choose its own emission limitations as long as the national standards are attained, was a question answered in *Train v. Natural Resources Defense Council*, 421 U.S. 60, 79 (1975). The U.S. Supreme Court answered yes. This rule allows states flexibility in fixing limitations and time schedules for compliance, which in turn allows them to grant variances based on economic and technological grounds, subject to EPA approval as a revision of the SIP.

The 1970 CAA was amended in various subsequent years. In 1977, as amended, the EPA was required to review and revise the NAAQS by December 31, 1980, and at 5-year intervals thereafter. Two other important amendments of note to the 1970 CAA were: (1) § 304 which provided for citizens' suits whereby any person may bring a civil action in a federal district court "against the Administrator (EPA) where there is an alleged failure of the Administrator to perform any act or duty under this chapter which is not discretionary with the Administrator"; and (2) § 307 of the act vested the U.S. Court of Appeals for the District of Columbia with exclusive jurisdiction to review a variety of rules promulgated and other final actions by the administrator. However, it should be noted that there are a range of other EPA actions that can be reviewed in the appropriate circuit of courts of appeals. The jurisdictional scheme for judicial review of EPA decision making is actually bifurcated. The act provides that suits to compel the administrator to perform nondiscretionary duties may be brought only in district courts, whereas petitions seeking review of the administrator's discretionary actions must be brought in the District of Columbia Court of Appeals.

Few changes have been made in the NAAQS since 1971. In 1978, EPA added lead to the list, specifying a limiting standard level of concentration of 1.5 $\mu g/m^3$ in a maximum quarterly average.

The addition of lead was a result of a citizens suit, *Natural Resources Defense Council v. Train*, 545 F.2d 320 (2d Cir. 1976). The EPA had evaluated lead, conceding that lead met its standards for pollutants that "endanger the public health or welfare," but refusing to include it on its NAAQS list. The Natural Resources Defense Council brought suit under § 304 of the act (citizens' suit provision) seeking to compel EPA to include lead on the NAAQS. The court ruled that because EPA found lead as endangering public health and welfare, it had a nondiscretionary duty to include lead on its NAAQS list of pollutants, and ordered the addition of lead.

In 1979, EPA changed the ozone standard from 0.08 to 0.12 ppm, not to be exceeded more than 1 day/year. Standards for hydrocarbons were revoked in 1983. The sole function of hydrocarbon standards was to ensure attainment for ozone oxidants, which were later found unreliable. This reduction of 50% was upheld in *American Petroleum Institute v. Costle*, 665 F.2d 1176 (U.S. App. D.C. Cir. 1981).

4.2.1   Contaminant Source Points

Man-made point sources of air pollutant emissions are generally divided between mobile (vehicular) and stationary sources, with the latter being subdivided as to existing and new point sources.

Table 4.1 reveals the four main man-made source points for emission of air pollutants, viz., transportation, stationary fuel combustion, industry processes, and solid waste. Table 4.2 shows the percentages emitted by each of the four main categories. Transportation, or internal combustion engines, has been and remains the most culpable air polluter.

From Table 4.1, the following observations and deductions may be made.

For these same years, the total pollutant emissions were reduced: 1970 to 1975, a 22% reduction; from 1975 to 1985, a reduction of 50%; from 1985 to 1986, a reduction of 11%. An overall reduction of air pollutants for the 16-year period was 65%. A remarkable improvement in air quality was made over the period from 1970 to 1986.

4.2.2   Urban Pollution

The Pollutant Standards Index (PSI) is used as an air quality indicator for checking and describing urban area trends (Table 4.3). PSI is calculated at about a dozen monitoring sites around the nation and computes particulate matter (PM_{10}), SO_2, CO, O_3, and NO.

4.2.3   State Implementation Plan

The principal legal instrument for attaining the national AAQS is the SIP. Section 107(a) of the 1970 act requires that each state will formulate a SIP that will "specify the manner in which the primary and secondary AAQS will be achieved and mandated within each air quality control region in each state." Section 107(b) of the act requires state authorities to designate **air quality control regions**. The EPA may designate interstate air quality regions under § 107(c). The United States has been

TABLE 4.1 U.S. Air Pollutants According to Source and Type for the Years 1970, 1975, 1986

Year and Pollutant	All Man-Made Sources	Trans-porta-tion	Station-ary Fuel Com-bustion	Indus-trial Pro-cesses	Solid Waste	Other
Emissions in 10^6 Metric Tons per Year						
1970 — particulate matter	18.5	1.2	4.6	10.5	1.1	1.1
1975 — particulate matter	10.6	1.3	2.8	5.2	0.6	0.7
1986 — particulate matter	6.8	1.4	1.8	2.5	0.3	0.8
1970 — sulfur oxides	28.4	0.6	21.3	6.4	0.3	0.1
1975 — sulfur oxides	26.0	0.7	20.4	5.0	0.3	0.1
1986 — sulfur oxides	21.2	0.9	17.2	3.1	0.3	0.1
1970 — nitrogen oxides	18.1	7.6	9.1	0.7	0.4	0.3
1975 — nitrogen oxides	19.1	8.9	9.3	0.7	0.1	0.1
1986 — nitrogen oxides	19.3	8.5	10.0	0.6	0.1	0.1
1970 — hydrocarbons	27.5	12.4	1.1	8.9	1.8	3.3
1975 — hydrocarbons	22.8	10.2	1.1	8.1	0.9	2.5
1986 — hydrocarbons	19.5	6.5	2.3	7.9	0.6	2.2
1970 — carbon monoxide	98.7	71.8	4.4	9.0	6.4	7.2
1975 — carbon monoxide	81.0	62.0	4.2	6.9	3.1	4.8
1986 — carbon monoxide	60.9	42.6	7.2	4.5	1.7	5.0
Emissions in 10^3 Metric Tons per Year						
1970 — lead	203.8	163.6	9.6	23.9	6.7	5.0
1975 — lead	147.0	122.6	9.3	10.3	4.8	5.0
1985 — lead	8.6	3.5	0.5	1.9	2.7	5.0

Note: μg = a micron, 0.001 of a millimeter, or 1/25,000 of an inch, 10 μg particles can be seen by the unaided eye.

From Office of Air Quality Planning and Standards, Technical Support Division National Air Data Branch; National Air Pollutant Emissions Estimates, 1940-1986, EPA 450/4-87-024, U.S. EPA, Research Triangle Park, NC, January 1988; Health, United States 1988, U.S. Department of Health and Human Services, PHS, CDC, Hyattsville, MD, March 1989, p. 61.

Table 4.2 Percent Share of Total Air Pollutant Emissions by Category for the Years 1970, 1975, 1985, 1986

Source Category		1970	1975	1985	1986
Transportation	Emitted	65%	67%	51%	47%
Industry	Emitted	15%	12%	14%	15%
Stationary fuel combustion (mainly electric generation)		13%	15%	26%	29%
Solid waste	Emitted	4%	3%	4%	4%
Other	Emitted	3%	3%	6%	6%
Totals	Emitted	100%	100%	101%	101%

TABLE 4.3 Pollutant Standards Index

Index Range	Descriptor Words (Health Effect)
0–50	Good
51–100	Moderate
101–199	Unhealthful
200–299	Very unhealthful
300+	Hazardous

divided into 247 air quality control regions. Regions often have varying pollutant-contributing sources. Division into regions allows authorities to regulate the varying pollutants and their generating sources more exactingly and expeditiously.

Section 110 of the act is a directive of what a SIP is to contain and its need to be kept current. Section 110(a)(2(A) of the 1990 amended act states that a SIP must include "enforceable emission limitations and other control measures, means, or techniques (including economic incentives such as fees, marketable permits, and auctions of emissions rights), as well as schedules and timetables for compliance, as may be necessary or appropriate to meet the applicable requirements of this Act."

Under the SIPs, a state is responsible for bringing existing stationary sources (as industrial and power generating plants) into compliance with the AAQS. In nonattainment areas, where the primary criteria pollutants exceeded the NAAQS, reasonable available control technology (RACT) is required under a permitting program to gradually bring the sources into compliance with the standards.

This issue was the basis for *Union Electric Co. v. EPA*, 427 U.S. 246 (1976). Missouri had set standards for emission of SO_2 more stringent than the federal standards, which is permissible under the CAA. Union Electric petitioned for a review of the EPA-approved SIP for Missouri on the grounds that it was economically and technologically infeasible to comply with. Union Electric argued that compliance with emissions limitations was impossible, that Missouri had proceeded more rapidly than required, and that the state could not force technology on its own. The question before the court of appeals and the U.S. Supreme Court was "can the EPA Administrator reject an SIP under the Clean Air Act because it is economically or technologically infeasible?"

The Supreme Court, in affirming the appeals court, said no. However, *Union Electric* did not mean that a state could not offer special treatment in their SIP, or that all sources unable to comply must shut down. As long as requirements will be met, individual plants may be allowed hardship variances and extensions. The Chief Justice and Justice Powell, both concurring in the decision, noted that with respect to public utilities necessary to public welfare, more flexibility and leniency should be expected.

4.2.4 New Source Performance Standards

Finally, in review of the older CAA of 1970 and its amendments up to 1977, New Source Performance Standards (NSPS) should be mentioned. Section 111 of the act requires the EPA administrator to promulgate standards of performance for new stationary sources. Under this section, the EPA has sole authority to set emission limitations, grant variances, and regulate new sources of pollution. It is an exception to a state authority. The federal EPA publishes lists of categories for new sources and performance standards that must be incorporated and implemented in the SIPs. A *new source* is a stationary source commencing construction or modification after the date of a proposal of an applicable NSPS.

In determining whether the NSPS program applies, § 111(a)(3) defines any stationary source as "any building, structure, facility, or installation which emits, or may emit, any air pollutant." The NSPS

> shall reflect the degree of emission limitation and the percentage reduction achievable through the application of the best technology system of continuous emission reduction which (taking into consideration the cost of achieving such emission reduction, and any non-air quality health and environmental impact and energy requirements) the Administrator determines has been adequately demonstrated.

Attention is called that NSPS are set for industrial categories and not pollutants.

An early case involving a new source point was *New Mexico Citizens for Clean Air and Water v. Train,* 6 ERC 2061. The recommended case brief style, previously mentioned in Chapter 1, Section 2.4, is used here as an illustration for engineers following the topical form of: (1) nature of litigation, (2) background facts, (3) issues, (4) arguments, (5) court decisions, and (6) conclusions.

A Case Brief for *New Mexico Citizens for Clean Air and Water v. Train,* 6 ERC 2061

Nature of Litigation: An action to review a state permit for the construction of a new plant.

Background Facts: During a period of critical national need for increased copper smelting capabilities, Phelps Dodge Company agreed to construct a plant in New Mexico. Based on an EIS, the New Mexico EPA granted Phelps Dodge a permit to begin construction. New Mexico Citizens for Clean Air and Water (hereafter, Citizens) sought to enjoin construction and to have the permit revoked. Citizens argued that the plant would violate existing national primary and secondary ambient air quality standards. Additionally, they argued that sulfur effluents released will result in a significant deterioration of existing air quality. Citizens argued for permit revocation because the EIA had not required installation of the most modern pollution controls. Citizens alleged standing under the Citizens Provision, § 304 of the CAA Citizens complained that the New Mexico EIA failed to follow its own rules which prescribed a 90% reduction in emissions. Citizens failed to produce evidence on this point and it was disregarded.

Issue: Whether the alleged violations of national ambient air standards confer jurisdiction under Section 304(a)(1) of the Act.

Court's Decision: No. The Citizens only have standing to bring such a complaint if founded on actual performance violative of promulgated standards. Congress has granted jurisdiction only for citizen suits to challenge emission standards and limitations violations. National ambient standards are not emission standards. They are overall air quality standards, the enforcement of which is reposed in the Administrator. Citizens' second ground, significant deterioration of existing air standards, can prevail only if a regulation of the EPA makes such conduct a violation of the Clean Air Act. No such regulation exists. Citizens allege that the Administrator is in violation of the Act for failing to require the state to correct the situation. § 113 requires the Administrator to notify violators that their actions violate the Act. This duty is mandatory; enforcement by the Administrator is discretionary. A court cannot order him to initiate a suit. The Administrator investigated the allegations and concluded no violation existed. The non-degradation policy under the Act is also discretionary and the Administrator cannot be forced to bring suit to effectuate New Mexico's policy. This claim was stayed pending the court-ordered adoption of non-degradation regulations by the EPA. Citizen's argument that Phelps Dodge's plant will violate existing federal and state regulations is speculative and not real at present. Unless an actual showing of irreparable harm is made, no prospective injunctive relief can be granted.

Conclusions: The EPA has a duty to check all charges of air pollution emissions. However, an allegation against a source point that is not founded on fact, but is only speculative as to the future, cannot be adjudicated. Under the CAA, citizens may only challenge standards set by the EPA, or allege actual violations taking place. The national ambient air standards are not emission standards. The state has a wide latitude over violations.

4.2.5 Nonattainment Areas

In 1977, the CAA was amended adding several relief provisions for stationary sources unable to meet statutory deadlines for compliance with emission limitations. With the original 1975 EPA deadline for NAAQS unattained, it was extended to 1982. Early in 1983, EPA designated more than 100 counties in the United States as probable noncompliance areas. An exception for carbon monoxide was made, and for ozone, which seemed closely linked to vehicular emissions. The deadline for attainment of the primary NAAQS was extended and reextended several times, eventually to December 31, 1987, but the SIPs were required to contain a number of provisions designed to achieve the goals as expeditiously as possible. The statute provided that each plan "require permits for the construction and operation of new or modified stationary sources in accordance with § 173" in nonattainment areas (i.e., those areas that have not attained NAAQS). *Modified* means any physical change in the method of operation of a stationary source that increases the amount of air pollutants emitted into the air.

The extent of the controls over a modified stationary source is illustrated by *U.S. v. City of Painesville*, 644 F.2d 1186 (6th Cir. 1981), where a contract for the purchase of a boiler after publication of regulations made it a new source despite planning activities prior to the regulation. In *Painesville*, even a subsequent court remand of the NSPS did not nullify the obligation for the defendant to comply.

Before issuing a permit, § 173 requires the state agency to determine that (1) there will be sufficient emissions reductions in the region to offset the emissions from the new source and also to allow for reasonable further progress toward attainment, or the increased emissions will not exceed an allowance for growth established pursuant to § 172(b)(5); (2) the applicant must certify that his other sources in the state are in compliance with the SIP; (3) the agency must determine that the applicable SIP is otherwise being implemented; and (4) the proposed source complies with the lowest achievable emission rate (LAER).

4.2.6 Hazardous Air Pollutants

Hazardous air pollutants for existing and new sources are subject to direct federal regulation under the National Emission Standards for Hazardous Air Pollutants (NESHAP). Section 112(a)(1) of the Act defines a hazardous pollutant as "one to which no ambient air quality standard is applicable, and which in the judgement of the Administrator causes, or contributes to, air pollution which may reasonably be anticipated to result in an increase in mortality or an increase in serious irreversible or incapacitating, reversible, illness."

For 20 years under the former § 112 of the CAA, the program resulted in the issue of only eight standards for hazardous air pollutants (i.e., asbestos, beryllium, mercury, vinyl chloride prior to 1977, and after the 1977 amendments, benzene, radionuclides, arsenic, and coke oven emissions). By 1984, only 19 states had any control programs in place over toxic air pollutants.

Several prominent court cases involving hazardous air pollutants and the EPA in the earlier years, viz., *Reserve Mining Co. v. EPA*, 514 F.2d 492 (8th Cir. 1975) concerning asbestos and iron mining wastes (this case will be reviewed later in Chapter 5 on water pollution); *Adamo Wrecking Co. v. U.S.*, 434 U.S. 275 (1978) concerning removal of asbestos; *Ethyl Corp. v. EPA*, 541 F.2d 1 (D.C. Cir. 1976) concerning gasoline additives (lead); *Lead Industries Association v. EPA*, 647 F.2d 1130 (D.C. Cir. 1980) concerning lead; *AFL–CIO v. American Petroleum Institute*, 448 U.S. 607 (1980) concerning benzene; *NRDC v. U.S. EPA*, 824 F.2d 1146 (D.C. Cir. 1987) concerning vinyl chloride.

4.2.7 Offsets and the Bubble Effect

In nonattainment areas, EPA gave consideration to the question whether a plant with multiple emission points should be considered a single source, or each emission point (e.g., a single smokestack) within an industrial plant or complex should be counted as a single source point (the latter concept is called **netting**). Under the plant-wide concept, new equipment that is less pollutant-emitting and will yield a reduction in total plant-wide pollutant emissions may be averaged with other existing equipment that is over-producing pollutants in its emissions so long as the average will yield a total reduction of pollutants into the air. This is called an **offset**. The use of offsets within the same source, or plant, is called the **bubble**.

The same concept of offset may be used on larger scales as in an SIP for a region or local area of several plants where there is nonattainment. Generally, *offsets* refer to intersource reductions of air pollutants, and *bubble* refers to intrasource reductions of air pollutants (i.e., multi-plant facilities under common ownership).

The question of the bubble concept had been at issue in three cases before the D.C. Circuit court. In two of them, the bubble policy was held invalid for new source standards, whereas in the third case, which dealt with prevention of significant deterioration (PSD) rules, the concept was upheld. On reaching the U.S. Supreme

Court in *Chevron U.S.A., Inc. v. NRDC*, 467 U.S. 837 (1984), the bubble policy was upheld (see case that follows).

An excerpt concerning the bubble concept from an earlier case, *Alabama Power Co. v. Costle,* 636 F.2d 323 (D.C. Cir. 1979) gives some insight with its relation to the PSD program. Circuit Judge Wilkey wrote this part of the court's opinion.

II B. EPA's Qualified Application of the "Bubble" Concept of PSD

An important issue under the Act arises from the problem of determining what types of industrial changes will be construed as "modifications" subject to PSD review requirements. Under the Act, the PSD permit and review process applies to construction and modification of major emitting facilities. As discussed in the previous section, the Act defines "modification" as any physical or operational change in a stationary source which "increases the amount of any air pollutant emitted by such source."[48] There are two possible ways to construe the term "increases." First, one can look at any change proposed for a plant, and decide whether the net effect of all the steps involved in that change is to increase the emission of any air pollutant — this is commonly termed the "bubble" concept. Second, one can inspect the individual units of a plant, which are affected by an operational change, and determine whether any of the units will consequently emit more of a pollutant. In its regulations, EPA has adopted a qualified form of the "bubble" concept for defining modifications subject to PSD review.

Congress did not, in any pertinent part of legislative history, specify which of these two constructions was to be controlling, but an analysis of the implications of the two possible interpretations shows the second to be unreasonable and contrary to the expressed purposes of the PSD provisions of the Act. It is important first to recognize that alterations of almost any plant occur continuously; whether to replace depreciated capital goods, to keep pace with technological advances, or to respond to changing consumer demands. This dynamic aspect of American industry was not disputed by the parties. To apply the second construction of "increases," however, would require PSD review for many such routine alterations of a plant; a new unit would contribute additional pollutants, these increases could not be set off against the decrease resulting from abandonment of the old unit, and thus the change would become a "modification" subject to PSD review. Not only would this result be extremely burdensome, it was never intended by Congress in enacting the Clean Air Act Amendments.

The intent of the relevant portion, Part C, of the Clean Air Act as amended in 1977, is succinctly stated by the title of that part: "Prevention of Significant Deterioration of Air Quality" — in areas that currently attain air quality standards. According to their stated purposes, the PSD provisions seek *"to* assure that any decision to permit *increased* air pollution *in any area to* which this section applies is made only after careful evaluation of all the consequences of such a decision and after adequate procedural opportunities for informed public participation in the decision making process."

Congress wished to apply the permit process, then, only where industrial changes might increase pollution in an area, not where an existing plant changed its operations in ways that produced no pollution increase. It is true that Congress intended to generate technological improvement in pollution control, but this approach focused upon "rapid adoption of improvements in technology as new sources are built," not as old ones were changed without pollution increase. The interpretation of "modification" as requiring a net increase is thus consistent with the purpose of the Act; while the other interpretation is not. The

EPA has properly exempted from best available control technology (BACT) and ambient air quality review those "modifications" of a source that do not produce a net increase in any pollutant. Within the terminology of the Act, of course, industrial changes meeting this standard are not "modifications" at all.

The "bubble" regulation for PSD must be compared with an earlier EPA regulation, which applied the bubble concept to the new source performance standards of the Act and which was struck down by this court in *ASARCO Inc. v. Environmental Protection Agency.* That regulation stated that a modification of a source for NSPS purposes "shall not be deemed to occur if an existing facility undergoes a physical or operational change where ... the total emission rate of any pollutant has not increased from all facilities within the stationary source...."

The *ASARCO* case struck down that regulation because it expanded the definition of "source," within which offsets were allowed, to include combinations of facilities, contrary to the statutory definition of "source." Here we start with the same premise as *ASARCO,* that the Agency may not define "source" to include a combination of facilities. Several factors prevent us, however, from drawing the same conclusion. First is a difference between the two regulations. The present EPA regulation allows offsets within a "source"; it does not, in light of our decision in this case, allow offsets within any "combination of facilities." Thus, it does not suffer from the defect on which the *ASARCO* decision turned. Second, *ASARCO* did not rule out the interpretation of "increases" in pollution as net increases. The case stated that a bubble concept would be contrary to the intent of the NSPS provisions, but such is clearly not the case with regard to the PSD provisions. Third, the PSD provisions express a purpose of ensuring that economic growth occurs in a manner consistent with preservation of clean air. The bubble concept is precisely suited to preserve air quality within a framework that allows cost-efficient, flexible planning for industrial expansion and improvement. Finally, it is relevant that EPA had its NSPS bubble concept in effect at the time Congress enacted the 1977 Clean Air Act Amendments. Though we are reluctant to assume that Congress expressly endorsed the specific bubble regulation, the Conference Committee approved the congressional policy as enacted at that time in existing EPA regulations. *ASARCO,* in short, dealt with a significantly different regulation and statutory purpose. Its holding is therefore not inconsistent with our decision today, upholding the bubble concept for the PSD regulations.

The Agency retains substantial discretion in applying the bubble concept. First, any offset changes claimed by industry must be substantially contemporaneous. The Agency has discretion, within reason, to define which changes are substantially contemporaneous. Second, the offsetting changes must be within the same source, as defined by EPA. In light of the statutory intent to treat modification the same as construction, EPA's definition of "statutory source" for the PSD provisions will govern both the definition of "modification" and the coverage of section 169(1).

---

[48] Clean Air Act § 111(a)(4), 42 U.S.C. § 7411(a)(4) (Supp. I 1977).

As previously described, the effect of the bubble concept was to make an allowance for an industrial firm to *offset* one source of pollution with a reduction in another source. As stated earlier, the bubble policy was upheld by the U.S. Supreme Court in *Chevron U.S.A., Inc. v. NRDC,* 467 U.S. 837 (1984). The most pertinent arguments of *Chevron* are given in the following longer reproduction of the court's decision. Certain, less pertinent parts have been omitted.

For contrast, a case brief of the same *Chevron* decision follows the longer version. Reading of the longer version is recommended because it furnishes a valuable description of the rules, regulations, and procedures under the CAA.

CHEVRON U.S.A., INC. v. NATURAL RESOURCES DEFENSE COUNCIL, INC.

Supreme Court of the United States, 1984
467 U.S. 837

Justice STEVENS delivered the opinion of the Court.

In the Clean Air Act Amendments of 1977, Pub. L. 95-95, 91 Star. 685, Congress enacted certain requirements applicable to States that had not achieved the national air quality standards established by the Environmental Protection Agency (EPA) pursuant to earlier legislation. The amended Clean Air Act required these "nonattainment" States to establish a permit program regulating "new or modified major stationary sources" of air pollution. Generally, a permit may not be issued for a new or modified major stationary source unless several stringent conditions are met. The EPA regulation promulgated to implement this permit requirement allows a State to adopt a plant-wide definition of the term "stationary source."[2] Under this definition, an existing plant that contains several pollution-emitting devices may install or modify one piece of equipment without meeting the permit conditions if the alteration will not increase the total emissions from the plant. The question presented by this case is whether EPA's decision to allow States to treat all of the pollution-emitting devices within the same industrial grouping as though they were encased within a single "bubble" is based on a reasonable construction of the statutory term "stationary source."

I

The EPA regulations containing the plant-wide definition of the term stationary source were promulgated on October 14, 1981. 46 Fed. Reg. 50766. Respondents filed a timely petition for review in the United States Court of Appeals for the District of Columbia Circuit pursuant to 42 U.S.C. § 7607(b)(1). The Court of Appeals set aside the regulations. *National Resources Defense Council, Inc. v. Gorsuch*, 222 U.S. App. D.C. 268, 685 F.2d 718 (1982).

The court observed that the relevant part of the amended Clean Air Act "does not explicitly define what Congress envisioned as a 'stationary source' to which the permit program ... should apply," and further stated that the precise issue was not "squarely addressed in the legislative history." *Id.*, at 273, 685 F.2d, at 723. In light of its conclusion that the legislative history bearing on the question was best contradictory, it reasoned that "the purposes of the nonattainment program should guide our decision here." *Id.*, at 276, n. 39, 685 F.2d, at 726, n. 39. Based on two of its precedents concerning the

---

[2] "(i) 'Stationary source' means any building, structure, facility, or installation which emits or may emit any air pollutant subject to regulation under the Act."

"(ii) 'Building, structure, facility, or installation' means all of the pollutant-emitting activities which belong to the same industrial grouping, are located on one or more contiguous or adjacent properties, and are under the control of the same person (or persons under common control) except the activities of any vessel." 40 CFR § 51.18(j)(1)(i) and (ii) (1983).

applicability of the bubble concept to certain Clean Air Act programs, the court stated that the bubble concept was "mandatory" in programs designed merely to maintain existing air quality, but held that it was "inappropriate" in programs enacted to improve air quality, *Id.*, at 276, 685 F.2d, at 726. Since the purpose of the permit program — its "Raison d'etre," in the court's view — was to improve air quality, the court held that the bubble concept was inapplicable in this case under its prior precedents. *Ibid.* It therefore set aside the regulations embodying the bubble concept as contrary to law. We granted certiorari to review that judgement, 461 U.S. 956, 103 S.Ct. 2427, 77 L.Ed.2d 1314(1983), and we now reverse.

In light of ... well-settled principles it is clear that the Court of Appeals misconceived the nature of its role in reviewing the regulations at issue. Once it determined, after its own examination of the legislation, that Congress did not actually have an intent regarding the applicability of the bubble concept to the permit program, the question before it was not whether in its view the concept is "inappropriate" in the general context of a program designed to improve air quality, but whether the Administrator's view that it is appropriate in the context of this particular program is a reasonable one. Based on the examination of the legislation and its history which follows, we agree with the Court of Appeals that Congress did not have a specific intention on the applicability of the bubble concept in these cases, and conclude that the EPA's use of that concept here is a reasonable policy choice for the agency to make.

The Clean Air Act Amendments of 1977 are a lengthy, detailed, technical, complex, and comprehensive response to a major social issue. A small portion of the statute — 91 Stat. 745–751 (Part D of Title I of the amended Act, 42 U.S.C. §§ 7501–7508) expressly deals with nonattainment areas. The focal point of this controversy is one phrase in that portion of the Amendments.

Basically, the statute required each State in a nonattainment area to prepare and obtain approval of a new SIP by July 1, 1979. In the interim those States were required to comply with the EPA's interpretative Ruling of December 21, 1976. 91 Stat. 745. The deadline for attainment of the primary NAAQS's was extended until December 31, 1982, and in some cases until December 31, 1987, but the SIP's were required to contain a number of provisions designed to achieve the goals as expeditiously as possible.

Most significantly for our purposes, the statute provided that each plan shall:

"6) require permits for the construction and operation of new or modified major stationary sources in accordance with section 173...." 91 Stat. 747.

Before issuing a permit, § 173 requires the state agency to determine that (1) there will be sufficient emissions reductions in the region to offset the emissions from the new source and also to allow for reasonable further progress toward attainment, or that the increased emissions will not exceed an allowance for growth established pursuant to § 172(b)(5); (2) the applicant must certify that his other sources in the State are in compliance with the SIP, (3) the agency must determine that the applicable SIP is otherwise being implemented, and (4) the proposed source complies with the lowest achievable emission rate (LAER).

The 1977 Amendments contain no specific reference to the "bubble concept." Nor do they contain a specific definition of the term "stationary source," though they did not disturb the definition of "stationary source" contained in § 111(a)(3), applicable by the terms of the Act to the NSPS program. Section 302(j), however, defines the term "major stationary source" as follows:

"(j) Except as otherwise expressly provided, the terms 'major stationary source' and 'major emitting facility' mean any stationary facility or source of air pollutants which directly emits, or has the potential to emit, one hundred tons per year or more of any air pollutant (including any major emitting facility or source of fugitive emissions of any such pollutant, as determined by rule by the Administrator)." 91 Stat. 770.

The legislative history of the portion of the 1977 Amendments dealing with nonattainment areas does not contain any specific comment on the "bubble concept" or the question whether a plant-wide definition of a stationary source is permissible under the permit program. It does, however, plainly disclose that in the permit program Congress sought to accommodate the conflict between the economic interest in permitting capital improvements to continue and the environmental interest in improving air quality. Indeed, the House Committee Report identified the economic interest as one of the "two main purposes" of this section of the bill. It stated:

"Section 117 of the bill, adopted during full committee markup establishes a new section 127 of the Clean Air Act. The section has two main purposes: (1) to allow reasonable economic growth to continue in an area while making reasonable further progress to assure attainment of the standards by a fixed date; and (2) to allow States greater flexibility for the former purpose than EPA's present interpretative regulations afford.

"The new provision allows States with nonattainment areas to pursue one of two options. First, the State may proceed under EPA's present 'tradeoff' or 'offset' ruling. The Administrator is authorized, moreover, to modify or amend that ruling in accordance with the intent and purposes of this section. Specifically, the controversy in this case involves the meaning of the term 'major stationary sources' in § 172(b)(6) of the Act, 42 U.S.C. § 7502(b)(6). The meaning of the term 'proposed source' in § 173(2) of the Act, 42 U.S.C. § 7503(2), is not at issue."

"The State's second option would be to revise its implementation plan in accordance with this new provision." H.R. Rep. No. 95-294, p. 211 (1977).

The portion of the Senate Committee Report dealing with nonattainment areas states generally that it was intended to "supersede the EPA administrative approach," and that expansion should be permitted if a State could "demonstrate that these facilities can be accommodated within its overall plan to provide for attainment of air quality standards." S. Rep. 95-127, p. 55 (1977). The Senate Report notes the value of "case-by-case review of each new or modified major source of pollution that seeks to locate in a region exceeding an ambient standard," explaining that such a review "requires matching reductions from existing sources against emissions expected from the new source in order to assure that introduction of the new source will not prevent attainment of the applicable standard by the statutory deadline." *Ibid.* This description of a case-by-case approach to plant additions, which emphasizes the net consequences of the construction or modification of a new source, as well as its impact on the overall achievement of the national standards, was not, however, addressed to the precise issue raised by this case.

***

As previously noted, prior to the 1977 Amendments, the EPA had adhered to a plant-wide definition of the term "source" under a NSPS program. After adoption of the 1977 Amendments, proposals for a plant-wide definition were considered in at least three formal proceedings.

In January 1979, the EPA considered the question whether the same restriction on new construction in nonattainment areas that had been included in its December 1976 ruling should be required in the revised SIPs that were scheduled to go into effect in July 1979. After noting that the 1976 ruling was ambiguous on the question "whether a plant with a number of different processes and emission points would be considered a single source,"

44 Fed. Reg. 3276 (1979), the EPA, in effect, provided a bifurcated answer to that question. In those areas that did not have a revised SIP in effect by July 1979, the EPA rejected the plant-wide definition; on the other hand, it expressly concluded that the plant-wide approach would be permissible in certain circumstances if authorized by an approved SIP. It stated:

"Where a state implementation plan is revised and implemented to satisfy the requirements of Part D, including the reasonable further progress requirement, the plan requirements for major modifications may exempt modifications of existing facilities that are accompanied by intra-source offsets so that there is no net increase in emissions. The agency endorses such exemptions, which would provide greater flexibility to sources to effectively manage their air emissions at least cost." *Ibid.*

In April, and again in September 1979, the EPA published additional comments in which it indicated that revised SIPs could adopt the plant-wide definition of source in nonattainment areas in certain circumstances. (See *id.*, at 20372, 20379, 51924, 51951, 51958.) On the latter occasion, the EPA made a formal rulemaking proposal that would have permitted the use of the "bubble concept" for new installations within a plant as well as for modifications of existing units. It explained:

"Bubble' Exemption: The use of offsets inside the same source is called the 'bubble.' EPA proposes use of the definition of 'source' (see above) to limit the use of the bubble under nonattainment requirements in the following respects:

"i. Part D SIPs that include all requirements needed to assure reasonable further progress and attainment by the deadline under section 172 and that are being carried out need not restrict the use of a plant-wide bubble, the same as under the PSD proposal.

"ii. Part D SIPs that do not meet the requirements specified must limit use of the bubble by including a definition of 'installation' as an identifiable piece of process equipment."

Significantly, the EPA expressly noted that the word "source" might be given a plant-wide definition for some purposes and a narrower definition for other purposes. It wrote:

"Source means any building, structure, facility, or installation which emits or may emit any regulated pollutant. 'Building, structure, facility or installation' means plant in PSD areas and in nonattainment areas except where the growth prohibitions would apply or where no adequate SIP exists or is being carried out." [*Id.*, at 51925.][28]

The EPA's summary of its proposed ruling discloses a flexible rather than rigid definition of the term "source" to implement various policies and programs:

"In summary, EPA is proposing two different ways to define source for different kinds of NSR programs:

"(1) For PSD and complete Part D SIPs, review would apply only to plants, with an unrestricted plant-wide bubble.

"(2) For the offset ruling, restrictions on construction, and incomplete Part D SIPs, review would apply to both plants and individual pieces of process equipment, causing the plant-wide bubble not to apply for new and modified major pieces of equipment.

"In addition, for the restrictions on construction, EPA is proposing to define 'major modification' so as to prohibit the bubble entirely. Finally, an alternative discussed but not favored is to have only pieces of process equipment reviewed, resulting in no plant-wide bubble and allowing minor pieces of equipment to escape NSR regardless of whether they are within a major plant." [*Id.* at 51934.]

---

[28] In its explanation of why the use of the bubble concept was especially appropriate in preventing significant deterioration (PSD) in clean air areas, the EPA stated: "In addition, application of the bubble on a plant-wide basis encourages voluntary upgrading of equipment, and growth in productive capacity." [*Id.*, at 51932.]

In August 1980, however, the EPA adopted a regulation that, in essence, applied the basic reasoning of the Court of Appeals in this case. The EPA took particular note of the two then-recent Court of Appeals decisions, which had created the bright-line rule that the bubble concept should be employed in a program designed to maintain air quality but not in one designed to enhance air quality. Relying heavily on those cases,[29] EPA adopted a dual definition of "source" for nonattainment areas that required a permit whenever a change in either the entire plant, or one of its components, would result in a significant increase in emissions even if the increase was completely offset by reductions elsewhere in the plant. The EPA expressed the opinion that this interpretation was "more consistent with congressional intent" than the plant-wide definition because it "would bring in more sources or modifications for review" 45 Fed. Reg. 52697 (1980), but its primary legal analysis was predicated on the two Court of Appeals decisions.

In 1981, a new administration took office and initiated a "Government-wide reexamination of regulatory burdens and complexities." 46 Fed. Reg. 16281. In the context of that review, the EPA reevaluated the various arguments that had been advanced in connection with the proper definition of the term "source" and concluded that the term should be given the same definition in both nonattainment areas and PSD areas.

In explaining its conclusion, the EPA first noted that the definitional issue was not squarely addressed in either the statute or its legislative history and therefore that the issue involved an agency "judgement as how to best carry out the Act." *Ibid.* It then set forth several reasons for concluding that the plant-wide definition was more appropriate. It pointed out that the dual definition "can act as a disincentive to new investment and modernization by discouraging modifications to existing facilities" and "can actually retard progress in air pollution control by discouraging replacement of older, dirtier processes or pieces of equipment with new, cleaner ones." *Ibid.* Moreover, the new definition "would simplify EPA's rules by using the same definition of 'source' for PSD, nonattainment new source review and the construction moratorium. This reduces confusion and inconsistency." *Ibid.* Finally, the agency explained that additional requirements that remained in place would accomplish the fundamental purposes of achieving attainment with NAAQs as expeditiously as possible. These conclusions were expressed in a proposed rulemaking in August 1981 that was formally promulgated in October. [See *id.* at 50766.]

Statutory Language

The definition of the term stationary source in § 111(a)(3) refers to "any building, structure, facility, or installation" which emits air pollution. This definition is applicable only to the NSPS program by the express terms of the statute; the text of the statute does not make this definition "applicable to the permit program." Petitioners therefore maintain that there is no statutory language even relevant to ascertaining the meaning of stationary source in the permit program aside from § 302(j), which defines the term major stationary source. See *supra*, at 12. We disagree with petitioners on this point.

---

[29] "The dual definition also is consistent with *Alabama Power* and *ASARCO. Alabama Power* held that EPA had broad discretion to define the constituent terms of 'source' so as best to effectuate the purposes of the statute. Different definitions of 'source' can therefore be used for different sections of the statute....
"Moreover, *Alabama Power* and *ASARCO* taken together suggest that there is a distinction between Clean Air Act programs designed to *enhance* air quality and those designed only to *maintain* air quality....
"Promulgation of the dual definition follows the mandate of *Alabama Power,* which held that, while EPA could not define 'source' as a combination of sources, EPA had broad discretion to define 'building,' 'structure,' 'facility,' and 'installation' so as to best accomplish the purposes of the Act." 45 Fed. Reg. 52697 (1980).

The definition in § 302(j) tells us what the word "major" means — a source must emit at least 100 tons of pollution to qualify — but it sheds virtually no light on the meaning of the term "stationary source." It does equate a source with a facility — a "major emitting facility" and a "major stationary source" are synonymous under § 302(j). The ordinary meaning of the term facility is some collection of integrated elements which has been designed and constructed to achieve some purpose. Moreover, it is certainly no affront to common English usage to take a reference to a major facility or a major source to connote an entire plant as opposed to its constituent parts. Basically, however, the language of § 302(j) simply does not compel any given interpretation of the term source.

Respondents recognize that, and hence point to § 111(a)(3). Although the definition in that section is not literally applicable to the permit program, it sheds as much light on the meaning of the word source as anything in the statute. As respondents point out, use of the words "building, structure, facility, or installation," as the definition of source, could be read to impose the permit conditions on an individual building that is a part of a plant. A "word may have a character of its own not to be submerged by its association." *Russell Motor Car Co. v. United States*, 261 U.S. 514, 519, 43 S. Ct. 428, 429, 67 L. Ed. 778 (1923). On the other hand, the meaning of a word must be ascertained in the context of achieving particular objectives, and the words associated with it may indicate that the true meaning of the series is to convey a common idea. The language may reasonably be interpreted to impose the requirement on any discrete, but integrated, operation which pollutes. This gives meaning to all of the term — a single building, not part of a larger operation, would be covered if it emits more than 100 tons of pollution, as would any facility, structure, or installation. Indeed, the language itself implies a bubble concept of sorts: each enumerated item would seem to be treated as if it were encased in a bubble. While respondents insist that each of these terms must be given a discrete meaning, they also argue that § 111(a)(3) defines "source" as that term is used in § 302(j). The latter section, however, equates a source with a facility, whereas the former defines source as a facility, among other items.

We are not persuaded that parsing of general terms in the text of the statute will reveal an actual intent of Congress. We know full well that this language is not dispositive; the terms are overlapping and the language is not precisely directed to the question of the applicability of a given term in the context of a larger operation. To the extent any congressional "intent" can be discerned from this language, it would appear that the listing of overlapping, illustrative terms was intended to enlarge, rather than to confine, the scope of the agency's power to regulate particular sources in order to effectuate the policies of the Act.

Significantly, it was not the agency in 1980, but rather the Court of Appeals that read the statute inflexibly to command a plant-wide definition for programs designed to maintain clean air and to forbid such a definition for programs designed to improve air quality. The distinction the court drew may well be a sensible one, but our labored review of the problem has surely disclosed that it is not a distinction that Congress ever articulated itself, or one that the EPA found in the statute before the courts began to review the legislative work product. We conclude that it was the Court of Appeals, rather than Congress or any of the decision makers who are authorized by Congress to administer this legislation, that was primarily responsible for the 1980 position taken by the agency.

Policy

Judges are not experts in the field, and are not part of either political branch of the Government. Courts must, in some cases, reconcile competing political interests, but not on the basis of the judges' personal policy preferences. In contrast, an agency to which Congress has delegated policymaking responsibilities may, within the limits of that delegation, properly rely upon the incumbent administration's views of wise policy to inform its judgements. While agencies are not directly accountable to the people, the Chief Executive is, and it is entirely appropriate for this political branch of the Government to make such policy choices — resolving the competing interests which Congress itself either inadvertently did not resolve, or intentionally left to be resolved by the agency charged with the administration of the statute in light of everyday realities.

When a challenge to an agency construction of a statutory provision, fairly conceptualized, really centers on the wisdom of the agency's policy, rather than whether it is a reasonable choice within a gap left open by Congress, the challenge must fail. In such a case, federal judges — who have no constituency — have a duty to respect legitimate policy choices made by those who do. The responsibilities for assessing the wisdom of such policy choices and resolving the struggle between competing views of the public interest are not judicial ones: "Our Constitution vests such responsibilities in the political branches." *TVA v. Hill*, 437 U.S. 153, 195, 98 S.Ct. 2279, 2302, 57 L. Ed.2d 117 (1978).

We hold that the EPA's definition of the term "source" is a permissible construction of the statute which seeks to accommodate progress in reducing air pollution with economic growth. "The Regulations which the Administrator has adopted provide what the agency could allowably view as ... [an] effective reconciliation of these twofold ends...." *United States v. Shimer*, 367 U.S., at 383, 81 S.Ct., at 1560.

The judgement of the Court of Appeals is reversed.

For contrast with the longer version, above, of the *Chevron* decision, read the much shorter case brief following.

Chevron, U.S.A., Inc. v. Natural Resources Defense Council, Inc.,

467 U.S. 837 (1984)

Nature of litigation: An appeal from decision setting aside EPA regulation.

Background Facts: The 1977 amendments to the Clean Air Act were designed to address the non-attainment problems posed by states which had failed to meet earlier clean air standards and schedules enacted by the EPA. Under the amended Act, these "nonattainment states" were required to establish a permit program as to new or modified stationary sources of air pollution which restricted the issuance of permits unless stringent requirements were met. The EPA promulgated regulations which provided for a plant-wide definition of the term "stationary source" that allowed states to treat all of the pollution emitting devices within the same industrial grouping as if they were encased within a single "bubble." After the regulations containing the plant-wide definition for the term "stationary source" were promulgated in 1981, the Natural Resources Defense Council (NRDC) petitioned for review. The court of appeals set aside the regulations which concluded that the adoption of the definition was inappropriate in the general context of the program designed to improve air quality. Chevron appealed.

Issue: Where Congress has implicitly delegated a particular question of statutory construction to the appropriate agency for determination, can a court substitute its own construction of the provision in place of a reasonable interpretation made by the administrator of the agency?

Court's Decision: No. Where Congress has implicitly delegated a particular question of statutory construction to the appropriate agency for determination, a court may not substitute its own construction of the provision in place of a reasonable interpretation made by the administrator of the agency. Considerable weight should be accorded to the construction given by the agency entrusted to administer the program in question in accordance with the principle of deference to administrative interpretation. Congress has not explicitly spoken on the issue before this Court. However, the legislative history demonstrates that in adopting the 1977 amendments to the Act, Congress sought to accommodate the conflict between economic interests and environmental concerns. Previously adopted EPA regulations created a bright-line rule that the "bubble" concept was inappropriate within the context of a program designed but upon earlier court of appeals decisions which inflexibly read the statute to create the distinction between programs designed to enhance air quality and those designed to maintain air quality. The record indicates that the plant-wide definition for the term 'stationary source' is fully consistent with one of the underlying concerns faced by Congress: allowance for reasonable economic growth. Courts are ill equipped to decide such questions of policy, and have a duty to respect legitimate policy choices made by those who have been entrusted with the authority to resolve such questions, as the EPA. Since the EPA's definition of the term at issue was based on a reasonable construction of the statutory provision, the court of appeals erred in substituting its own choice in light of that reasonable construction.

The decision of the Court of Appeals is reversed.

Having become familiarized with the fundamentals of the older air pollution controls, we progress to an outline of the 1990 amended CAA, which is presently in force.

4.3 CAA AMENDMENTS — 1990

Sweeping amendments of 1990 revamped the CAA (1977) [and as amended November 15, 1990 Public Law 101-549, 104 Stat. 2399 (1990), U.S.C. § 7401–7642.] Three major areas of the former CAA that were thought to require overhauling were (1) the nonattainment program for air quality standards (i.e., the NAAQS, Title I) (2) the hazardous air pollutant program, Title III, and (3) acid rain pollutants under Title IV. The amended act was said to be one of the most significant pieces of environmental statutory legislation ever devised. "Its very size dwarfs the former CAA" and the costs to implement it were estimated at the time to range annually from $20 billion to $100 billion.

4.3.1 Outline of the 1990 CAA Amendments

The various titles of the amended 1990 CAA are summarized as follows. Notations on the more important provisions are given below each title summarization.

4.3.1.1 Title I — Attainment and Maintenance of National Ambient Air Quality Standards

The amended act covered and revised the nonattainment provisions of the NAAQS, dividing those areas into categories, with particular attention to those not attaining the standards for ozone, carbon monoxide, and particulate matter (PM_{10}). New designations for lead were also required. Sanctions by EPA were provided for states and cities not reducing overall emissions and not attaining compliance.

§ 102 — Non-attainment area designations: The states must designate regions as attainment, non-attainment, or unclassifiable with regard to the NAAQS.

§§ 103–104 — Submission and approval of SIPs and implementation plans for NAAQS by reasonable available control technology (RACT); permitting requirements for new and modified sources require analyses for alternative site, processes and control techniques.

§ 105 — Non-attainment areas: Classification of ozone, CO, and particulate matter (PM_{10}). Nearly 50% of Americans live in areas of violation of NAAQS for ozone.

Area classifications for ozone attainment deadlines are:

Area Classification	Design Value (ppm)	Primary Standards Date (by Nov. 15 of)
Marginal	0.121–0.138	3 yrs. after enactment
Moderate	0.138–0.160	6 yrs. after enactment
Serious	0.160–0.180	9 yrs. after enactment
Severe	0.180–0.280	15 yrs. after enactment
Severe	0.190–0.280	17 yrs. after enactment
Extreme	0.280 and above	20 yrs. after enactment

Only the metropolitan Los Angeles area is currently classified as extreme; 8 U.S. areas are severe; 16 areas are serious. SIPs must include current emissions inventories, and be updated every three years. Stationary sources for NO_x and volatile organic compounds (VOCs) emissions must be reported annually (but waived for sources with less than 25 tons per annum). Deadlines are established for attainment; sanctions and consequences are provided for nonattainment.

Interstate pollution: An ozone transport region is established consisting of 11 Northeastern and Middle Atlantic states which is subject to additional control measures, monitoring and modeling.

PM_{10} — Initially, all nonattainment areas have been designated as "moderate." Stringent controls are provided.

NOTE: Concerning National Ambient Air Quality Standards (NAAQS), see § 4.4, *infra*, Legislative and Litigation Updates, for an important U.S. Supreme Court case, *Whitman (EPA) v. American Trucking Associations, Inc. et al*, 531 U.S. 457 (2001).

4.3.1.2 Title II — Provisions Relating to Mobile Sources

This covers mobile sources of air pollutants, providing requirements for more strict tailpipe emissions and reformulated gasoline resulting in cleaner fuel.

The clean fuels program establishes means to develop the technology and infrastructure to make available new fuels and vehicles that can utilize them. Designated are pilot programs to be tried in California beginning in 1996. A more strict program is scheduled for 2001. A fleet program is scheduled for the year 2001. The fleet program is designed for centrally fueled vehicle fleets operating in the 25 worst ozone and CO nonattainment areas. Other states may choose to adopt the California fuel vehicle standards.

4.3.1.3 Title III — Hazardous Air Pollutants

This covers the regulation and reduction of 189 hazardous air pollutants (expanded list from former eight pollutants under CAA, 1972, 1977) by controlled technology-based standards. The provisions for PSD (prevention of significant deterioration) do not apply to the new list of hazardous pollutants under § 112(b)(6).

Sources: EPA must develop and publish, within 12 months, a list of sources by categories and subcategories of major and area sources of hazardous air pollutants. (An **area source** is defined as "any stationary source of hazardous air pollutants that is not a major source; and, does not include motor vehicles".)

Area source program: In achieving a reduction of hazardous emissions, an equivalent reduction in public health risk must be made for sources associated with hazardous pollutants, including a 75% reduction in the occurrence of cancer attributable to such sources.

Emission standards: EPA must establish and promulgate emission standards for not fewer than 40 categories or subcategories by November 14, 1992 and emission standards for coke oven batteries by December 31, 1992. Additionally, another 25% of categories must be established and promulgated by November 14, 1994; and, another 25% of the categories by November 14, 1997; all categories and subcategories must have emission standards established by November 14, 2000.

Atmospheric deposition to Great Lakes and coastal waters: EPA must conduct an assessment study and establish a monitoring network to examine the deposition of hazardous pollutants into the Great Lakes, Chesapeake Bay, Lake Champlain and the coastal waters within three years after the enactment, and every two years thereafter.

Specific studies: EPA is charged with making studies of electric utility generating units; coke oven production technology; POTWs; hydrogen sulfide; and, hydrofluoric acid.

A Chemical Safety and Hazardous Investigation Board is established with members appointed by the President.

4.3.1.4 Title IV — Acid Deposition Control

This covers acid rain deposition control and will especially impact plants burning fossil fuels. Public electric generating utilities will receive the greatest cost impact along with consumers. Provisions are made for reduction of nitrogen oxides, and a two-phase program is established for reduction of SO_2 emissions, both of which apply only to steam generating electric utilities. Other provisions are for an EPA sale of allowances, continuous monitoring, record keeping, and reporting by all operators generating SO_2 and NO_x.

4.3.1.5 Title V — Permits

This establishes a new permitting system for all major stationary sources and others. It designates minimum criteria for state 5-year permitting programs, deadlines for permitting, the permitting process, and enforcement.

4.3.1.6 Title VI — Stratospheric Ozone Protection

This regulates the protection of the stratosphere and global warming by phasing out of some 25 ozone-depleting chemicals.

4.3.1.7 Title VII — Enforcement

This covers enforcement of regulation through both civil and criminal penalties. Bounties for citizen reporting violations are established.

4.3.1.8 Title VIII — Miscellaneous Provisions

This contains 22 miscellaneous provisions, such as air pollution from outer continental shelf activities, visibility impairment, and international border areas.

4.3.1.9 Title IX — Clean Air Research

This authorizes a clean air research program and continuation of investigations under the Acid Precipitation Act of 1980.

4.3.1.10 Title X — Disadvantaged Business Concerns

This establishes availability of at least 10% of research funding to disadvantaged business concerns.

4.3.1.11 Title XI — Clean Air Employment Transition Assistance

This provides funds to assist workers whose work was terminated as a result of compliance with the CAA requirements.

4.4 CAA LEGISLATIVE AND LITIGATION UPDATES — 1990–2001*

4.4.1 Title I — Section 107, Nonattainment

States must designate regions for nonattainment for NAAQS. Section 172(b) sets designations for ozone, carbon monoxide, and fine particulate matter (PM_{10} less than 10 microns, aerodynamic diameter). Daily emissions standards have been set, viz., primary for 24-hour average at 150 $\mu g/m^3$ and half that (75 $\mu g/m^3$) for secondary; annual limit for primary and secondary is 50 $\mu g/m^3$.

In *Virginia v. EPA*, 108 F.3d 1397 (D.C. Cir. 1997), the Appeals Court struck down the EPA's final rule requiring 12 northeastern states and the District of Columbia to revise their state implementation plans (SIPs) to reduce oxides of nitrogen (NO_X) and volatile organic compounds (VOC). The court found that under Section 110, each state has the liberty to "adopt whatever mix of emissions limitations it deems best-suited to its particular situation" (Mansfield, 1997, p. 166).

In *Whitman (EPA Administrator) v. American Trucking Associations, Inc., et al.*, combined with *American Trucking Associations, Inc. v. Whitman (EPA Administrator)*, 531 U.S. 457, 121 S. Ct. 903, 149 L. Ed.2d 1 (2001), the U.S. Supreme Court granted *certiorari* to the appeals of both parties from the lower court, the D.C. Circuit Court of Appeals, concerning the EPA-revised NAAQS for ozone and particulate matter.

In the D.C. Circuit holding [175 F.3d 1027 (D.C. Cir. 1999)], the court concluded that the EPA had failed to articulate any "intelligent principle" for selecting a revised standard and the EPA's interpretation of the act constituted an unconstitutional delegation of legislative power. The D.C. court also held that: (1) the EPA could not consider cost in setting the NAAQS; (2) the EPA must consider health benfits of ozone; (3) the EPA's promulgation of the revised standards did not violate the NEPA, the Unfunded Mandates Reform Act; or (4) the Regulatory Flexibility Act; the 1990 amendments to the act limit the EPA's options to enforce new ozone NAAQS; and (5) the EPA's use of PM_{10} as a surrogate for $PM_{2.5}$ was arbitrary and capricious (Mansfield, 1999, p. 198). Both parties appealed for *certiorari* to the U.S. Supreme Court. Following is a condensed form of the Supreme Court's *certiorari* decision.

CHRISTINE TODD WHITMAN, ADMINISTRATOR OF ENVIRONMENTAL
PROTECTION AGENCY, ET AL., PETITIONERS v. AMERICAN TRUCKING
ASSOCIATIONS, INC., ET AL.
AMERICAN TRUCKING ASSOCIATIONS, INC., ET AL., PETITIONERS v.
CHRISTINE TODD WHITMAN, ADMINISTRATOR OF ENVIRONMENTAL
PROTECTION AGENCY, ET AL.
Nos. 99-1257 and 99-1426

SUPREME COURT OF THE UNITED STATES
531 U.S. 457; 121 S. Ct. 903; 2001 U.S. LEXIS 1952; 149 L. Ed.2d 1; 69 U.S.L.W. 4136;
51 ERC (BNA) 2089 (2001)
November 7, 2000, Argued, February 27, 2001, Decided*

---

* *Note:* Though much of the CAA-EPA litigation through 1994 was in reference to actions that had been filed before the 1990 amended CAA and rules under the new act had been promulgated, only a few of the more important cases of significance are included.

SYLLABUS:

Section 109(a) of the Clean Air Act (CAA) requires the Environmental Protection Agency (EPA) Administrator to promulgate national ambient air quality standards (NAAQS) for each air pollutant for which "air quality criteria" have been issued under § 108. Pursuant to § 109(d)(1), the Administrator in 1997 revised the ozone and particulate matter NAAQS. Respondents in No. 99-1257, private parties and several States (hereinafter respondents), challenged the revised NAAQS on several grounds. The District of Columbia Circuit found that, under the Administrator's interpretation, § 109(b)(1) — which instructs the EPA to set standards "the attainment and maintenance of which … are requisite to protect the public health" with "an adequate margin of safety" — delegated legislative power to the Administrator in contravention of the Federal Constitution, and it remanded the NAAQS to the EPA. The Court of Appeals also declined to depart from its rule that the EPA may not consider implementation costs in setting the NAAQS. And it held that, although certain implementation provisions for the ozone NAAQS contained in Part D, Subpart 2, of Title I of the CAA did not prevent the EPA from revising the ozone standard and designating certain areas as "nonattainment areas," those provisions, rather than more general provisions contained in Subpart 1, constrained the implementation of the new ozone NAAQS. The court rejected the EPA's argument that it lacked jurisdiction to reach the implementation question because there had been no "final" implementation action.

Held:

1. Section 109(b) does not permit the Administrator to consider implementation costs in setting NAAQS. Because the CAA often expressly grants the EPA the authority to consider implementation costs, a provision for costs will not be inferred from its ambiguous provisions. *Union Elec. Co. v. EPA*, 427 U.S. 246, 257, 49 L. Ed.2d 474, 96 S. Ct. 2518, and n. 5. And since § 109(b)(1) is the engine that drives nearly all of Title I of the CAA, the textual commitment of costs must be clear; Congress does not alter a regulatory scheme's fundamental details in vague terms or ancillary provisions, see *MCI Telecommunications Corp. v. American Telephone & Telegraph Co.*, 512 U.S. 218, 231, 129 L. Ed. 2d 182, 114 S. Ct. 2223. Respondents' arguments founder upon this principle. It is implausible that § 109(b)(1)'s modest words "adequate margin" and "requisite" give the EPA the power to determine whether implementation costs should moderate national air quality standards. *Cf. ibid.* And the cost factor is both so indirectly related to public health and so full of potential for canceling the conclusions drawn from direct health effects that it would have been expressly mentioned in §§ 108 and 109 had Congress meant it to be considered. Other CAA provisions, which do require cost data, have no bearing upon whether costs are to be taken into account in setting the NAAQS. Because the text of § 109(b)(1) in its context is clear, the canon of construing texts to avoid serious constitutional problems is not applicable. See, e.g., *Miller v. French*, 530 U.S. 327, 341, pp. 4–11, 147 L. Ed.2d 326, 120 S. Ct. 2246.

2. Section 109(b)(1) does not delegate legislative power to the EPA. When conferring decisionmaking authority upon agencies, Congress must lay down an intelligible principle to which the person or body authorized to act is directed to conform. *J. W. Hampton, Jr., & Co. v. U.S.*, 276 U.S. 394, 409, 72 L. Ed. 624, 48 S. Ct. 348. An agency cannot cure an unlawful delegation of legislative power by adopting in its discretion a limiting con-

---

*Together with No. 99-1426, *American Trucking Associations, Inc, et al. v. Whitman, Administrator of Environmental Protection Agency, et al.*, also on certiorari to the same court.

struction of the statute. The limits that § 109(b)(1) imposes on the EPA's discretion are strikingly similar to the ones approved in, e.g., *Touby v. United States*, 500 U.S. 160, 114 L. Ed.2d 219, 111 S. Ct. 1752, and the scope of discretion that § 109(b)(1) allows is well within the outer limits of the Court's nondelegation precedents, see, e.g., *Panama Refining Co. v. Ryan*, 293 U.S. 388, 79 L. Ed. 446, 55 S. Ct. 241. Statutes need not provide a determinate criterion for saying how much of a regulated harm is too much to avoid delegating legislative power. pp. 11–15.

3. The Court of Appeals had jurisdiction to consider the implementation issue under § 307 of the CAA. The implementation policy constitutes final agency action under § 307 of the CAA because it marked the consummation of the EPA's decision making process, see *Bennett v. Spear*, 520 U.S. 154, 137 L. Ed.2d 281, 117 S. Ct. 1154. The decision is also ripe for review. The question is purely one of statutory interpretation that would not benefit from further factual development, see *Ohio Forestry Assn., Inc. v. Sierra Club*, 523 U.S. 726, 733, 140 L. Ed.2d 921, 118 S. Ct. 1665; review will not interfere with further administrative development; and the hardship on respondent States in developing state implementation plans satisfies the CAA's special judicial-review provision permitting preenforcement review, see *id.* at 737. The implementation issue was also fairly included within the challenges to the final ozone rule that were before the Court of Appeals, which all parties agree is final agency action ripe for review. pp. 16–20.

4. The implementation policy is unlawful. Under *Chevron U.S.A. Inc. v. Natural Resources Defense Council, Inc.*, 467 U.S. 837, 81 L. Ed.2d 694, 104 S. Ct. 2778, if the statute resolves the question whether Subpart 1 or Subpart 2 applies to revised ozone NAAQS, that ends the matter; but if the statute is ambiguous, the Court must defer to a reasonable agency interpretation. Here, the statute is ambiguous concerning the inter-action between Subpart 1 and Subpart 2, but the Court cannot defer to the EPA's interpretation, which would render Subpart 2's carefully designed restrictions on EPA discretion nugatory once a new ozone NAAQS has been promulgated. The principal distinction between the subparts is that Subpart 2 eliminates regulatory discretion allowed by Subpart 1. The EPA may not construe the statute in a way that completely nullifies textually applicable provisions meant to limit its discretion. In addition, although Subpart 2 was obviously written to govern implementation for some time into the future, nothing in the EPA's interpretation would have prevented the agency from aborting the subpart the day after it was enacted. It is left to the EPA to develop a reasonable interpretation of the nonattainment implementation provisions insofar as they apply to revised ozone NAAQS. pp. 20–25.

175 F.3d 1027 and 195 F.3d 4, affirmed in part, reversed in part, and remanded.

Future arguments by industry will be based on the premise that ambient air quality standards for $PM_{2.5}$ and ozone should be based on quantitative and comparative risk analyses.

4.4.1.1 State Implementation Plans

In *Pan American Grain Manufacturing, Inc. v. EPA*, 95 F.3d 101 (1st Cir. 1996), plaintiff contested EPA's approval of Puerto Rico's SIP revision banning "the use of clamshell devices in grain removal operations to reduce particulate matter emissions." The challenge failed because it "was time-barred" having been filed "more than sixty days after EPA's final designation action upon publication in the Federal Register." The court also found that the rule "was connected to compliance with the PM_{10} NAAQS" (Mansfield, 1996, p. 123).

In *Missouri Limestone Producers Assn. v. Browner*, 165 F.3d 619 (8th Cir. 1999), "the Missouri business associations petitioned for judicial review of the EPA's decision to approve Missouri's revision of its fugitive dust provisions under the CAA. The Missouri Department of Natural Resources (MDNR) submitted the proposed SIP revision to the EPA, after which petitioners appealed the SIP revision to the Missouri Air Conservation Commission (MACC), a sub-agency of MDNR. The EPA approved the SIP revision notwithstanding the pending appeal. The Eighth Circuit held that despite petitions pending appeal to MACC, MDNR had discretion to submit SIP revision to the EPA for approval. The court also held that MDNR's decision to submit the revision to the EPA was final and enforceable unless and until it was invalidated by either MACC or a Missouri state court. The court reasoned that holding an otherwise valid regulation to be non-final based solely on a challenge by a private party would greatly impinge on a state's ability to enact and enforce regulations. In addition, the Eighth Circuit held that the MDNR was not required to hold a public hearing prior to submitting the SIP revision to the EPA, noting that adequate hearings were held prior to submitting the SIP MACC's adoption of the state rule. Finally, the court held that the EPA was not required to provide detailed factual data when acting under the APA; the EPA must provide merely the terms or substance of the proposed rule, or a description of the subjects and issues involved so that an interested party may make an informed criticism and comments" (Mansfield, 1999, p. 200).

The full text of the short opinion of the Eighth Circuit Court of Appeals in *Missouri Limestone Producers Association v. Browner*, 165 F.3d 619 (8th Cir. 1999) is reproduced, following, for elaboration of details.

UNITED STATES COURT OF APPEALS FOR THE EIGHTH CIRCUIT
165 F.3d 619; 1999 U.S. App. LEXIS 367; 47 ERC (BNA) 2025; 29 ELR 20455
November 18, 1998, Submitted; January 13, 1999, Filed

JUDGES: Before McMILLIAN, WOLLMAN, and HANSEN, Circuit Judges.
OPINION BY: WOLLMAN, Circuit Judge.

The Missouri Limestone Producers Association and the Missouri Ag Industries Council (Petitioners) seek review of a final action of the federal Environmental Protection Agency (EPA), which approved Missouri's revision of its state implementation plan (SIP) under the Clean Air Act, 42 U.S.C. §§ 7401–7671q. We deny the petition for review.

I. Missouri adopted four area-specific airborne, or "fugitive," dust regulations as part of its original SIP in 1972. The regulations were federally enforceable because the SIP was formally approved by the EPA. In 1990, Missouri revised the regulations. It replaced the four area-specific rules with one consistent, statewide rule. See Mo. Code Regs. tit. 10, § 10-6.170 (1990). The EPA did not approve the new rule, however, because the Missouri

Department of Natural Resources (MDNR) submitted only that the EPA remove the four area-specific rules, rather than replace them with the new statewide rule. This would have virtually eliminated federally enforceable fugitive-dust regulations in Missouri, which was unacceptable because the EPA may not approve a SIP revision that lessens air-quality standards. As a result, Missouri's state-enforced fugitive-dust rule differed from the federally enforced ones.

To eliminate the discrepancy, in February 1997 the MDNR submitted that the EPA replace the 1972 rules with the current statewide rule. Petitioners brought an appeal to the Missouri Air Conservation Commission (MACC), a sub-agency of the MDNR, challenging the decision to submit the SIP revision to the EPA. After notice and a public comment period, the EPA formally approved the SIP revision despite the pending appeal. [See Approval and Promulgation of Implementation Plans; State of Missouri, 63 Fed. Reg. 3037 (1998) (to be codified at 40 C.F.R. Part 52).]

Petitioners seek review of the EPA action under 42 U.S.C. § 7607(b)(1). They allege that their appeal to the MACC deprived the EPA of authority to approve the SIP revision under 40 C.F.R. Part 51, Appendix V. In the alternative, Petitioners claim that the action was invalid because the MDNR did not hold public hearings on the decision to submit the SIP revision and because the EPA did not disclose sufficient factual information about the revision.

II. Petitioners contend that our review of the EPA action falls within 42 U.S.C. § 7607(d)(9). Section 7607 applies when the EPA promulgates federal implementation plans, however, not when it approves state implementation plans. See 42 U.S.C. § 7607(d)(1)(B) (citing 42 U.S.C. § 7410(c)) (describing proper EPA procedure for promulgating federal implementation plans). Here, the EPA acted pursuant to its general authority under the Administrative Procedure Act (APA), 5 U.S.C. § 553. Accordingly, the deferential standards of the APA govern our review. See *Western States Petroleum Ass'n v. EPA*, 87 F.3d 280, 283 (9th Cir. 1996); *National Steel Corp. v. Gorsuch*, 700 F.2d 314, 320 (6th Cir. 1983); *South Terminal Corp. v. EPA*, 504 F.2d 646, 655 (1st Cir. 1974).

We will set aside an agency action only if it was "arbitrary, capricious, an abuse of discretion, or otherwise not in accordance with law." 5 U.S.C. § 706(2). This standard of review gives agency decisions "a high degree of judicial deference." *Dubois v. Thomas*, 820 F.2d 943, 948–49 (8th Cir. 1987); see also *Mueller v. EPA*, 993 F.2d 1354, 1356 (8th Cir. 1993); *Mausolf v. Babbitt*, 125 F.3d 661, 667 (8th Cir. 1997), cert. denied, 141 L. Ed.2d 735, 118 S. Ct. 2366 (1998). We also consider whether any errors committed by the agency were prejudicial. See 5 U.S.C. § 706.

First, Petitioners argue that the EPA failed to comply with the Clean Air Act regulations that govern approval of SIP revisions. The regulations require that the EPA receive a formal request from the proper designee of the state before approving a SIP revision. See 40 C.F.R. Part 51, Appendix V, § 2.1(a). The request must demonstrate that the revision has been adopted "in final form." *Id.* § 2.1(b).

The MACC has the authority to adopt and amend Missouri's air-quality rules. See Mo. Code Regs. tit. 10, § 10-1.010(2)(A) (1998). The MDNR, however, has the specific responsibility to "submit revisions of the State Implementation Plan to the United States Environmental Protection Agency for approval." *Id.* § 10-1.010(2)(B). Therefore, the MDNR was the proper agency to submit Missouri's SIP revision to the EPA. Once the MACC adopted the new statewide fugitive-dust rule, it was squarely within the MDNR's discretion to submit the revision to the EPA.

Furthermore, even assuming that the MDNR's decision to submit the revision was appealable, there is no authority for Petitioners' assertion that the appeal rendered the MDNR decision "non-final." The case cited by Petitioners to support this argument, *State ex rel. Lake Lotawana Development Company v. Missouri DNR*, 752 S.W.2d 497, 497–98 (Mo. Ct. App. 1988), simply requires plaintiffs to exhaust administrative remedies before seeking judicial review. It does not hold that an otherwise valid regulation becomes non-final once a private party challenges its validity. See *id*. To hold as much would essentially allow a party to enjoin the enforcement of a state regulation by filing an appeal rather than obtaining an injunction, which would greatly impinge on a state's ability to enact and enforce regulations. Accordingly, we hold that the MDNR decision is final and enforceable, unless and until it is invalidated by the MACC or a Missouri state court. See *Sierra Club v. Indiana-Kentucky Elec. Corp.*, 716 F.2d 1145, 1151 (7th Cir. 1983) (recognizing that the EPA may not enforce a regulation that the state has found invalid). Thus, the EPA complied with the Clean Air Act regulations.

Next, Petitioners argue that the MDNR was required to hold public hearings on whether to submit the SIP revision to the EPA. Assuming that this argument was sufficiently raised during the public comment period to avoid waiver, it is nonetheless unavailing. Petitioners acknowledge that adequate hearings were held prior to the MACC's adoption of the new statewide fugitive-dust rule. They also admit that they were allowed to raise their objections to submission of the revision at a special MACC meeting on that topic. They contend that another hearing was required, however, before the MDNR could submit the revision to the EPA. To support their argument, Petitioners cite the regulation that requires states to hold "one or more public hearings" before "adoption and submission" of a SIP revision. See 40 C.F.R. § 51.104.

Nothing in the text of the Clean Air Act or the regulations requires states to hold a hearing on the submission of a properly adopted SIP revision. In fact, Petitioners' interpretation of the regulations contradicts the language of the Act itself, which provides that "each revision to an implementation plan submitted by a State … shall be adopted by such State after reasonable notice and public hearing." 42 U.S.C. § 7410(l). This language indicates that a hearing is required before adoption of a revision, not before the revision is submitted to the EPA.

Petitioners have provided no evidence that additional hearings on whether to submit a SIP revision have ever been held in Missouri, or in any other state. Since the EPA's original approval of Missouri's SIP in 1972, the state has made many SIP revisions without holding separate hearings on the decision to submit. The EPA, which is entrusted with interpretation and enforcement of the Clean Air Act, has never required hearings on whether properly adopted SIP revisions should be submitted. In light of our deferential review of an agency's interpretation of its own regulations, we hold that the MDNR was not required to hold an additional hearing before submitting the SIP revision to the EPA.

Finally, Petitioners contend that the EPA did not provide sufficient factual information about the proposed SIP revision. The EPA is not required to provide detailed factual data, however, when it acts under the APA. See 5 U.S.C. § 553(b). It must provide public notice of "either the terms or substance of the proposed rule or a description of the subjects and issues involved." *Id*. § 553(b)(3). We have held that an agency's notice is sufficient if it allows interested parties to offer "'informed criticism and comments.'" [*Northwest Airlines, Inc. v. Goldschmidt*, 645 F.2d 1309, 1319–20 (8th Cir. 1981) (quoting *Ethyl Corp. v. EPA*, 176 U.S. App. D.C. 373, 541 F.2d 1, 48 (D.C. Cir. 1976)]. The EPA included a lengthy description of the substance and purpose of the SIP revision in its public notice. This was sufficient to allow for informed public comment, as Petitioners demonstrated by their extensive remarks during the public comment period.

The petition for review is denied, as is Petitioners' motion for stay pending disposition of the appeal before the Missouri Air Conservation Commission.

4.4.1.2 Section 111 — New Source Performance Standards

Source performance standards for nonmetallic mines and mineral processing plants are (1) no greater than 0.05 g/dscf from any source, and (2) no greater than 7% opacity from any source.

In *Davis County Solid Waste Management v. EPA*, 108 F.3d 1454 (D.C. Cir. 1997), the court of appeals "reconsidered its initial decision to vacate EPA's emissions standards for municipal waste combustor (MWC) units in their entirety." The court agreed with EPA that only the emissions standard as applied to small MWC units and cement kilns should be vacated, and that NSPS for new and large units should be retained" (Mansfield, 1997, p. 167).

In *Lignite Energy Council, et al. v. U.S. EPA*, 339 U.S. App. D.C. 183; 198 F.3d 930 (D.C. Cir. 1999), Lignite Council challenged EPA's new source performance standards for nitrogen oxides, emissions from utility and industrial boilers. The D.C. Court of Appeals held that EPA did not exceed its discretion under § 111 of the CAA in promulgating these standards, and therefore denied the petitions. The court opinion stated:

Fossil-fuel fired steam generating units ("boilers") emit nitrogen oxides (NO_x), air pollutants that can cause deleterious health effects and contribute to the formation of acid rain. Section 111 of the Clean Air Act requires EPA to establish performance standards for the emission of NO_x from newly constructed boilers; these "new source performance standards" are to be set at a level that reflects the degree of emission limitation achievable through the application of the best system of emission reduction which (taking into account the cost of achieving such reduction and any non-air quality health and environmental impact and energy requirements) the Administrator determines has been adequately demonstrated.

In its 1990 Clean Air Act Amendments, Congress specifically directed EPA to exercise its Section 111 authority and establish new NO_x standards that incorporate "improvements in methods for the reduction of emissions of oxides of nitrogen."

In response to these statutory mandates, EPA promulgated a rule lowering its NO_x new source performance standards to .15 lb/MMBtu (pounds of NO_x emitted per million BTU burned) for utility boilers[1] and .20 lb/MMBtu for industrial boilers. These standards reflect

---

[1] To be precise, the emission standard for utility boilers is an output-based standard of 1.6 pounds of NO_x emitted per megawatthour of electricity generated. However, as this output-based standard was intended by EPA to correlate with a .15 lb/MMBtu input-based standard, we refer to its input-based equivalent for simplicity's sake throughout this opinion. We reject petitioners' argument that EPA's decision to shift to an output-based standard for utility boilers unfairly "penalizes" the use of low-energy coals, like lignite; it would seem just as easy to argue that an input-based standard "penalizes" high-energy fuels.

the level of NOx emissions achievable by what EPA considers to be the "best demonstrated system" of emissions reduction: the use of selective catalytic reduction (SCR) in combination with combustion control technologies.[2]

Petitioners' central claim is that EPA selected SCR as the basis for its NO_x standards without properly balancing the factors that Section 111 requires it to "take into account." Because Section 111 does not set forth the weight that should be assigned to each of these factors, we have granted the agency a great degree of discretion in balancing them, see, e.g., *New York v. Reilly*, 297 U.S. App. D.C. 147, 969 F.2d 1147, 1150 (D.C. Cir. 1992); EPA's choice will be sustained unless the environmental or economic costs of using the technology are exorbitant.

Petitioners argue that SCR is not the "best demonstrated system" under Section 111 because the incremental cost of reducing NO_x emissions is considerably higher with SCR than with combustion controls. Recent improvements in combustion controls will enable many boilers to attain emissions levels close to EPA's SCR-based standards; accordingly, petitioners assert that EPA should have based its standards on these less expensive technologies. However, in light of EPA's unchallenged findings showing that the new standards will only modestly increase the cost of producing electricity in newly constructed boilers. We do not think that EPA exceeded its considerable discretion under Section 111. Moreover, petitioners' argument stressing the comparable environmental merits of advanced combustion controls is to a certain extent self-defeating, since the new source performance standards set by EPA are not technology-forcing, and continuing advances in combustion control technologies will reduce the amount of NO_x reduction that must be captured by the more expensive SCR technology.

It was also within EPA's discretion to issue uniform standards for all utility boilers, rather than adhering to its past practice of setting a range of standards based on boiler and fuel type. Petitioners recognize that EPA is not required by law to subcategorize — Section 111 merely states that "the Administrator may distinguish among classes, types, and sizes within categories of new sources," but argue that it was arbitrary and capricious for EPA to decline to do so. EPA explains that its change to uniform standards is justified by SCR's performance characteristics: Unlike the technologies on which past new source performance standards were based, flue gas treatment technologies like SCR limit NO_x emissions after combustion, and the effectiveness of SCR is thus far less dependent upon boiler design or fuel type. Petitioners respond that there are reasons to expect SCR to perform less adequately on boilers burning high-sulfur coals, but EPA collected continuous emissions monitoring data on two high-sulfur coal-fired utility boilers that showed that the .15 lb/MMBtu standard was achievable, and supplemented this study with similar evidence from foreign utility boilers. EPA also considered petitioners' concerns about the impact of alkaline metals on the performance of the catalyst used in the SCR process, and concluded that such "catalyst poisoning" is not a significant problem in coal-fired boilers. Mindful of the high degree of deference we must show to EPA's scientific judgment, see, e.g., *Appalachian Power Co. v. EPA*, 328 U.S. App. D.C. 379, 135 F.3d 791, 801–02 (D.C. Cir. 1998), we accept these determinations and sustain EPA's uniform standard for utility boilers.

Petitioners offer a broader challenge to EPA's .20 lb/MMBtu standard for industrial boilers, claiming that SCR is not "adequately demonstrated" for any coal-fired industrial boilers. EPA was unable to collect emissions data for the application of SCR to these

---

[2] SCR is a "flue gas treatment technology"; it reduces NO_x after combustion by injecting ammonia into the flue gas in the presence of a catalyst, breaking down NO_x and producing nitrogen and water. In setting past standards, EPA had focused solely on combustion control technologies, which instead reduce NO_x by suppressing its formation during the combustion process.

boilers, but this absence of data is not surprising for a new technology like SCR, nor does it in and of itself defeat EPA's standard. Because it applies only to new sources, we have recognized that Section 111 "looks toward what may fairly be projected for the regulated future, rather than the state of the art at present." *Portland Cement Ass'n v. Ruckelshaus*, 158 U.S. App. D.C. 308, 486 F.2d 375, 391 (D.C. Cir. 1973). Of course, where data are unavailable, EPA may not base its determination that a technology is adequately demonstrated or that a standard is achievable on mere speculation or conjecture, but EPA may compensate for a shortage of data through the use of other qualitative methods, including the reasonable extrapolation of a technology's performance in other industries.

EPA has done precisely that here, concluding from its study of utility boilers that SCR is "adequately demonstrated" and the .20 lb/MMBtu standard is "achievable" for coal-fired industrial boilers as well. Utility and industrial boilers are similar in design and both categories of boilers can attain similar levels of NO_x emissions reduction through combustion controls, which means that SCR will be required to capture comparable quantities of NO_x for both boiler types. While petitioners argue that SCR is less likely to be effective on industrial boilers because they have widely fluctuating load cycles, EPA has shown that SCR can be successfully applied to coal-fired utility boilers under a "wide range of operating conditions" including those analogous to the load cycles of industrial boilers. We think that it was reasonable for EPA to extrapolate from its studies of utility boilers in setting an SCR-based new source performance standard for coal-fired industrial boilers.[3]

We also sustain EPA's application of the .20 lb/MMBtu standard to combination boilers, which simultaneously combust a mixture of fuels. The preexisting NO_x emissions standards established a range of values for combustion boilers that varied by fuel type: while combination boilers burning natural gas with non-coal solid fuels (e.g., wood) were subject to a .30 lb/MMBtu standard, the performance standards for combination boilers combusting coal with oil or natural gas were determined based upon the proportion of the boiler's total heat input provided by each fuel. It is difficult to understand petitioners' objection to the application of the industrial boiler standard to boilers burning natural gas and wood. A reduction of that standard from .30 to .20 lb/MMBtu is perfectly reasonable in light of the significant advances in NO_x emissions technology since 1986; indeed, EPA studies show that wood-fired boilers can reach emissions levels far lower than .20 lb/MMBtu through the application of flue gas treatment technologies. And our conclusion that the .20 lb/MMBtu standard is achievable for boilers burning only coal necessarily defeats petitioners' objection that the industrial boiler standard is unreasonable as applied to combination boilers burning coal simultaneously with other fuels with lower NO_x emissions characteristics.

Petitioners' final objection is to EPA's valuation of steam energy produced by "cogeneration facilities." EPA's adoption of an output-based standard for utility boilers raised the question of how to calculate the energy produced by these units, which generate thermal steam energy in addition to electrical energy. Steam energy produced by cogeneration facilities is exported for several different industrial uses; however, because of inefficiencies in transporting and converting steam, only a fraction of steam energy produced by cogeneration facilities is actually used in the industrial process. EPA resolved this problem by

---

[3] For similar reasons, we do not think that EPA's lack of data on domestic SCR applications to boilers burning lignite renders its standards unlawful. In assessing a new technology like SCR, EPA is not required to provide evidence of its application to boilers burning every type of coal from every geographical location. It is acceptable for EPA to extrapolate from the successful applications of SCR to domestic high-sulfur coal-fired boilers and to foreign boilers burning lignite.

assigning a 50% credit for steam energy when determining a cogeneration unit's output. Petitioners describe this credit as an arbitrary and capricious "discounting" of steam energy's value, but it just as easily could be called a subsidy: The maximum efficiency for the conversion of steam to electrical energy is only 38%, and EPA's final rule justifies the 50% credit on the ground that it will encourage cogeneration. *Id.* In light of the difficulties that would attend calculating the useful energy of steam heat produced by cogeneration facilities on a unit-by-unit basis, we conclude that EPA's resolution of this issue was acceptable.

The petitions for review are denied. So ordered.

In the year 2000 case of *Bowers v. Pollution Control Hearings Board*, 13 P.3d 1076 (Wash. App., Div. 1, 2000), appellant Gregory Bowers sought to overturn an order issued by the Southwest Air Pollution Control Authority (SWAPCA) and affirmed by the Pollution Control Hearings Board (PCHB). The order directed the Centralia Power Plant to reduce its emissions of four pollutants, including sulfur dioxide and nitrogen oxide. SWAPCA issued the order pursuant to the Washington Clean Air Act, which requires all existing sources of air pollutants in the state to control their emissions by using reasonably available control technology (RACT).

Bowers raised a host of challenges before the PCHB. His main argument was that SWAPCA did not conduct its RACT review properly, resulting in emission levels that were too high and control technologies that did not adequately protect human health. The PCHB rejected Bowers' claims and affirmed SWAPCA's decision.

The court held that Bowers did not meet his burden in challenging the PCHB's findings of fact and conclusions of law. The PCHB did not erroneously interpret the Washington Clean Air Act as applied to the requirements for setting RACT. Substantial evidence supported its findings that addressed the scientific underpinnings of the emissions limits and the cost effectiveness and impact of the various control technologies. PCHB's decision to affirm the SWAPCA order was not arbitrary or capricious, and was affirmed by the court of appeals.

4.4.2   Title II — Mobile Sources

More stringent standards are required for tailpipe standards to control exhaust pollutants from cars and trucks; improvements in evaporative emissions; required use of reformulated gasoline; and diesel to reduce VOC and toxic air pollutants starting in 1995.

In *American Automobile Manufacturing Assn. v. Cahill*, 973 F. Supp. 288 (N.D. N.Y. 1997), the plaintiff challenged New York from enforcing legislation "requiring the sale of zero emission vehicles (ZEV) beginning in the model year 1998." The court held that under §§ 177 and 209 of the act, the ZEV sales mandate was an enforcement mechanism and not an emissions standard, consequently, not subject to the state adopted emissions limitations" (Mansfield, 1997, p. 168).

In *Sierra Club, Illinois Chapter v. U.S. Dept. of Transportation*, 962 F. Supp. 1037 (N.D. Ill. 1997), "the environmental group alleged that U.S. DoT failed to comply with the NEPA and § 4(f) of the Transportation Act in the planning of a

12.5-mile toll road in Illinois," and that U.S. DoT had not adequately analyzed "the air pollution impact of the proposed toll road because it did not quantify the potential ozone-producing effect from the tollroad's" motor vehicles. The DoT study was found to be "inadequate in scope and the pertinent emissions information was not incorporated into the final EIS." DoT was directed "to produce a study justifying their analysis that the toll road would not have any adverse ozone-producing effect, or why it was not possible"(Mansfield, 1997, p. 169).

In *Sierra Club v. EPA*, 167 F.3d 658 (D.C. Cir. 1999), the D.C. Circuit "reviewed the EPA's rule establishing standards for medical waste incinerators (MWIs). The court held that the EPA use of state permit and regulatory data rather than performance data, in setting MACT floors for emissions standards for existing MWIs was not forbidden by the CAA. The court also held that the EPA was not required to compel MWIs to use pollution prevention measures to reduce mercury and chlorinated plastics in their waste streams. However, because the EPA did not adequately explain its use of a combination of state regulatory data and uncontrolled values in setting such floors, and it did not adequately explain its method of setting floors for newly constructed MWIs, the case was remanded" (Mansfield, 1999, p. 198).

4.4.3 Title III — Hazardous Air Pollutants

Under § 112(b), a list of hazardous air pollutants must be created for organic chemicals and metals common to industrial processes. A source categories list must be published by the EPA in 12 months from November 1990. A major source is any stationary source emitting 10 tons/year or more of any listed hazardous pollutant. Existing sources can avoid for 6 years otherwise applicable control technology-based standard through early reductions.

In *Fried v. Sungard Recovery Services, Inc.*, 925 F. Supp. 364 (E.D. Pa. 1996), "a citizen's suit alleged violations of the asbestos National Emission Standards for Hazardous Air Pollutants (NESHAP). The alleged violations arose from Sungard's renovation of two floors of a commercial building over an extended period. In denying Sungard's motion for summary judgement, the federal District Court stated that: (1) under NESHAP requirements, planned renovation operation includes all renovation activities planned together and cannot be limited to discrete projects individually priced, scheduled, and budgeted as independent activities; (2) if the minimum amount of regulated asbestos is disturbed in any one year, the entire renovation operation, even if occurring over more than one year, is covered by the asbestos NESHAP requirements; (3) the activity of chipping tile presented a question of fact as to whether this activity made the asbestos friable; and (4) each particular NESHAP requirement is its own 'parameter' for purposes of determining whether alleged violations are continuing or repeated, and the evidence created a genuine issue of material fact as to whether violations were continuing" (Mansfield, 1996, p. 124).

In *National Lime Association v. EPA*, 233 F.3d 625 (D.C. Cir. 2000), the court considered petitions by National Lime, a trade association, and the Sierra Club, an environmental organization, challenging the EPA's hazardous air pollutant emission regulations for cement manufacturing.

With respect to the Sierra Club petition, the court held: "(1) we reject its challenge to the emission standards for hazardous metals and dioxin/furan; (2) we find the Agency's failure to set standards for hydrogen chloride, mercury, and total hydrocarbons contrary to the Clean Air Act's plain language; (3) we direct EPA to consider the health impacts of potentially stricter standards for hazardous metals; and (4) we sustain the regulation's monitoring requirements.

"Concluding that the National Lime Association has associational standing, we (1) reject its argument that EPA's use of particulate matter as a surrogate for non-volatile metal hazardous air pollutants violates the Clean Air Act and is arbitrary and capricious; and (2) reject its challenge to the testing method EPA adopted for determining whether a manufacturer qualifies as a 'major source' of hazardous air pollutants."

The court remanded the rule to the EPA to allow the agency (1) to set MACT floor standards for HCl, mercury, and total hydrocarbons; (2) to consider setting beyond-the-floor standards for hazardous air pollutant (HAP) metals; and (3) to respond to comments suggesting improvements to Method 26/26A for measuring HCl emissions. With respect to all other issues discussed herein, the petitions were denied.

4.4.4   Title IV — Acid Deposition Control

Mineral-producing plants using fossil fuels in kilns, roasters, dryers, etc. must be aware of this title and reduction standards for sulfur dioxide emissions. Excess emission penalties are provided for at $2000 per ton for SO_2.

4.4.4.1   Section 404 — Acid Rain

American Municipal Power of Ohio v. EPA, 98 F.3d 1372 (2d Cir. 1996) involved a utility's challenge to the EPA's interpretation of the term *thermal energy* in § 410(f) of the act. Under § 410(f), utility units that opt into the Acid Rain Program may not transfer or bank Title IV emission allowances on reduced utilization or shutdown of units unless the transfer falls within an exception for thermal energy. "The EPA defined the term 'thermal energy' as 'the thermal output produced by a combustion source used solely as a part of a manufacturing process but not used to produce electricity.' Petitioners argued that the EPA's interpretation was unreasonable because small utilities would not qualify as producing thermal energy. The Court rejected this argument, holding that the EPA's interpretation was consistent with legislative history indicating that Congress intended that the 'thermal energy' exception be narrow in its scope, and was consistent with the principal goal of the Acid Rain Program, namely, to reduce sulfur dioxide emissions" (Mansfield, 1996, p. 129).

In *Ormet Corp. v. Ohio Power Co.*, 98 F.3d 799 (4th Cir. 1996), "Ormet claimed a right to 89% of the Title IV emission allowances the EPA had issued to Ohio Power's Kammer Generating Station. Ormet argued its contract with Ohio Power required it to pay a proportional share of the Kammer Station's operation and maintenance cost, that it had paid 89% of those costs, and that therefore, pursuant to § 408(I) of the Act, it was entitled to 89% of the emission allowances allotted for the Kammer plant. § 408(I) provides that where there are multiple owners of a fossil-fired unit, 'allowances and the proceeds of transactions involving allowances will be deemed to be held or distributed in proportion to each holder's legal, equitable, leasehold, or other contractual reservation or entitlement....'

"During trial, the lower, district court had dismissed Ormet's complaint on the grounds that the EPA's issuance of allotments was final agency action, reviewable only in the court of appeals. The court of appeals held that there was no final agency action because the Clean Air Act does not authorize the EPA to resolve private disputes over a unit's allowances. The Court rejected Ormet's contention that § 408(I) creates a private cause of action, but held that the district court had jurisdiction because resolution of Ormet's claim involved interpretation and application of the Clean Air Act to the contractual arrangement between the parties" (Mansfield, 1996, p. 130).

4.4.5   Title V — Permits, Penalties, and Enforcement

The permitting program is to be eventually turned over to the states after approval of their programs by EPA. However, EPA still retains overview powers. States had 3 years from November 1990 to develop and submit to EPA for approval state permitting programs.

An Administrative Penalty Scheme (§ 113[d]) provides that the EPA administrator may assess penalties for violations not exceeding $200,000 and that the proceeding must be brought within 1 year of the first alleged date of violation. Field inspectors may assess violations at $5000/day for appropriate minor violations. Section 113(b)(2) of the 1990 amendment allows for actions for violations of the new sections where an operator has violated the act with civil penalties up to *$25,000/day for each violation*, not per day of violation. Citizens awards of $10,000 may be made for information and services in reporting violators that lead to a conviction. Increases in larger assessments for violations are being evidenced (e.g., a phosphate mining operator in North Carolina was assessed a $5.7 million civil penalty for air pollution violations in 1987).

Compliance for open pit air permits will call for keeping pit, crusher, plant, stockpiles, and road areas watered down and dust free. Rule of thumb estimates for quarry dust emissions are 30 to 50% due to crushing, screening, and processing; 20 to 40% for roadways; and lesser amounts from stockpiles wind erosion. Adding and replacing equipment will likely require permitting permission. Letters of notification to the state EPA are usually required before any new construction starts. Plant equipment under federal NSPS must meet strict standards for emissions. Equipment manufactured after August 31, 1983 is subject to controlled specifications (e.g., screens, conveyors, and bins, 10% opacity limit; crushers, 15% opacity limit). Equipment manufactured before August 31, 1983 is exempt from the stricter opacity limits, but limited to a range of 20 to 40%.

EPA mine permitting requires a list of emission sources, calculation of annual uncontrolled emissions, specification of controlled apparatus, estimate of control efficiency, and calculation of annual controlled emissions. Simply stated, if dust can be seen, there is a violation.

Some illustrative litigated cases for air pollution are: In *Clean Air Implementation Project v. EPA*, No. 96-1224, 1886 WL 393118 (D.C. Cir. June 28, 1996), the court of appeals granted petitioners' motion for summary *vacatur* (to vacate the complaint) and remand of the EPA "potential to emit" definition in its Title V regulations in light of the decisions in *National Mining Assn. v. EPA*, 59 F.3d 1351 (D.C. Cir. 1996) (holding the EPA had "failed to satisfactorily explain why its § 112 "potential to emit" rule eliminated consideration of non-federally enforceable controls").

However, later, in *Engine Manufacturers Assn. v. EPA*, 88 F.3d 1075 (D.C. Cir. 1996), the court of appeals "rejected the National Mining Association's challenge to the EPA's decision, pursuant to § 213 of the Act, to regulate very large (greater than 750 hp) engines used in mining equipment, as well as the EPA's decision to regulate smoke from large compression–ignition engines." The court held that EPA had properly considered the costs its regulations would impose on manufacturers and users of very large mining engines and equipment and that the EPA properly promulgated smoke regulations for very large compression–ignition engines on the basis of scientific data (Mansfield, 1996, p. 131).

Referring to *Engine Manufacturers Assn.*, above, "the Court of Appeals for D.C. Circuit considered petitions challenging two rules on non-road engines. First, the court granted, in part, Engine Manufacturers' challenge to EPA's interpretation of § 209(e) which preempts states from adopting requirements relating to emissions from specific categories of 'new' non-road engines. The court held that: (1) the EPA had reasonably construed 'new' to mean that the engine or vehicle had not been sold to the ultimate purchaser or put into use (the "showroom-new" definition); (2) the EPA reasonably interpreted § 209 not to preempt states from adopting in-use regulations for non-road sources; and (3) the EPA improperly interpreted § 209(e)(2) as an implied preemption provision of the Act that authorizes California to adopt emissions standards for non-road sources that are not expressly preempted, to apply only to new non-road sources — because the plain language indicated that the implied preemption provision covers both new and used non-road sources" (Mansfield, 1996, pp. 130–131).

4.4.5.1 EPA Enforcement — Section 113

The United States brought an enforcement action under § 113(b) in *U.S. v. Hawaiian Cement*, No. 97-01204 ACK (D. Hawaii September 16, 1997 — proposed consent decree). "At its Portland cement manufacturing plant, Hawaiian Cement failed to achieve emission limitations for particulate standards established by the Hawaiian SIP and also violated applicable new source performance standards. The consent decree requires the defendant (Hawaiian Cement) to pay a civil penalty of $1,162,500 and undertake a comprehensive program that will ensure defendant complies with CAA requirements" (Mansfield, 1997, p. 171).

In *U.S. v. Shell Oil Co.*, No. 97-539-WDS (S.D. Ill. June 20, 1997 — proposed consent decree), "defendants agreed to resolve the U.S. civil claims against them by paying a $678,000 civil penalty and performing injunctive relief, which included installing an enhanced biodegradation unit for controlling benzene emissions from water extracted from groundwater production wells at Shell's facility" (Mansfield, 1997, p. 172).

In *U.S. v. J & D Enterprises*, 955 F. Supp. 1153 (D. Minn. 1997), the U.S. sought civil penalties against a contractor with the City of St. Paul razing an abandoned, municipally owned warehouse, which included the removal of asbestos before demolition. "The government alleged that the defendant (J & D) violated a number of the asbestos NESHAP provisions governing removal of asbestos. ... the defendant sought to file a third-party complaint against St. Paul seeking indemnity for all of the civil penalties that may be assessed. The court examined the legislative intent of the CAA and concluded that under Minnesota law, indemnification for civil penalties under the CAA was violative of public policy" (Mansfield, 1997, p. 173).

In another asbestos removal penalty case, *U.S. v. Burrell*, No. 95-385-2, 1996 WL 230039 (E.D. Pa. May 3, 1996), defendant Burrell moved the court to set aside his conviction for failure to file a written notice of asbestos removal and violation of asbestos removal work practice standards on the basis that "as an employee of the business supervising the removal, Congress did not intend that his activities would make him criminally liable as an 'operator' under § 113(c) of the Act. The court noted that Congress distinguished those in charge or in control of an asbestos removal operation from those who are employees following a supervisor's orders. To clarify the difference between an employee and an independent contractor, the court referred to factors enumerated in the restatement (Second) of Agency, section 220(2). The court held that in this instance, the defendant was clearly 'in charge or control' of the removal operations and denied the defendant's motion" (Mansfield, 1996, pp. 126–127).

Also, see *U.S. v. Fern, infra*, in Chapter 8 on expert witnessing, Section 8.10 — False Statements.

4.4.6 Additional Recent CAA Cases for Reference and Study

1. *Appalachian Power Co. v. EPA*, 208 F.3d 1015 (D.C. Cir. 2000). Re: Title V Permitting.
2. *Allied Local & Regional Manufacturers Caucus v. EPA*, 215 F.3d 61 (D.C. Cir. 2000). Re: Title I — General Issues — VOCs — architectural coatings, paints.

CHAPTER 5

Water Pollution and the Clean Water Act*

It isn't pollution that's harming the environment. It's the impurities in our air and water that are doing it.

Vice President Al Gore, 1998

5.1 INTRODUCTION — HISTORICAL COMMENT ON WATER POLLUTION

Fresh, clean drinking water is one of the basic, raw materials that man draws from the Earth, one that he cannot exist without. Man consumes about 3 to 4 lb. of water daily. Fortunately, water is a replenishable raw material, but its reserves are easily polluted and its contaminations must be guarded against.

Water pollution has existed in the world with the beginning of life forms, but man only became greatly concerned with its universal deterioration in the last hundred years. Unfortunately, man had previously viewed the earth's water bodies as limitless depositories with an innate ability to cleanse and absorb his deposited wastes. At various times in history, human populations have been plagued and ravaged by outbreaks of typhoid and waterborne disease epidemics. Only then did man concern himself with the reason, discovering and finally accepting that careless disposal of wastes, particularly human and animal excreta, was the cause of polluting sources of drinking water. This discovery ultimately led to local sanitation laws loosely regulating privies and their location, along with establishment of municipal sewer systems throughout the country. Nevertheless, collected raw sewage continued to be dumped into rivers, lakes, and seas. The polluting activity continues in some places. Sanitation and water treatment gradually became a field of civil engineering; and along with geological studies, the earth's hydrologic cycle became recognized as an interchange between the waters of the ocean, inland surface water bodies, and groundwater systems.

---

* The Clean Water Act (CWA) (33 U.S.C. § 1251–1376 officially first enacted as the Federal Water Pollution Control Act in 1956.

5.2 BRIEF REVIEW OF MAN'S WATER POLLUTION VS. NATURE'S TREATMENT PROCESS

Man has long felt that wastes dumped into bodies of water were dissipated by washing and dilution. That is a truism within limits. Reference is frequently made to the quotation from Samuel Taylor Coleridge's "Rhyme of the Ancient Mariner," "... water, water everywhere, nor any drop to drink." Of the earth's water 97% is in the oceans, leaving only 3% potentially potable from land sources.

Man-made water pollution is caused (1) when the volume of organic wastes exceeds the available oxygen in the water that is required for the natural bacterial action for decomposing the organic wastes; and (2) because inorganic matter and chemicals dumped into waters do not react with the natural bacterial process and in fact may even destroy the bacterial process for organic wastes.

The amount of oxygen in a body of water is a measure of its ecological health. Decomposition of organic wastes deposited in bodies of water is measured in biochemical oxygen demand (BOD) and chemical oxygen demand (COD). The effect of overtaxing the natural limits of oxygen reserves in a body of water not only pollutes but also affects the fish population in number and type. As waters become more polluted with dissolved oxygen and BODs consumed for the organic wastes deposited by sewers and other sources, popular and sporting species of fish requiring more oxygen are driven off or die and are replaced by a lower order of fish and water life that can survive in lesser oxygen content.

Geology teaches us that lakes are temporary features on the earth's surface (eutrophication). Pollution, particularly the introduction of heated waters, for example, from the water cooling process of energy-producing plants, and other industrial processes dumping wastewaters into lakes and rivers, hastens the natural, biological process by accelerating chemical reactions (such as the growth of algae and aquatic plants), thereby shortening the life of the lake many-fold. Raising the water temperature of lakes and streams reduces the amount of dissolved oxygen in the water, thereby affecting the fish population and even their reproductivity. With less oxygen, the body of water supports fewer fish.

5.3 EARLY REMEDIES AT LAW FOR WATER POLLUTION

The earliest legal concerns for pollution and contamination of water had been mainly for drinking and household water supplies. Recourse for damages by contamination of one's water supply was largely left to individuals to bring an action in common law against the alleged polluter. (**Common law** is that body of law that originated and developed in England, and became the law of the American colonies and other British-founded colonies. It serves as basic law in the United States, but is distinguished from statutory law as enacted by legislatures.) The complaints were brought in private nuisance and trespass with claims for damages to one's water supply. Riparian water law supported such claims in the eastern United States, and appropriation water rights law in the West. Governmental or statutory protection of one's household water purity from upstream or nearby polluters was nonexistent. Gradually, municipalities regulated against pollution of their water supplies, but individual users in agricultural and rural areas were left to pursue their own common law remedies.

A few examples of older case decisions and law follow and illustrate how individuals were left to take action against polluters of their water rights. A mine operator may have been liable to one whose well flow had been destroyed, or whose well had been contaminated by the proximate causes of mining. However, in *Bayer v. Nello Teer Co.*, 256 N.C. 509 (1962), a rock quarry, operating with the best practices of open-pit mining, which pumped no more run-in and percolating waters from its pit than necessary, was found not liable in damages to an adjoining landowner for contamination of his water supply from waters percolating into his well.

In *Gilmore v. Royal Salt Co.*, 84 Kan. 729 (1911), a salt mining company deposited a large quantity of refuse salt upon its land. By action of rain upon the salt, the water underlying an adjacent tract was impregnated with salt water through percolation, making it unfit for use and harmful for vegetation in irrigation. The mine operator was liable to the adjacent landowner. Additionally, in *Sunray DX Oil Co. v. Thurman*, 384 S.W. 2d 482 (1964), where a pit constructed to hold salt water overflowed onto the land and killed vegetation and timber, the lessee was held liable for the damages from flooding.

Also, in *Freel v. Ozark Mahoning Co.*, 208 F. Supp. 93 (1962), an action for damages was brought against a fluorite mining company in Colorado for injury to private property and health resort from contamination and pollution of the stream and the flooding of such property by failure of the company to contain its mill tailings ponds. The court found that the plaintiffs were entitled to recover damages without proof that the conduct of the defendant mining company was negligent or intentional. Evidence demonstrated, however, that the defendant's conduct was negligent, willful, wanton, and reckless.

The mining company had built several large tailing ponds containing harmful and noxious chemicals from the plant flotation process, and later had failed to repair a small breach of the pond's containing walls. The downstream initial harm and injury to Freel was caused by leaking of the chemicals into the nearby stream. Greater injury was exacerbated on two occasions when the containing walls were breached and flooding from the mill ponds severely damaged Freel's health resort buildings and polluted its mineral springs.

For an Arizona open-pit copper mining operation, the court found that the right under the Arizona statutes to use the water of the public streams for mining purposes did not give such user any right to send tailings and waste material from his reduction works down the stream to the destruction or substantial injury of the riparian rights of a user below for irrigation purposes [*Arizona Copper Co. v. Gillespie*, 230 U.S. 46 (1913)].

Similarly, in *Montgomery Limestone Co. v. Bearden*, 256 Ala. 269 (1951), a cause of action for a nuisance was made where the quarry operated a sump pump for the purpose of keeping the pit dewatered from run-in waters, resulted in the deposit of debris in the river, which then flowed through the complainant's premises in a polluted condition.

After World War II's industrial expansion and with the advent of environmentalism accelerating in the 1960s, in addition to the historical type of damage claims from mining and industrial operations — such as machinery noise and vibrations, and dust from blasting — new types of claims were entered in the field of litigation (i.e., damages to public waters, stream pollution, fish, and wildlife). An example from 1963 was the case of *People (of California) v. New Penn Mines, Inc.*, 28 Cal. Rptr. 337 (Cal. App. 1963), which was an action by the state's attorney general in the name of the state for abatement of an alleged public nuisance caused by drainage of toxic mine and mill wastes into a river resulting in damage to fish life and contaminated water quality.

The state's complaint alleged the Penn Mine, once an extensive producer of copper and zinc, had been inactive in recent decades. During its operation, fluid ore tailings and mill wastes were placed in settling ponds, and mine waste rock piled in dump areas. The rock dumps and tailings ponds are rich in mineral salts. During the rainy season, surface waters flow over the dumps and ponds, picking up concentrations of minerals that drain into the Mokelumne River. The river is a seasonal spawning ground of the king salmon and steelhead trout. The mineral pollutants are extremely harmful to the fish life and have resulted in kills of salmon and trout. Injunctive relief was sought.

The court held that although the owner of an inactive mine was not discharging industrial waste within the meaning of the water pollution act, a condition of pollution or nuisance, actual or threatened, may occur when the surface water or some other mechanism causes drainage of accumulated mine wastes into a public stream.

The New Penn Mines correctly objected on the grounds that the state in the person of the attorney general lacked jurisdiction to bring the suit. The court upheld Penn Mine's position that the injunction must fail on the ground that such an action must be brought by the appropriate regional water pollution control board acting under the provisions of the Dickey Water Pollution Act. (*Note:* However, for later litigation on the Penn Mine as a point source site of water pollution, see Chapter 7, Section 7.3.4, on the Penn Zinc and Copper mine, Calaveras County, California.)

A water pollution case involving downstream riparian rights that was decided in 1975, well after the Clean Water Act (CWA) was enacted, was *Springer v. Joseph Schlitz Brewing Co.* for liability of an industry to downstream riparian owners when its industrial wastes overloaded a city treatment system and caused water pollution. The water damage claim was initiated in North Carolina in 1969 and 1970, about the time the CWA was enacted. The case is of interest and noteworthy because it treats the claim under riparian rights and also involved a publicly owned treatment works (POTW), which became an important part of the CWA. The reason that there is no reference made to the CWA requirements in the decision is explained following the case.

Springer v. Joseph Schlitz Brewing Co.

United States Court of Appeals, Fourth Circuit, 1975
510 F.2d 468

BUTZNER, Circuit Judge.

This North Carolina diversity case raises the question of the liability of an industry to downstream riparian owners when its wastes overload a city's treatment facilities and cause water pollution. The plaintiffs, David and Diana Springer, own an interest in a large farm on the Yadkin River. Seeking an injunction and compensatory and punitive damages, they contend that, beginning in late 1969, wastes from a new brewery owned by Joseph B. Schlitz Brewing Company in Winston-Salem, North Carolina, overloaded the city's sewage treatment plant, causing it to pollute the Yadkin and interfere with their riparian rights. The Springers introduced evidence, which, viewed in the light most favorable to them, showed that Schlitz knew, or in the exercise of reasonable care should have known, that the city sewage plant lacked the capacity to treat the brewery's waste; that in discussions with the city, Schlitz underestimated the quantity and harmfulness of the waste; and that the company violated the city sewage ordinance. The evidence also established that in the Spring and Summer of 1970, after Schlitz reached full production, inadequately treated sewage from the overloaded plant caused six unprecedented fish kills and otherwise impaired the quality of the Yadkin River.

At the close of the Springers' case, the court directed a verdict for Schlitz on the ground that North Carolina absolves the user of a municipal sewer system of liability for the city's failure to adequately treat its sewage. We reverse because we believe the case is controlled by exceptions to North Carolina's rule of immunity.

I.

In North Carolina, a riparian landowner has a right to the agricultural, recreational, and scenic use and enjoyment of the stream bordering his land, subject, however, to the rights of upstream riparian owners to make reasonable use of the water without excessively diminishing its quality. Though he does not own the fish in the stream, the riparian owner's rights include the opportunity to catch them. Interference with riparian rights is an actionable tort, and a riparian owner may join several polluters as joint tortfeasors.

Nevertheless, an industry that uses a municipal sewage system to dispose of its waste is not liable to a riparian landowner for the pollution caused by the city's failure to provide adequate treatment. Quoting from 43 CJ. 158, the North Carolina Supreme Court stated in *Hampton v. Spindale*, 210 N.C. 546, 548, 187 S.E. 775, 776 (1936):

"[T]he inhabitants of a city who invoke its power to construct and control a sewer, and who use the sewer ... for the purpose and in the way prescribed by law, are not liable jointly with the city for the damages which result to third persons from the negligence of the city in the construction, management, or operation of the sewer."

In the only other North Carolina case to consider the point, the court justified the rule by emphasizing the inability of a private sewer user to control a city's treatment of its wastes after they entered the system. *Clinard v. Town of Kernersville*, 215 N.C. 745, 748, 3 S.E.2d 267, 270 (1939). The district court considered itself bound by these decisions to enter a directed verdict for Schlitz.

The Springers claim that their proof is sufficient to invoke exceptions to the general rule of immunity. Specifically, they contend that Schlitz should be held liable if it violated the city sewage ordinance, or if Schlitz knew, or should have known, of the inability of the city to adequately treat the brewery's wastes. Since no North Carolina court has considered these exceptions, we must determine the common law of the state by examining the rationale for the established rule, developments on this point in other states, and analogous areas of the state's common law.... We will treat the exceptions on which the Springers rely in Parts II and III of this opinion.

II.

In February 1970, the City of Winston-Salem enacted a comprehensive sewage ordinance to take effect in May. The ordinance requires every user of industrial sewers to have a discharge permit. Users may discharge only wastes containing 2500 ppm BOD or less, and they are forbidden to release sewage containing a wide variety of dangerous or difficult-to-treat substances. The ordinance imposes surcharges for BOD pound loadings caused by a concentration above 300 ppm. Originally, it allowed the city to furnish advice and technical assistance, but it did not provide for variances or exemptions. By state law, its violation is a misdemeanor.

Schlitz's effluent contained more than 2500 ppm BOD until April 1971. Beginning in May 1970, the city billed, and Schlitz paid, all BOD surcharges. The brewery and other industries, however, were allowed to operate in violation of the ordinance and without permits as long as they submitted schedules for compliance and conformed to them. It received its permit, one of the first issued to any industry, in May 1971.

The violation of a municipal sewage ordinance which is intended to protect downstream riparian owners can subject an industrial sewage source to private civil liability. *Hampton v. Spindale*, 210 N.C. 546, 548, 187 S.E. 775, 776 (1936), expressly restricts freedom from liability to those persons who use the sewers "in the way prescribed by law." Although this is dictum it is consistent with the rationale for the private user's immunity. When an industry turns over the control of its sewage to the city, it can reasonably expect that the city will safeguard riparian property by effective treatment. But it is not reasonable for an industry to expect a city to safely treat prohibited sewage. Consequently, the reason for granting immunity does not then apply.

This reading of *Hampton* conforms to North Carolina's general law regarding the effect of regulatory legislation on civil liability. The state is firmly committed to the proposition that the "violation of a statute designed to protect persons or property is a negligent act, and if such negligence proximately causes injury, the violator is liable." *Murray v. Bensen Aircraft Corp.*, 259 N.C. 638, 131 S.E.2d 367 (1963) (federal aircraft safety statute); accord, *Bell v. Page*, 271 N.C. 396, 156 S.E.2d 711 (1967) (town ordinance regulating private swimming pools). The statute or ordinance, serving as a legislative declaration of a standard of care, creates a private right not to be harmed by its violation.

Schlitz's failure to obtain a permit until May 1971 does not afford the Springers a ground for recovery. The permit does not protect riparian owners. It is only an instrument of the city's enforcement program, and its absence does not pollute the stream. Moreover, the ordinance does not explicitly forbid the discharge of wastes containing more than 2500 ppm BOD. The ordinance is a criminal statute which must be construed in the defendant's favor in civil proceedings as well as in criminal ones. Since the discharge of more than 2500 ppm BOD is not a crime, it does not constitute negligence per se.

In contrast, the discharge of sewage prohibited by Section 23-2(2) [of the Winston-Salem Code of Ordinances] is a crime, and under North Carolina law it is actionable if it proximately causes damage to riparian property. Section 23-2(2) states in part:

"(2) Except as hereinafter provided, it shall be unlawful for any person to discharge or cause to be discharged any of the following described materials, waters, liquids, or wastes into any public sanitary sewer:

"(g) Liquid wastes containing any toxic or poisonous substances in sufficient quantities to (i) interfere with the biological processes used in a sewage treatment plant, or (ii) which, in combination with other liquid wastes, upon passing through a sewage treatment plant will be harmful to persons, livestock, or aquatic life utilizing the receiving streams into which water from a sewage treatment plant is discharged."

In order to establish that Schlitz's violation of the ordinance was negligence, the Springers would have to prove that the brewery wastes had the characteristics forbidden by Section 23-2(2). Viewed in the light most favorable to them, the evidence showed that the brewery's wastes had a toxic or poisonous effect on bacteria that are essential to the sewage treatment process. The evidence also showed that after passing through the plant, the wastes were harmful to aquatic life in the receiving stream. The jury should, therefore, be allowed to determine whether or not the discharge violated the ordinance. If it did, this was negligence, and the jury should then decide whether it proximately caused damage to the Springers' property.

Schlitz argues that its discharge of brewery wastes is not actionable because the city's officials did not require compliance with the ordinance until May 1971, except for the payment of surcharges. This contention lacks merit. Section 23(d) of the ordinance authorized the water and sewer officials to:

"consult with and furnish technical assistance and advice to industrial users of the city's sewerage system in order to assist them in devising procedures and constructing equipment to reduce or eliminate from industrial wastes objectionable characteristics or properties which may not otherwise be discharged into the public sanitary sewers under Section 23-2."

While this provision recognized that compliance was not instantly attainable and allowed city officials to furnish technical assistance and advice, its terms gave them no power to affect the rights of third parties. A statute protecting life or property creates a private right not to be harmed by its violation unless the legislature specifically provides otherwise. Public officials have no more power to alter these private rights than the legislature chooses to give them.

It is argued, however, that the ordinance should not be read literally because the only reasonable construction is one which would allow city officials to dispense with its requirements. While it is true that the city's director of water and sewers believed that immediate, rigorous enforcement would be undesirable because it would "shut down just about every industry in Winston-Salem," there is no conflict between the public interest in keeping a factory open and the private right to recover damages for pollution. The two may be reconciled by requiring the source of pollution to pay damages while allowing it to operate. See *Boomer v. Atlantic Cement Co.*, 26 N.Y.2d 219, 309 N.Y.S.2d 312, 257 N.E.2d 870 (1970). Since the plain language of the ordinance does not authorize city officials to dispense with its requirements, their actions could not affect the Springers' rights.

In addition, paying effluent surcharges did not relieve Schlitz from having to obey the rest of the ordinance. The city levies these charges "to cover some of the cost to the city of treating such wastes." They begin not at 2500 ppm but at 300 ppm, the approximate dividing line between domestic and industrial strength sewage. The charges are enumerated in the ordinance section which sets other fees, not in the one which describes prohibited uses. They are revenue measures rather than substitutes for civil liability.

III.

As an alternative ground of recovery, the Springers argue that Schlitz should be held liable if it knew, or in the exercise of reasonable care should have known, that the city could not adequately treat the brewery's waste. The validity of this claim depends on an analysis of the underlying reasons for the general rule of immunity.

In absolving the user of a municipal sewage system from liability, North Carolina follows the leading case of *Carmichael v. Texarkana*, 116 F.845 (8th Cir. 1902).... *Carmichael* involved a suit by a riparian owner against a city and its inhabitants. The court affirmed the dismissal of the complaint against the inhabitants, holding that the city alone was responsible for the nuisance created by its sewer system. It analogized the problem to a more familiar area of vicarious responsibility — the employer's liability for the torts of an independent contractor.

In contrast, a manufacturing company is in a position to know enough about its own operations to understand the nature of its wastes and the problems of treating them. If it has not yet selected a plant site, it may have a choice among cities and disposal systems. It may also be large enough to pre-treat or properly dispose of its own sewage. When the company represents a desirable source of jobs and tax revenues, the local authorities can be expected to provide it with technical information about the municipal sewers. In brief, a large industrial sewer user can make an informed decision whether to use a city sewer system to render its wastes nuisance free.

According to the Springers' evidence, Schlitz, which has more than a century of experience in brewing beer, considered several cities in the Southeast before selecting Winston-Salem for a brewery site. As part of its investigation, the company requested information about the city's sewage facilities, and its representatives toured the treatment plant. Had the company inquired, it would have learned that the plant was operating at or over its daily capacity of 18 million gallons of sewage containing 76,000 pounds of BOD. The city assured Schlitz that it would adequately treat its wastes after Schlitz advised the city to expect 15,000 pounds of BOD per day. Before the brewery had been open a year, however, it was discharging 56,000 pounds of BOD per day into a system which had previously reached its 76,000 pound capacity. Brewery wastes are known by sanitary engineers to be difficult to treat and to interfere with the treatment of other wastes because of their high concentration of BOD and fluctuating alkalinity.

IV.

The record discloses that after the 1970 fish kills, Schlitz expended more than $1,300,000 for sewage treatment facilities, and it is now in compliance with the effluent quality standards of the city ordinance. The city has doubled the size of its treatment plant, and it can now properly treat the waste from the brewery. In view of this evidence, it is unlikely that an injunction is appropriate under North Carolina law. [Citations omitted]

On remand, the court should instruct the jury that the Springers cannot recover if the evidence discloses only that Schlitz's waste, after passing through the Winston-Salem treatment plant, polluted the Yadkin, killed fish, and caused other damage to the Springers' riparian property. Under North Carolina's general rule of immunity, Schlitz cannot be held liable if the evidence discloses no more than those facts.

The jury should be told that in order for the Springers to recover they must prove by a preponderance of the evidence that Schlitz violated Winston-Salem's sewage ordinance by discharging sewage prohibited by § 23 2(2) or that Schlitz knew, or in the exercise of reasonable care should have ascertained, that the city's treatment plant could not adequately treat its sewage. In either event, the Springers must also prove that the brewery's waste proximately caused their injury; that is, that it was foreseeable that Schlitz's waste, alone, or in conjunction with other waste, would damage the Springers' riparian property. This outline of the elements of the plaintiffs' claim, of course, is not intended to preclude the court from charging the jury about defenses Schlitz may present when the case is fully tried.

These facts disclose that Schlitz knew the characteristics of its own sewage and that it exercised control over its selection of a site and of the sewage system which would dispose of its waste. The evidence also indicates that Schlitz could not rely on the city's acceptance of its sewage, because it had not furnished the city accurate information. Finally, the proof supports an inference that Schlitz knew, or in the exercise of reasonable care should have ascertained, that the city could not adequately treat its brewery wastes. The jury might reasonably have decided that Schlitz negligently selected the city of Winston-Salem to treat its sewage. Drawing upon North Carolina's common law principles governing the liability of an employer of an independent contractor, we conclude that Schlitz is not, as a matter of law, immune from liability.

Accordingly, on remand, the district court should submit the issue of Schlitz's liability on the Springers' alternative theory.

Before the NEPA era, **the federal court systems had proclaimed that there was no federal common law**. Common law claims were reserved to state legal systems (e.g., nuisance, trespass). However, in the years before federal clean water laws became effective enough to regulate interstate water pollution, the *U.S. Supreme Court established an interim recognition of the existence of a federal common law* in *Illinois v. City of Milwaukee*, 406 U.S. 91 (1972) that would support a claim of nuisance caused by interstate water pollution. The doctrine was established in support of an action by the state of Illinois seeking to compel the city of Milwaukee to control sewage and storm water flows into Lake Michigan. The waste discharges allegedly carried disease pathogens by lake currents to the Illinois shores. Adoption of the common law doctrine was deemed necessary to allow federal courts to control interstate nuisance claims where the water pollution laws under the EPA were as yet ineffective.

The doctrine was continued by the federal courts in other interstate claims until 1981 when the U.S. Supreme Court reversed itself in *City of Milwaukee II,* **holding that there was no federal common law doctrine remedy to downstream states**. The court's reason was that the common law nuisance claim had been preempted by the Federal Water Pollution Control Act passed 5 months after *City of Milwaukee I* (1972).

In 1987, federal common law downstream nuisance claims against upstream polluters in interstate waterways was laid to rest in *International Paper Co. v. Ouellette*, 479 U.S. 481 (1987). In *Ouellette*, Vermont property owners brought damage claims for a continuing nuisance under Vermont laws in a Vermont federal court against the New York paper mill. The federal district court found for the Vermonters' claim (in *International Paper Co. v. Ouellete*, 602 F. Supp. 264 [D. Vt. 1985]) holding that an action for interstate pollution could be maintained in the state where the injury occurred. The holding was affirmed by the Second Circuit Court of Appeals, and appealed to the U.S. Supreme Court. The Supreme Court reversed finding that Vermont law did not govern. If allowed, surely followed by similar subsequent claims, such a holding would interfere with the design and purpose of the CWA intended by Congress.

For more recent arguments and treatment of pollution of interstate waterways, see *Oklahoma v. EPA,* 908 F.2d 595 (10th Cir. 1990) and also *Arkansas v. Oklahoma,* 112 S. Ct. 1046 (1992), *infra,* at Section 5.8.10.1.

5.4 POLLUTED WATERS — HEALTH HAZARD — WASTEWATER TREATMENT

In areas where human excreta disposal and sewage treatment facilities are lacking or even where facilities are available, lack of or insufficient proper pretreatment leads to one of the major factors threatening human health. Various chemicals and organisms present in discharges containing fecal matter are known to cause bacillary dysentery (shigellosis), infectious hepatitis, diarrhea, salmonella infection, and many other types of disease. Extermination of the organisms must be completed in sewage treatment. If not exterminated in the sewage plant, they will live another day to destroy human health by being retained in the sewage sludge produced by a POTW. Sludges are frequently used as fertilizers and spread on the earth's surface. Insufficient treatment allows certain of the organisms to survive and be retained in the sewage sludge. It has been proved that some pathogenic organisms will survive in certain soil conditions. Dry soil is unfavorable for pathogen survival, but some may survive as long as 2 years in freezing moist soil.

Salvato reports that studies of enteric pathogens survival in wastewater treatment plant effluents show that up to 50% of coliform and pathogenic bacteria is removed by primary sedimentation, but is "relatively ineffectual in removing viruses and protozoa. Activated sludge or trickling filter treatment removes about 90% of the coliform or pathogenic bacteria remaining after primary sedimentation. Viruses, although reduced, survive activated sludge, and, especially, trickling filter treatment. Chemical coagulation, flocculation, sedimentation, and filtration will remove nearly all bacteria, viruses, protozoa, and helminths, particularly if supplemented by chlorination" (Salvato, 1982, p. 377).

5.5 EVOLUTION OF WATER POLLUTION REGULATION

Sanitation and pollution of public waterways had long been recognized as a matter for local concern and regulation. The earliest federal attempt to control pollution of public waters was the enactment of the Rivers and Harbors Act of 1899, known as the Refuse Act, Title 33 U.S.C. § 403 et seq. Its concern for dumping and filling in the public waters was more with obstructions to water craft than for pollution. However, its § 407 made unlawful the depositing or discharging of "any refuse matter of any kind or description whatever other than that flowing from streets and sewers and passing therefrom in a liquid state, into any navigable water of the United States, or into any tributary of any navigable water from which the same shall float or be washed into such navigable water. Section 411 provided penalties for fines and imprisonment for convicted violators.

In tracing the history of the federal government's involvement in controlling water pollution, Columbia School of Law Professor Frank Grad's note on water pollution relates: "Later federal legislation dealt mainly with water pollution as a vector in the spread of communicable diseases. The Public Health Services Act of 1912 authorized the investigation of the effect of pollution in navigable lakes and streams on public health. Subsequent cooperation between the Public Health Service and state agencies resulted in the voluntary adoption of nationwide standards for the treatment of drinking water. The adoption of these standards has almost entirely eliminated waterbourne diseases" (Grad, 1978, p. 2–73).

Professor Grad's Historical Note continues:

The Federal Government had involved itself in environmental controls rather gradually, with reliance on the commerce power in the direct control of pollution being a fairly recent phenomenon of federal regulation; initially, it operated indirectly through the Federal Government's power to tax and spend for the general welfare (U.S.Constitution, Art. 1, Sect. 8, Cl. 3).... federal involvement (began) through sponsorship of ... grant-in-aid programs, with federal standards ... being imposed as a condition of receipt of federal funds for environmental control. This traditional, congressional approach to the pollution control problem (was) exemplified in the policy declaration of the Federal Water Pollution Act of 1948, the first major federal legislation on the subject: "It is hereby declared to be a policy of Congress to recognize, preserve and protect the primary responsibilities and rights of the states in controlling water pollution." 33 U.S.C. Sect. 466 (1964)

Under early water pollution and air pollution control acts, the federal function was to (cooperate) with state and interstate agencies as well as with local and municipal governments. When pollution problems worsened rapidly, the wisdom of this secondary federal role was questioned. The new direction was first reflected in the declaration of the policy of the Water Quality Control Act of 1965: "The purpose of this act is to enhance the quality and value of our water resources and to **establish a national policy** for the prevention, control, and abatement of water pollution" [33 U.S.C. Sect. 466 (1964 Supp.V)] (Grad, 1978, p. 2–72) (emphasis by Aston).

Thus, with the acquiescence of the states over water pollution regulation and the declaration in 1965 of the federal government's national policy for water quality control, followed by a series of congressional water quality acts, viz., the Water Quality Act of 1965, the Clean Water Restoration Act of 1966, and rather closely by the passage of NEPA in 1969, the federal government assumed command not only for control over clean water but also for all environmental quality in the nation.

The Federal Water Pollution Control Act Amendment (FWPCA) of 1972, better known as the CWA, is a command-and-control type of legislative act. It states its objectives: "to restore and maintain the chemical, physical, and biological integrity of the Nation's waters" [FWPCA, amend. 1972, 33 U.S.C. § 1251(a)]. Other goals set in the amended act were "wherever attainable, an interim goal of water quality which provides for the protection and propagation of fish, shellfish, and wildlife, and provides for recreation in and on the water be achieved by July 1, 1983"; and the national goals "that the discharge of pollutants into navigable waters be eliminated by 1985...," and "the prohibition of the discharge of toxic pollutants in toxic amounts" [FWPCA, § 1251(a) 2, § 1251 (a)(1), § 1251(a)(3)]. Ambient water quality standards were to be supplemented by discharge standards in the form of effluent limitations applicable to all point sources.

5.6 SOURCES OF WATER POLLUTION

The National Water Commission in its Final Report (1973) to the President and to Congress on "Water Policies for the Future" stated that sources of water pollution are of two types: (1) waste discharges from identifiable points (point sources); and (2) diffused wastes reaching water through land runoff, washout from the atmosphere, or other means (nonpoint sources), each differing in their amenability to control.

The point sources were listed as municipal sewerage systems, storm water runoff, industrial wastes, and animal wastes from commercial feedlots.

The nonpoint sources listed were sediment, agricultural chemical, mine drainage, and spills of oil and other hazardous substances; and miscellaneous nonpoint sources such as salting of highways for ice control and discharge of waste from vessels.

5.7 EFFLUENT DISCHARGE CONTROL UNDER THE CLEAN WATER ACT

The statutory authority for setting effluent standards and objectives is found in §§ 301 and 304 of CWA.

CLEAN WATER ACT § 301 [§ 1311]

(b) In order to carry out the objective of this Act there shall be achieved —

(1)(A) not later than July 1, 1977, effluent limitations for point sources, other than publicly owned treatment works, (i) which shall require the application of the best practicable control technology currently available as defined by the Administrator pursuant to section 304(b) of this Act, or (ii) in the ease of a discharge into a publicly owned treatment works which meets the requirements of subparagraph (B) of this paragraph, which shall require compliance with any applicable pretreatment requirements and any requirements under section 307 of this Act; and

(B) for publicly owned treatment works in existence on July 1, 1977, or approved pursuant to section 203 of this Act prior to June 30, 1974 (for which construction must be completed within four years of approval), effluent limitations based upon secondary treatment as defined by the Administrator pursuant to section 304(d)(1) of this Act; or,

(C) not later than July 1, 1977, any more stringent limitation, including those necessary to meet water quality standards, treatment standards, or schedule of compliance, established pursuant to any State law or regulations (under authority preserved by section 510), or any other Federal law or regulation, or required to implement any applicable water quality standard established pursuant to this Act. (2)(A) for pollutants identified in subparagraphs (C), (D), and (F) of this paragraph effluent limitations for categories and classes of point sources, other than publicly owned treatment works, which (i) shall require application of the best available technology economically achievable for such category or class, which will result in reasonable further progress toward the national goal of eliminating the discharge of all pollutants, as determined in accordance with regulations issued by the Administrator pursuant to section 304(b)(2) of this Act, which such effluent limitations shall require the elimination of discharges of all pollutants if the Administrator finds, on the basis of information available to him (including information developed pursuant to section 315), that such elimination is technologically and economically achievable for category or class of point sources as determined in accordance with regulations issued by the Administrator pursuant to section 304(b)(2) of this Act or (ii) in the case of the introduction of a pollutant into a publicly owned treatment works which meets the requirements of subparagraph (B) of this paragraph, shall require compliance with any applicable pretreatment requirements and any other requirement under section 307 of this Act.

CLEAN WATER ACT § 304 [§ 1314]

(b) For the purpose of adopting or revising effluent limitations under this Act the Administrator shall, after consultation with appropriate Federal and State agencies and other interested persons, publish within one year of enactment of this title, regulations, providing guidelines for effluent limitations, and, at least annually thereafter, revise, if appropriate, such regulations. Such regulations shall —

(1)(A) identify, in terms of amounts of constituents and chemical, physical, and biological characteristics of pollutants, the degree of effluent reduction attainable through the application of the best practicable control technology currently available for classes and categories of point sources (other than publicly owned treatment works); and

(B) specify factors to be taken into account in determining the control measures and practices to be applicable to point sources (other than publicly owned treatment works) within such categories or classes. Factors relating to the assessment of best practicable control technology currently available to comply with subsection (b)(l) of section 301 of this Act shall include consideration of the total cost of application of technology in relation to the effluent reduction benefits to be achieved from such application, and shall also take into account the age of equipment and facilities involved, the process employed, the engineering aspects of the application of various types of control techniques, process changes, non-water quality environmental impact (including energy requirements), and such other factors as the Administrator deems appropriate;

(2)(A) identify, in terms of amounts of constituents and chemical, physical, and biological characteristics of pollutants, the degree of effluent reduction attainable through the application of the best control measures and practices achievable including treatment techniques, process and procedure innovations, operating methods, and other alternatives for classes and categories of point sources (other than publicly owned treatment works); and

(B) specify factors to be taken into account in determining the best measures and practices available to comply with subsection (b)(2) of section 301 of this Act to be applicable to any point source (other than publicly owned treatment works) within such categories of classes. Factors relating to the assessment of best available technology shall take into account the age of equipment and facilities involved, the process employed, the engineering aspects of the application of various types of control techniques, process changes, the cost of achieving such effluent reduction, non-water quality environmental impact (including energy requirements), and such other factors as the Administrator deems appropriate.

In spite of the idealistic goals, the "no discharge into navigable waters" was not reached by 1985. However, EPA had early-on instituted a two-phased set of standards for existing industrial point sources to effectuate the goal as quickly as possible. CWA, § 304(1) dealt with individual control strategies (ICS) for industries. Industrial dischargers were to employ the *best practicable control technology* (BPT) *currently available* [CWA, § 301(b)(1)(A)] to achieve effluent limitations by 1977. From 1977 to 1983, the best available technology (BAT) economically achievable was to be used in accomplishing set effluent standards [CWA, § 301(b)(2)(A)].

Similar effluent standards were set for POTWs [§ 307(b)]. Section 307(c) set pretreatment effluent standards for point sources discharging into POTWs. New point sources were regulated by new source performance standards using the best available demonstrated technology.

5.7.1   National Pollutant Discharge Elimination System

Control of effluent discharges for all point sources was to be effectuated through the National Pollutant Discharge Elimination System (NPDES) permit (CWA, § 402), either by EPA or an EPA–state approved program. Without an NPDES permit, the CWA made it unlawful for any point source to discharge pollutants into a body of water.

Permits had to incorporate applicable effluent limitations established under §§ 301, 302, 306, and 307 including enforceable schedules of compliance meeting the 1977 and 1983 deadlines. In 1987, a § 402 amendment provided that the renewal of a facility permit cannot be less stringent.

The NPDES program has four elements: (1) industrial and municipal permits, (2) federal facilities permitting, (3) pretreatment program, and (4) general permits. (General permits allow the issue of one permit of uniform limitations for a specified class of dischargers in a defined geographic area.) The usual permit is issued for a 5-year period and may be renewed.

Reading the following case of *Westvaco Corp. v. U.S. EPA* will be informative of the NPDES program and several other EPA program procedures.

WESTVACO Corporation v. United States EPA

United States Court of Appeals, Fourth Circuit, 1990, 899 F.2d 1383.

PHILLIPS, Circuit Judge.

This matter is before the court on the motion of the United States Environmental Protection Agency (EPA) to dismiss consolidated petitions for review filed by Westvaco Corporation (Westvaco) challenging certain agency actions taken by EPA Region III on June 2, 1989. Specifically, on that date EPA proposed to partially disapprove lists of "impaired waters" submitted by the states of Maryland and Virginia, respectively, pursuant to § 304(*l*) of the Clean Water Act (CWA). 33 U.S.C.§ 1314(*l*). EPA contends that this court lacks jurisdiction over these petitions. The parties have submitted extensive legal memoranda in support of and opposition to the motion to dismiss. We agree with the EPA's contentions and will dismiss the petitions for lack of jurisdiction in this court to review the challenged actions at this time.

The general legal background and procedural history of this case is given as agreed to by the parties.

As a primary means of achieving its ultimate goals, the CWA prohibits the discharge from any point source into protected national waters of any pollutant unless that discharge complies with specific requirements of the CWA. Section 301(a), 33 U.S.C. § 1311(a); § 502(12); 33 U.S.C. § 1362(12). Compliance may be achieved by obtaining a permit issued pursuant to § 402. 33 U.S.C. § 1342.

Section 402 establishes the National Pollutant Discharge Elimination System (NPDES) permit program. 33 U.S.C. § 1342. See *Environmental Protection Agency v. California*, 426 U.S. 200, 205, 96 S. Ct. 2022, 2025, 48 L. Ed. 2d 578 (1975). NPDES permits are issued by EPA or, in those states in which EPA has authorized a state agency to administer the NPDES program, by that agency subject to EPA review. 33 U.S.C. § 1342(a)–(d). EPA has approved 39 states to issue NPDES permits, including Maryland and Virginia. See 33 U.S.C. § 1342(b).

NPDES permits may be issued for terms up to five years. Section 402(b)(1)(B), 33 U.S.C. § 1342(b)(1)(B). Permits must incorporate technology-based controls, *i.e.*, limitations based on the degree of effluent control which can be achieved by point sources using various levels of pollution control technology. See §§ 301, 304, 33 U.S.C. §§ 1311, 1314. See also *E.I. DuPont de Nemours & Co. v. Train*, 430 U.S. 112, 126-36, 97 S. Ct. 965, 974, 51 L. Ed. 2d 204 (1977). In addition to technology-based controls, permits must contain any more stringent limitations that are necessary to meet water quality standards developed by the states pursuant to § 303. 33 U.S.C. § 1313. If standards are not established by a state, EPA must establish the water quality standards for the waters in that state. Section 301(b)(1)(C), 33 U.S.C. § 1311(b)(1)(C). Water quality standards consist of: (i) a designated "use" for the waters in question (*e.g.*, public water supply), and (ii) "water quality criteria" specifying the amount of various pollution which may be present in those waters and still achieve the designated use(s). 40 C.F.R. §§ 131.2, 131.3 (1988). The state "water quality criteria" may be expressed as numerical concentration limits or in narrative form. 40 C.F.R.§ 131.3(b).

Unlike technology-based limitations, water quality standards are not developed based on an evaluation of the capability of pollution control technologies but on the physical attributes of the water segment necessary to support the designated uses. Once water quality standards have been set, NPDES permit limitations must be established to assure compliance, regardless of the availability or effectiveness of treatment technologies.

The CWA requires that approved states' NPDES permitting programs be consistent with minimum federal requirements. 33 U.S.C. § 1314(i). Accordingly, EPA is given authority to review every state-issued NPDES permit. Section 402(b), 33 U.S.C. § 1342(b). To facilitate EPA's task, the CWA requires the state permitting authority to provide EPA with a copy of each permit application and provide notice of developments during the permitting process. Section 402(d)(1), 33 U.S.C. § 1342(d)(1); 40 C.F.R. §§ 123.43(a)(1), 123.43(a)(2)(1988). Section 402(d)(2) prohibits issuance of a permit by a state if the Administrator of EPA objects within 90 days. 33 U.S.C. § 1342(d)(2). If the state fails to submit a revised permit satisfying EPA's objections, EPA is authorized to issue a federal NPDES permit. Section 402(d)(4), 33 U.S.C. § 1342(d)(4) ; 40 C.F.R. § 123.44 (1988).

Where EPA assumes permit-issuing authority pursuant to § 402(d) and 40 C.F.R. § 123.44(h), EPA's regulations provide a comprehensive process for issuance of a final permit. See 40 C.F.R. §§ 124.6–124.15. If EPA intends to issue the permit, the Agency publishes a public notice of, and solicits public comment on, the draft permit. 40 C.F.R. § 124.10. After the close of the public comment period, EPA issues a final permit decision. 40 C.F.R. § 124.15(a). Any interested person may request an evidentiary hearing on EPA's final permit decision, with review before the Administrator. 40 C.F.R. §§ 124.74(a), 124.91. Final agency action on a permit does not occur until administrative remedies have been exhausted. 40 C.F.R. § 124.60(g). The Administrator's action in issuing or denying a permit is reviewable in the Courts of Appeals. Section 509(b)(1)(F), 33 U.S.C. § 1369(b)(1)(F).

In 1987, Congress passed the Water Quality Act (WQA) which placed greater emphasis on attaining state water quality standards. In particular, § 308 of the WQA amendments made several changes to the provisions of the CWA to focus attention on attaining water quality standards for toxic pollutants.

The first component of the WQA § 308 water quality program for toxic pollutants was the establishment of the § 304(*l*) program, entitled "Individual Control Strategies for Toxic Pollutants." 33 U.S.C. § 1314(*l*). Section 304(*l*)(1)(D) requires the states, within two years after February 4, 1987, to establish individual control strategies (ICS) which will reduce point source discharges of toxic pollutants sufficient to attain the water quality standards within three years. 33 U.S.C. § 1314(*l*)(1)(D). Section 304(*l*) requires EPA to approve or disapprove the state submissions of ICSs by June 4, 1989. 33 U.S.C. § 1314(*l*)(2). In the event of a state's failure to submit the lists and ICSs, or if EPA disapproves an ICS, EPA is required to develop the lists and the ICSs in cooperation with the state. Section 304(*l*), 33 U.S.C. § 1314(*l*)(3). These deadlines have effectively required the states and EPA to place high priority on identifying and controlling certain "toxic hot spots."

In order to identify these "toxic hot spots," § 304(*l*)(1)(B) requires the states to list those waters that are not expected to achieve applicable water quality standards, after application of technology based controls, due to discharges from point sources of toxic pollutants. 33 U.S.C. § 1314(*l*)(1)(B). This list is commonly referred to as the "B list." For each water segment listed on the B list, § 304(*l*)(1)(C) requires the states to identify the point sources responsible for the discharges of the toxic pollutants. 33 U.S.C. § 1314(*l*)(1)(C). This list is commonly referred to as the "C list."

For each point source on the C list, § 304(*l*)(1)(D) requires the states to develop the individual control strategy (ICS) discussed above.

Section 304(*l*) did not change the basic requirements of the CWA; rather it simply established a mandatory schedule for the completion of a toxic pollutant subset of the water quality-related activities that the CWA already imposed. Thus, before 1987, § 303(g) already had required states — without any deadline to evaluate their waters and identify those which needed controls beyond technology-based controls. 33 U.S.C. § 1313(d). Section 301(b)(1)(C) already had required limitations in permits to meet water quality standards for all pollutants. 33 U.S.C. § 13H(b)(1)(C).

EPA has now promulgated final regulations interpreting and implementing § 304(*l*). See 54 Fed. Reg. 246-58 (Jan. 4, 1989) and 23,86899 (June 2, 1989) (to be codified at 40 C.F.R. §§ 130.10, 123.46). The regulations establish the procedure for review of a state's lists of ICSs. In the event of a preliminary disapproval of the lists or ICSs, the procedure is similar to agency rulemaking, in that public participation is sought. On the other hand, if EPA approves a state's lists and ICSs, the § 304(*l*) listing process is complete and, therefore, constitutes final agency action. (Codified at 40 C.F.R. § 130.10(d)(8)). The regulations specify that EPA will make preliminary disapprovals by June 6, 1989. (To be codified at 40 C.F.R. § 130.10(d)(8)). The regulations require EPA to solicit and respond to public comment before any final disapproval decision on the lists or ICSs is made. (To be codified at 40 C.F.R. § 130.10(d)(10)). The public comment period is 120 days from the date of the region's preliminary disapproval. (To be codified at 40 C.F.R. § 130.10(d)(10)(vi)). Once the public comment period ends, EPA must make a final decision by June 1990, taking into account the public *comments*. (To be codified at 40 C.F.R..§ 130.10(d2(11,),). The regulations contemplate that EPA's preliminary disapproval of a state's lists and associated ICSs may be modified (*i.e.*, waters and point sources may be added or deleted) based on the additional data or information EPA receives during the public comment period. (To be codified at 40 C.F.R. § 130.10(dXll)(i)).

Exercising its agency discretion, EPA has defined an ICS to be a draft or final NPDES permit, with supporting documentation showing that effluent limits are sufficient to meet the applicable water quality standards. (To be codified at 40 C.F.R. § 123.46(c)). Therefore, in order to fulfill the mandate of § 304(*l*) that ICSs be developed by June 1990 for all listed point sources, some unexpired NPDES permits will have to be re-evaluated and possibly modified using existing NPDES permit issuance procedures.

Westvaco's petition for review in No. 89-2180 challenges EPA's preliminary disapproval of Maryland's § 304(*l*) lists and associated ICSs developed by the Maryland Department of the Environment. Westvaco owns and operates a bleached kraft pulp and paper mill in Luke, Maryland.

On February 3, 1989, the Maryland Department of the Environment had submitted its § 304(*l*) lists and ICSs to EPA Region III (the lists were subsequently modified by Maryland on May 10, 1989). Maryland did not include the North Branch of the Potomac River on its B list, or Westvaco's Luke Mill (which is located on that water segment) on its C list, and did not submit an ICS for this point source of dioxin. EPA Region III reviewed and proposed to disapprove Maryland's lists and ICSs on June 12, 1989. Among the reasons for the preliminary disapproval was Maryland's failure to list Westvaco's Luke Mill.

EPA Region III solicited public comment on its June 2 preliminary partial disapproval of Maryland's lists and ICSs by notice dated June 5, 1989. The notice was published in two daily newspapers of general circulation in Maryland. EPA Region III also directly solicited comment from Westvaco in a letter dated June 6, 1989. Westvaco and others submitted comments during the public comment period, which closed on October 5, 1989.

Pursuant to § 304(*l*)(3), EPA is required to take final action on the Maryland lists and respond to the public comments by June 1990. In making its final decision, EPA must consider the additional data and information received during the public comment period. EPA has discretion to add or delete waters and point sources prior to making its final decision.

Westvaco's petition for review in No. 89-2181 challenges EPA's preliminary disapproval of Virginia's § 304(*l*) lists and associated ICSs developed by the Virginia Water Control Board. Westvaco owns and operates a bleached kraft pulp and paper mill in Covington, Virginia.

On February 3, 1989, the Virginia Water Control Board had submitted its § 304(*l*) lists and ICSs to EPA Region III (the lists were subsequently modified by Virginia on February 16, 1989). Virginia did not place the Jackson River on its B list, or Westvaco's Covington Mill (which is located on that water segment) on its C list, and did not submit an ICS for this point source of dioxin. EPA Region III reviewed and proposed to disapprove Virginia's lists and ICSs on June 1, 1989. Among the reasons for the preliminary disapproval was Virginia's failure to list Westvaco's Covington Mill.

EPA Region III solicited public comment on its June 2 preliminary partial disapproval of Virginia's lists and ICSs by notice dated June 5, 1989. The notice was published in four daily newspapers of general circulation in Virginia. EPA Region III also directly solicited comment from Westvaco in a letter dated June 6, 1989. Westvaco and others submitted comments during the public comment period, which closed on October 5, 1989.

Pursuant to § 304(*l*)(3), EPA is required to take final action on the Virginia lists and respond to the public comments by June 1990. In making its final decision, EPA must consider the additional data and information received during the public comment period. EPA has discretion to add or delete waters and point sources prior to making its final decision.

The court concluded that it did not have jurisdiction under § 509 to review the challenged actions. It stated that the EPA's objection to the state's proposed "B" and "C" lists was inadequate to vest it with subject matter jurisdiction. Once a permit issued to the individual point sources, judicial review would then be available.

To implement the goals of the CWA, the EPA as the responsible agency had to formulate rules and guidelines for point sources of pollution to carry out the statutory command to attain an ultimate goal of zero water pollution.

As outlined at Chapter 2, Sections 2.2.7 to 2.2.8, under the procedure for administrative law rule making, EPA had to establish the scope of rule making for the various source categories to manage, control, and regulate effluent discharges for point sources. Preparatory information about industrial point source discharges was first obtained through industry survey questionnaires and accompanied by supporting on-site wastewater sampling. Industrial discharges were then characterized as to contained pollutants.

5.7.2   Introduction of EPA's Technology Standards

Schoenbaum describes the procedure for effluent assessment site visits for water pollutant control technology:

1) Sampling is generally conducted by contractors selected by the strict standards of the government contracting process. The logistics of coordinating the sampling can be extensive;

2) Successful site visits require the presence of knowledgeable plant personnel to answer pertinent questions and to assist the sampling team in various ways;

3) Site visits are useful only if plants are operating under "normal" conditions, therefore, visits must be scheduled to avoid "downtime" periods for maintenance or other interruptions; and

4) Scheduling of a site visit may depend on plant production schedules, if a plant produces numerous products or changes its product mix as part of a production cycle (Schoenbaum and Rosenburg, 1991, pp. 790–791).

Schoenbaum continues with an example of the CWA design of point source control for effluent discharges: CWA adopts a system of increasingly strict pollution control requirements theoretically leading toward the total elimination of point source discharges. The degree of source control for an industry depends on the level of technological achievement that has been attained.

Usually these discharge or effluent standards are described as "mass" based limits which establish the maximum quality of pollutants which may be discharged for each unit of production. For instance, the BAT limitations in the primary nickel and cobalt subcategory of the Non-Ferrous Metals Point Source Category state that a source may not discharge, on a daily average, more than 16.25 pounds of the pollutant copper for each million pounds of nickel produced. See 40 C.F.R. Sect. 421.233(c)(1988). In theory, as control technology improves, the amounts of waste copper to be disposed of should approach zero for each million pounds of nickel produced. Sometimes effluent standards are expressed in terms of permissible pollutant concentrations in wastewater. These concentration-based effluent standards are stated in terms of milligrams of a pollutant per liter of wastewater regardless of the amount of production. As such, they focus upon the amount of pollutants discharged and not those produced as do mass-based effluent standards. Although such limitations are acceptable to the EPA in limited circumstances, they run afoul of the basic anti-pollution policy of the Act. For example, a source could meet such a concentration-based standard by diluting its waste stream with intake water in order to meet the restriction (Schoenbaum and Rosenburg, 1991, pp. 788–789).

Industrial effluent discharge samples are reviewed by EPA's work group. Treatment controls with alternative options are prepared in formulating proposed guidelines and submitted to the administrator for water. After a proposal package is written, a final "red border" review is made within the EPA before signing by the administrator and submitting the approved guideline proposal for publication in the *Federal Register* for the public comment period. The comment periods are generally for 60 to 90 days, but may be longer depending on the complexity. At the end of the public comment period, the agency group considers and responds to the comments. It may make modifications or adjustments accordingly, and the proposed rule making undergoes another internal review before publishing the final rule. Implementation of the final rule after it becomes effective takes place and will stand unless challenged in a U.S. Court of Appeals (Schoenbaum and Rosenburg, 1991, pp. 792–793).

5.7.3 EPA's Effluent Standards

EPA's effluent standards include BPT (best practicable control technology), BCT (best cost-reasonable technology), and BAT (best available technology). Although § 301 and § 304 allow discretion for EPA in setting limitations on effluent dischargers in different categories, judicial interpretation has placed constraint on EPA's application of the standards.

5.7.3.1 Best Practicable Control Technology

Factors to be considered in determining BPT are (1) total cost of application of technology in relation to effluent reduction benefits, (2) age of equipment and plant facilities, (3) production process employed, (4) engineering aspects of control techniques, (5) process changes, (6) nonwater quality environmental impacts, and (7) any others deemed appropriate to the particular category.

The court in *American Meat Institute v. EPA,* 526 F.2d 442 (7th Cir. 1975) approved the EPA standard for BPT as "the average of the best existing performance by plants of various sizes, ages, and unit processes within each industrial category." An exception is where existing practices are uniformly inadequate.

An example where industrial inadequacy was found is in *Weyerhaeuser Co. v. Costle,* 590 F.2d 1011 (D.C. Cir. 1978) where the pulp and paper manufacturers challenged EPA's BPT regulations. Some of the industry's mills discharged effluents into the Pacific Ocean. In *Weyerhaeuser,* the pulpwood company argued that the receiving body of water was so great and, by comparison, the discharge so small that they should not have to greatly increase their manufacturing costs on added treatment equipment and lowering of effluent levels for wastes that could be absorbed and diluted. The *Weyerhaeuser* court instead rejected this argument, finding that the assimilative capacity of receiving waters was not intended to be considered regardless of the burden of costs placed on industry for an unacceptable process. In *EPA v. National Crushed Stone Assn.* (1980), discussed later, the court upheld the EPA by holding that individual economic inability to comply with effluent discharge was not a basis for a variance.

5.7.3.2 Best Cost-Reasonable Technology

In 1977, amended § 304 outlined the criteria for setting BCT effluent standards: "consideration of the reasonableness of the relationship between the costs of attaining a reduction in effluents and the effluent reduction benefits derived" and "the comparison of the cost and level of reduction of such pollutants from the discharge from publicly owned treatment works (POTWs) to the cost and level of reduction of such pollutants from a class or category of industrial sources."

Although the 1972 CWA reportedly failed to make its goals for waste discharges using BAT and BPT, 80% of the nation's industrial dischargers met the BPT standards. For the 1977 CWA amendment, Congress tightened the no-waste discharge scheme by initiating a four-part treatment plan dealing with (1) conventional pollutants, (2) toxic pollutants, (3) heat, and (4) nonconventional pollutants as follows:

1. Conventional pollutants became subject to a new standard, BCT, best conventional pollutant control technology, which was extended to 1989. The EPA has identified this group of pollutants under § 304(b)(4)(B) or § 1314(b)(4)(B), and usually associated with POTWs (e.g., suspended solids, fecal, coliform, pH, oil, and grease). Attainment of BCT standards was extended by the 1987 amendments to 1989.
2. The emphasis of the 1977 CWA amendment on toxic pollutants was shifted from the former health-based regulation (§ 307) after the 1976 industry case (*NRDC Inc. v. Train*), which resulted in the court's "Flannery decree." New legislation codified the litigation results, ordering EPA to use the BAT approach in regulating 65 toxic pollutants.
3. The 1974 EPA guidelines for the steam-generating electric industry barred thermal discharges except from cooling ponds and towers. That regulation was struck down by a Fourth Circuit case (*Appalachian Power v. Train*, 545 F.2d 1351) in 1976. Although considered pollutants, thermal water discharge permits are issued with limits under the variance provisions of § 316.
4. Nonconventional pollutants are all those pollutants other than conventional, toxic, and heat (e.g., iron, nitrates, chlorides, and ammonia). Special BAT modifications are available under CWA § 301(g).

5.7.3.3 Best Available Technology

Factors to be considered were similar to those for BPT. Concerning variances allowed for BAT standards, the economic and environmental impacts are given consideration. BAT variances are discussed later in Section 5.7.5 covering EPA's effluent limitations and evolution of variances.

5.7.3.4 New Sources

New sources of water pollution must comply with special national standards of performance that reflect the best available demonstrated control technology, processes, operating methods, or other alternatives, including where practicable, a standard permitting no discharge of pollutants.

By 1979, EPA had promulgated rules that were challenged in *American Paper Institute v. EPA*, 660 F.2d 954 (4th Cir. 1981). The Paper Institute argued that § 304 called for a two-part cost-reasonableness test, whereas the EPA maintained that it was not required to consider industry cost-effectiveness, but only to mandate a POTW cost comparison standard. The court rejected EPA's arguments and invalidated all BCT regulations.

In 1986, EPA adopted new BCT rules for industries where long-term performance data were available. BCT was to be more stringent than BPT requirements under a two-part costs test. If either test was failed, BCT would be equal to BPT for that industrial category.

Industries continued to object to the rising high costs to reach perfection, or zero discharge of conventional pollutants. A break point for doubly increased costs of pollutant removal occurred after 96 to 99% according to the chemical industry. In *Chemical Manufacturers Assn. v. EPA*, 870 F.2d 177 (5th Cir. 1989), the manufacturers argued that extremely high costs for compliance with EPA's rules to attain zero discharge were beyond reasonability and impermissible. The EPA counterargued that the break point or "knee-of-the-curve" test could be applied only to assess limitations beyond BPT. The Fifth Circuit Court of Appeals in *Chemical Manufacturers Assn.* stated:

> The BCT provisions were intended to establish an intermediate level between BPT and the stricter BAT limitations for conventional pollutants by adding a cost-effectiveness test for incremental technology requirements that exceed BPT technology. Under BCT, additional limitations on conventional pollutants that are more stringent than BPT can be imposed only to the extent that the increased cost of treatment (would) be reasonable in terms of the degree of environmental benefits.

5.7.4 Publicly Owned Treatment Works — Indirect Dischargers

The second most significant point source discharger of water pollutants is the POTWs. EPA had estimated that more than 60,000 industrial plants in 34 primary industrial groups discharge into POTWs (Schoenbaum and Rosenburg 1991, p. 801). These effluent generators are called *indirect dischargers*. Indirect dischargers have been criticized because they have allegedly shirked responsibility for pretreating their polluted discharges and shifted the responsibility to the municipal POTW for the discharge of their industrial wastes.

General pretreatment standards for POTWs are those "applicable to all indirect dischargers and generally prohibit discharges that 'pass through' or 'interfere' with the operation of the POTW.... Five specific kinds of pollutant discharge are banned by the EPA's general regulations, namely: (i) pollutants which cause a fire of an explosion hazard in the POTW; (ii) pollutants which will cause corrosive structural damage to the POTW (but, in no case discharges with a pH lower than 5.0, unless the works is specifically designed to accommodate such discharges); (iii) solid or viscous pollutants in amounts which will cause obstruction to the flow in the POTW resulting in interference; (iv) any pollutant, including oxygen-demanding pollutants (BOD, *etc.*) released in a discharge at a flow rate and/or pollutant concentration which will cause interference with the POTW; and (v) heat in amounts which will inhibit biological activity in the POTW resulting in interference, but in no case heat in such quantities that the temperature at the POTW treatment plant exceeds 40°C (104°F) unless the Approval Authority, upon request of the POTW, approves alternate temperature limits. (40 CFR § 403.5(b)(1)–(5) (1990)" (Schoenbaum and Rosenburg, 1991, pp. 859–860).

5.7.5   EPA's Effluent Limitations and Evolution of Variances

Several early challenges to EPA's authority to issue nationwide standards for effluent discharges were made. In *American Iron and Steel Institute v. EPA*, 626 F.2d 1029 (3d Cir. 1975), the appeals court decided against EPA and remanded the promulgated rules for restudy and rewriting. It found that EPA rules did not meet the requirements of § 301 (minimum control and maximum permissible effluent) and § 304 (prescribed guidelines for issuance of permits). The institute challenged the EPA's failure to consider engineering aspects and related costs. The court responded that engineering factors did not require the balancing of costs with achievement potentials; that compliance is the primary aim instead of costs. The court did allow that exceptions should be made for effluent discharges where intake water is already polluted.

An early U.S. Supreme Court case in which EPA's setting of standards for industrial effluent discharges was challenged was *E.I. DuPont de Nemours & Co. v. Train*, 430 U.S. 112 (1977). Eight chemical companies filed for a review of EPA regulations. At issue were "three important questions of statutory construction: (1) Whether EPA has the authority under Section 301 of the Act to issue industry-wide regulations limiting discharges by existing plants; (2) whether the Court of Appeals, which admittedly is authorized to review the standards for new sources, also has jurisdiction under Section 509 to review regulations concerning existing plants; and (3) whether new source standards issued under Section 306 must allow variances for individual plants."

The Supreme Court held that: (1) § 301 authorized the EPA to promulgate effluent limitations for classes and categories of existing point sources under § 301; (2) under § 509, the court of appeals for the federal judicial district in which the industrial plant is located may properly review EPA action promulgating effluent limitations for existing plants under § 301 and for new source points under § 306;

and (3) that portion of the judgment of the court of appeals requiring the EPA to provide a variance procedure for new sources is reversed.

With regard to holding (3), reversing the court of appeals' decision to allow variances for new point sources, Justice Stevens wrote, "The question, however, is not what a court thinks is generally appropriate to the regulatory process; it is what Congress intended for these regulations.... Congress intended these regulations to be absolute prohibitions."

5.8 CWA LEGISLATIVE AND LITIGATION UPDATES — 1990–2000

During the 1990s, the CWA underwent numerous amendments, of which the more prominent changes were:

5.8.1 Oil Pollution Act — 1990

This act amended subsection (b) of § 311 of the CWA to impose new requirements for the cleanup of oil spills.

5.8.2 Parts of Amended CWA

Certain parts of the CWA, as amended, are of particular interest, viz. § 301, § 304, and § 306 (effluent limitations guidelines and new source performance standards), and § 402 (NPDES permits, more commonly known as storm water runoff permits).

5.8.3 Section 301 — Effluent Limitations

This section covers secondary treatment for POTWs.

In *Maier v. U.S. EPA*, 114 F.3d 1032 (10th Cir. 1997), "the Tenth Circuit upheld the EPA's decision not to initiate rule-making regarding secondary treatment at POTWs. Citizens' groups had requested new secondary treatment standards that would address nitrogenous biochemical oxygen demand (NOD) in addition to carbonaceous biochemical oxygen demand (CBOD), but the court concluded that the EPA's decision to continue to regulate NOD on a case-by-case, permit-by-permit basis was reasonable and authorized under the statute" (Mansfield, 1997, p. 309).

5.8.4 Section 301(f) — Discharge of Chemical Warfare Agents

In *Chemical Weapons Working Group, Inc. v. U.S. Department of Army*, 111 F.3d 1485 (10th Cir. 1997), "the Tenth Circuit held that § 301(f)'s prohibition on the discharge of chemical warfare agents into navigable waters does not apply to emissions from an incinerator used to destroy chemical warfare agents. Significantly, the court emphasized that the CWA should not be used as a vehicle to regulate emissions that are already regulated by the Clean Air Act" (Mansfield, 1997, p. 309).

5.8.5   Section 303 — Water Quality Standards

The federal district court of Alaska held invalid EPA's regulation that extended the validity of state water quality standards during the interim period between EPA's disapproval of the state standards and the issuance of substitute standards [*Alaska Clean Water Alliance v. Clark*, 45 Env't. Rep. CAS (BNA) 1664 (W.D. Wash. 1997)] (Mansfield, 1997, p. 310).

In *Idaho Mining Assn. v. Browner*, 90 F. Supp. 2d 1078 (D. Idaho 2000), after the EPA had promulgated a rule establishing revised water quality standards for three water body segments in northern Idaho, the plaintiff, Idaho Mining Association, challenged the revised standards.

The Idaho Mining Association is a nonprofit corporation whose members included industrial facilities that conduct mining activities in the State of Idaho. Plaintiff's members held NPDES permits and were authorized to discharge certain amounts of industrial wastewater into particular waters in northern Idaho. On October 2, 1998, plaintiff filed a complaint against the U.S. EPA. Plaintiff alleged that the EPA failed to comply with the requirements of the Administration Procedure Act (APA) when it promulgated revised water quality standards for certain Idaho waters in 1997. The revised standards establish new designated uses for certain stream segments in northern Idaho and impose more stringent water quality criteria to protect the new uses. Plaintiff alleged that the new standards would significantly affect the ability of plaintiff's members to discharge mining pollutants into the affected waters pursuant to their NPDES permits and would negatively impact the economic viability of the mining industry in Idaho. Plaintiff sought an order vacating that portion of the EPA rule that established the new designated uses on the grounds that the EPA rule making for the revised standards was arbitrary and capricious, an abuse of discretion and otherwise not in accordance with the law.

EPA's disapproval of state standards developed for the three water segments at issue was based on its claim that the state standards did not protect aquatic life. The court held that EPA's reliance on a rebuttable presumption that fishable uses are attainable and must be protected by water quality standards. However, in view of contradictory evidence, the court found that EPA had acted arbitrarily and capriciously in relying on its presumption. The court granted summary judgment in part for both paries, denied in part for both parties, and remanded the part in question to the EPA for further consideration in accordance with the findings of the court.

5.8.6   Section 303(d) — Total Maximum Daily Loads — Web Site

This section of the CWA requires each state to identify water quality limited segments and to develop total maximum daily loads (TMDLs). A state-by-state summary of litigation can be found on the EPA TMDL Web site (see Environmental Protection Agency, *TMDL Litigation by State,* revised January 26, 1998: http://www.epa.gov/owow/tmdl/lawsuitl.html).

Reauthorization by Congress of the CWA and legislative activities concerning water had been stalemated for several years by a lack of a party majority in both houses of Congress. However, in 1999 and 2000, some Congressional activity centered on the national concern for TMDLs. A rider on an appropriations bill in 2000 prohibited EPA from finalizing and implementing the TMDL regulations until the end of 2001 fiscal year, and until scientific studies by EPA could be made.

The Section of Environment, Energy, and Resources of the American Bar Association reported in 1999 that in "at least thirty states, citizens have brought suits asking courts to require the EPA to develop TMDLs, alleging that state authorities have failed to do so in a timely manner."

Illustrative of the several citizen suits filed protesting the lack of action by the states and EPA to set TMDLs for impaired waters is *Heal the Bay v. Browner* (1999) in the federal district court of California and *American Canoe Assn., Inc. and American Littoral Society v. EPA,* 54 F. Supp. 2d 621 (E.D. Va. 1999). In *American Canoe,* the court summarized the plaintiffs' complaints:

> ... that EPA has failed to perform its duties under the CWA to identify Virginia's most heavily polluted waters and restore the chemical, physical, and biological integrity of those waters. Central to plaintiffs' allegation that EPA has failed to perform its duties under the CWA is their contention that EPA has a duty to establish total maximum daily loads (TMDLs) of pollutants for Virginia waters that it has failed to fulfill. The CWA compels states to establish TMDLs of pollutants for those waters within their boundaries that do not meet, or are not expected to meet, water quality standards even after the imposition of various enumerated controls and treatments. A TMDL represents the highest level at which a pollutant may be "loaded" into a water body without violating water quality standards. Thus, TMDLs must be established "at a level necessary to implement the applicable water quality standards with seasonal variations and a margin of safety which takes into account any lack of knowledge concerning the relationship between effluent limitations and water quality" for all pollutants that prevent or are expected to prevent the attainment of water quality standards.

> The gravamen of this complex action is the plaintiffs' allegation that they and their members have been harmed in their attempts to make aesthetic and recreational use of Virginia's rivers, streams, and coastlines because defendant, the United States Environmental Protection Agency (EPA), has failed to perform certain discretionary and nondiscretionary duties imposed on it by the Clean Water Act (CWA) and the Endangered Species Act (ESA), in conjunction with the Administrative Procedure Act (APA).

The court called attention to the facts that:

> According to the CWA, Virginia was to have submitted initial TMDLs to EPA by June 26, 1979, and thereafter from "time to time." Federal regulation states that the deadlines for these subsequent submissions are to be determined by the EPA regional administrator and the state. n7 [*624] When a state submits a TMDL, EPA must approve or disapprove the submission within thirty days, and in the event a TMDL is disapproved, EPA has thirty days from the date of disapproval to establish an appropriate TMDL. 33 U.S.C. § 1313(d)(2). In the nearly twenty years that have elapsed since the initial 1979 deadline, Virginia either has submitted no TMDLs or has submitted a single TMDL for one small tributary in the state, and EPA has never

established any TMDL for any of Virginia's waters. In ruling on defendants' motion to dismiss, this Court previously held that Virginia's twenty-year failure to submit TMDLs for EPA approval could properly be construed as a constructive submission that no TMDLs were necessary, triggering EPA's duty to approve or disapprove the constructive submission of "no TMDLs." (*American Canoe*, 623–624)

The consent decree proposed by plaintiffs and EPA, and approved by this court, seeks to rectify this long inaction by specifying an eleven-year schedule for the establishment of TMDLs for several hundred enumerated waters in Virginia. According to the decree, EPA expects Virginia either (i) to develop and submit TMDLs for the identified waters in accordance with the consent decree's schedule or (ii) to provide data and information showing that TMDLs are unnecessary according to that schedule. Should Virginia fail to meet the specified schedule, the decree sets deadlines by which EPA will establish TMDLs for various categories of waters. (*American Canoe*, *Id.*, 624)

The schedule for TMDL submissions is divided into four parts, with separate schedules for the creation of TMDLs for each of four categories of Virginia waters. The waters are categorized according to the sources or characteristics of their pollution, and each category currently includes 200 or more specific waters. According to the schedule, TMDLs for all "Category 1*," "Category 3," and "Category 4" waters will be established by May 1, 2011, and TMDLs for all "Category 2" waters will be established by May 1, 2006. The dates set forth in the consent decree for establishing TMDLs allot sufficient time for public notice of the TMDLs, consideration of public comment, revision of the TMDLs as necessary, and EPA final action on the TMDLs.

*NOTE: "Category 1" waters are those waters primarily affected by nonpoint source pollution (Additionally, Virginia committed to a goal of submitting TMDLs by 2003 for all "Category 2" waters which are waters listed because point sources discharging to these waters have water quality-based effluent limitations that are not stringent enough to achieve applicable water quality standards. Under the consent decree, some state submissions for "Category 2" waters are not required until 2005. The consent decree sets interim deadlines for TMDL submissions for other "Category 2" waters in 2002 and 2003. Finally, Virginia committed to setting TMDLs for 200 miscellaneous waters known as "Category 4" waters by 2010. Again, the consent decree establishes interim deadlines for submission of these TMDLs, requiring that 25% of these TMDLs be submitted by 2004, 50% by 2006, 75% by 2008, and 100% by the final 2010 deadline.

The joint motion to enter the consent decree must be granted. The consent decree represents a fair and reasonable solution to a difficult problem and evidences a substantial commitment to enforcement of the CWA's requirements in Virginia. As such, it is a proper and welcome resolution of this action. An appropriate order is entered.

s/ T. S. Ellis, III, United States District Judge, Alexandria, Virginia, July 12, 1999. (Id.)

In *Pronsolino v. EPA*, 91 F. Supp. 2d 1337 (N.D. Cal. 2000), the issue was whether § 303(d) of the Federal Water Pollution Control Act Amendments of 1972 (later, the CWA), authorized the EPA to determine TMDLs for rivers and waters polluted only by logging and agricultural runoff and/or other nonpoint sources rather than any municipal sewer and/or industrial point sources (33 U.S.C. 1313[d]). The issue gathers importance from the fact that "nonpoint source pollution has become the dominant water quality problem in the United States, dwarfing all other sources of volume...." According to the EPA, 54% of California's substandard rivers and waters are impaired by nonpoint sources only and another 45% are impaired by a combination of both point and nonpoint sources (EPA Tab 23).

The court in *Pronsolino* found authority for the EPA to establish TMDLs for waters impaired by nonpoint sources.

5.8.6.1 Concentrated Animal Feeding Operations Required to Have NPDES Permits

Community Assn. for Restoration of the Environment v. Henry Bosma Dairy, 65 F. Supp. 2d 1129 (E.D. of Washington, 1999), concerned livestock (animal) waste causing harm to surface and groundwaters. Although agricultural stormwater discharges and return flows from irrigation are not within the statutory definition of a point source and excluded from it, the court found that unpermitted discharges into irrigation ditches are in violation of the CWA.

In *Save the Valley, Inc. v. EPA*, 99 F. Supp. 2d 981(S.D. Ind. 2000), plaintiffs, Save the Valley, Inc., filed for injunctive relief against EPA under the citizen suit provision of the CWA to compel the EPA to assume enforcement of NPDES permits in the State of Indiana, and to initiate proceedings to withdraw the state authority to enforce permits. Plaintiffs allege that the EPA administrator must address Indiana's failure to require industrial hog farms (confined or concentrated animal feeding operations [CAFOs]) to acquire NPDES permits, in contravention of the CWA. The court denied the EPA motion for dismissal.

5.8.7 Sections 301, 304, 306 — Effluent Limitations — New Source Performance Standards, Navigable Waters Pollution — Mining Discharge

In *Rybachek v. U.S. EPA*, 904 F.2d 1276 (9th Cir. 1990), an Alaskan mine operator, Rybachek, and the Alaska Miners' Association challenged EPA's regulations that dealt with treatment required for discharges of untreated dredged soil and rock directly into navigable streams. The EPA interprets dredged soil and rock as pollutants and requires settling pond treatment before discharge.

The court found that EPA's classification of settable soils as nonconventional pollutant and subject to BAT standards was both a reasonable and permissible construction of the CWA. Congress had not designated settable solids as either a conventional or a toxic pollutant. As to its determination of economic achievability of technology, the EPA must consider the cost of meeting BAT limitations, but need not compare such costs with benefits of effluent reduction in promulgating regulations under the CWA. EPA was held to have considerable discretion in weighing technology costs that are less important factors than in setting BPT limitations.

The court upheld EPA's determination that settling ponds are the BPT currently available. With regard to new source placer mines, the court upheld the criteria, concluding that the case-by-case approach for determining whether a new mine is a new source is not overly broad, and was within EPA's authority under the CWA.

However, in *Waste Action Project v. Dawn Mining Corp.; Newmont Gold*, 137 F.3d 1426 (9th Cir. 1998), a citizens' action group, Waste Action Project (WAP) brought a CWA suit against Dawn Mining. WAP alleged that Dawn was discharging pollutant wastes containing uranium, silica, heavy metals, sulfates, phosphates, chlorides, and other chemicals into Chamokane Creek (Washington) without an NPDES permit in violation of the CWA. WAP alleged that the pollutants leaked from Dawn's tailings disposal areas (TDAs) into the groundwater and eventually to Chamokane Creek.

The mining companies moved for summary judgment. The district court ruled that the uranium mill tailings and associated wastes identified by WAP are by-product material as defined in § 11(e)(2) of the Atomic Energy Act (AEA), 42 U.S.C. § 2014(e)(2), and hence, are not pollutants under the CWA. WAP appealed.

The appeals court found that where issues of interpretation of the CWA are involved, the U.S. Supreme Court opinion in *Train v. Colorado Public Interest Research Group*, 426 U.S. 1, 48 L. Ed. 2d 434, 96 S. Ct. 1938 (1975), governed. In *Train*, the Supreme Court addressed an issue virtually identical to *Dawn*, specifically, whether the EPA had authority under the FWPCA, now known as the CWA, to regulate discharge of nuclear waste materials subject to regulation by the Atomic Energy Commission and its successors under the Atomic Energy Act. The court held unanimously that Congress did not intend for materials governed by the Atomic Energy Act to be included in the category of pollutants subject to regulation by the EPA under the FWPCA (*Train*, p. 25). The court of appeals held that uranium mill tailings are not pollutants for purposes of the CWA and are not subject to the EPA NPDES permitting requirements. The district court's grant of summary judgment was affirmed.

As a sequel to *Rybachek*, effluent limitations and water quality standards (§§ 301, 303) were the concerns in *Ackels v. EPA*, 7 F.3d 862 (9th Cir. 1993). The court upheld the incorporation of Alaska's turbidity limits by EPA in issuing NPDES permits to gold placer mining operations. The court found that the turbidity limit was an economically feasible control.

Strict liability for releasing a contaminating discharge into public streams was addressed in *Cities Service Co. v. State of Florida*, 312 So.2d 799 (Fla. App. 1975), in which the operator of a phosphate rock mining pit was held **strictly liable** when a dam break occurred in one of its settling ponds, releasing contaminating waters.

EPA actions with regard to CWA permits are not subject to agency-hearing review within the time frames of wastewater pretreatment § 307(a)(2) CWA where EPA promulgates *technology-based* standards pursuant to CWA §§ 301, 304, and 306. (See *Rybachek v. EPA*, cited earlier.) The requirements of § 307 apply only when EPA promulgates *health-based* standards for specific toxic pollutants. Judicial review of EPA NPDES permitting action is unavailable by a federal court where an ICS is involved under § 304(1). (See *Glatfelter, Westvaco,* and *Natural Resources Defense Council* cases later.)

5.8.8 Section 304(l) — Individual Control Strategies — Cases Concerning State-Issued NPDES Permit with EPA Approval

In *P.H. Glatfelter Co. v. EPA,* No. 90-1488, slip opinion (4th Cir., December 10, 1990), the Fourth Circuit, U.S. Court of Appeals held that judicial review was unavailable for EPA's conditional approval of a state-issued NPDES permit as an ICS under CWA § 304(1). The court noted that it had jurisdiction only if the EPA had itself promulgated an ICS under § 509(b)(1)(G) of the CWA.

In *Westvaco Corp. v. EPA,* 899 F.2d 1383 (4th Cir. 1990), the Fourth Circuit U.S. Court of Appeals held that judicial review was unavailable for EPA's preliminary disapproval of a 304(1) list of "impaired waters" submitted by the states of Virginia and Maryland. (Also, see *American Paper Institute, Inc. v. EPA,* 726 F.2d 1256 [S.D. Ala. 1989] where a federal district court of Alabama held that an EPA policy statement giving guidance to states within Region IV concerning discharge of toxic pollutants into rivers and streams was not a final action under § 304[1]).

In *Natural Resources Defense Council v. EPA,* 915 F.2d 1314 (9th Cir. 1990), NRDC sought review of a final rule issued by EPA for some of the polluted waters listed under § 304(1) that states must identify responsible point sources and develop ICSs for such sources.

The Ninth Circuit Court of Appeals held that (1) EPA lacked authority to restrict the statutory scope of § 304(1); (2) states must identify point sources for all listed waters under §§ 304(1)(A) and (B); and (3) on remand, EPA must reconsider whether ICSs are required for each point source pursuant to § 304(1)(D).

5.8.9 Section 307(b) — Wastewater Pretreatment

In *International Union (UAW) v. Amerace Corp., Inc.,* 740 F. Supp. 1072 (D. N.J, 1990), a citizen's suit alleged violations of pretreatment and reporting standards. The court held that: (1) federal pretreatment, discharge, and reporting requirements apply to an indirect discharger even if they have not been included in a permit issued by a publicly owned water treatment (POWT) facility; (2) discharger must report all monitoring results even if monitoring was not required; and (3) an NPDES permit violation is not required to hold an industrial user liable for violating categorical standards.

5.8.10 Section 402 — NPDES Permit Violations

In August 1990, EPA promulgated new rules for § 402, the NPDES permits. NPDES permits are required for mining operations.

It was reported (by an unknown source) that drainage from 11 millions acres of active and abandoned mines from 20 different minerals pour harmful, even toxic, matter into streams and lakes in 31 states. Coal mine acid waters are the chief culprit. The unknown source estimated the number of abandoned mines to be about 90,000. Whether all these mine sites are polluting public streams is highly questionable. A reliable figure is that of these total estimated abandoned mines, only 50 of the sites (0.00055%) are on the U.S. Government Superfund cleanup list because of environmentally hazardous conditions.

Under the amended NPDES, state laws controlling mine and plant wastewaters were to require a closed-loop, no-discharge system. Compliance with the mine and plant's water permit is critical for uninterrupted operation. Self-monitoring is essential in maintaining discharge limits into public streams and waters. Sedimentation and turbidity limits for discharges were to be closely monitored by government inspectors.

In *Thompson & Phillips Clay Co. v. Department of Environmental Resources (DER)*, 582 A.2d 1162 (Pa. Commw. 1990), T&P's clay pit was downhill of an inactive coal mine formerly operated by another company. T&P's mine permit included conditions regulating its discharges and specified treatment thereof. DER denied the operator's application for a second-stage release of its reclamation bond of its mine site for acid mine drainage being discharged from their pit into a nearby stream. The acid water contained excessively high concentrations of metals and sulfates not allowable under Pennsylvania's Clean Streams Law, Act of 1937, P.L. 1987. The Pennsylvania Coal Association joined the litigation as *amicus curiae* (friend of the court; a party, or bystander, who has no right to appear in the suit, but who interposes and volunteers information, argument, or evidence, on some matter in law with regard to the suit to protect his interests, i.e., here, other coal mining operations).

T&P argued that it did not cause the pollution. It maintained that the acid drainage was flowing through forces of gravity from the abandoned coal mine uphill to its clay mine and, therefore, it was not responsible for treating the drainage. It further argued that before liability could attach, the element of causation must be proven by DER.

The court responded that "the statutory prohibition against unauthorized discharges into waters of the Commonwealth did not require a causal link between the clay mine's activities and acid mine drainage allegedly originating from another's mine." The clay mine operator was held liable and denial of release for the reclamation bond was upheld until corrective measures were made for the contaminated discharge by the clay pit operator. (However, for contrast, see the court's holding in *Friends of Santa Fe County v. Lac Minerals, Inc.*, 1995, Section 5.8.12.)

In *U.S. v. City of Toledo*, 867 F. Supp. 588 (N.D. Ohio 1994), the court held that the city's allegation of inaccuracy of effluent discharge monitoring measurement devices that yielded an inaccurate reading on one day, and not on all the days on which violations were alleged was insufficient to create a genuine issue of material fact to overcome the EPA's claim of violation.

In *Friends of the Earth, Inc. v. Laidlaw Environmental Services (TOC), Inc.*, 528 U.S. 167, 120 S. Ct. 693, 145 L. Ed. 2d 610 (2000), the owner of a South Carolina hazardous waste incinerator facility that included a wastewater treatment plant was granted an NPDES permit by a South Carolina state agency acting under 402(a)(1) of the CWA, 33 USCS 1342(a)(1). The permit authorized the company to discharge treated water into a nearby river, but placed limits on the discharge of several pollutants. In June 1992, two environmental organizations (subsequently joined by a third), alleging that the owner's discharges of pollutants had exceeded the limits set by the permit, notified the owner of an intention to file a citizen suit under 505(a) of the CWA — 33 USCS 1365(a) — after the expiration of a 60-day notice period. Meanwhile, at the owner's own request, the state agency agreed to file a separate lawsuit against the owner. On the last day before the 60-day notice period expired, the owner and the state agency reached a settlement with respect to the separate lawsuit; the settlement required the owner to pay civil penalties and to make every effort to comply with the permit obligations. Nevertheless, the environmental organizations filed a citizen suit against the owner in the U.S. District Court for the District of South Carolina shortly afterward, in which suit the organizations alleged noncompliance with the NPDES permit and sought declaratory and injunctive relief and an award of civil penalties. The owner moved for summary judgment on the ground that the organizations had failed to present evidence demonstrating injury in fact and therefore lacked standing under Article III of the Federal Constitution to bring the suit. In opposition to this motion, the organizations submitted affidavits and deposition testimony from some members of the organizations who lived near the river and who alleged that the owner's discharges (1) had curtailed those members' recreational use of the river, and (2) would subject those members to other economic and aesthetic harms. In 1993, the district court denied the summary judgment motion on finding that the organizations had standing. In 1997, the district court issued a judgment consisting of findings of fact, conclusions of law, and an order in which it was concluded that (1) although there was no demonstrated proof of harm to the environment from the discharge violations, the owner had gained an economic benefit as a result of noncompliance with the permit; (2) a civil penalty of $405,800 was appropriate; and (3) injunctive relief was inappropriate, because the owner had been in substantial compliance with the permit since at least August 1992 (956 F. Supp. 588). The organizations appealed the district court's civil penalty judgment on the ground that the penalty was inadequate, but they did not appeal the denial of declaratory or injunctive relief. In a cross-appeal, the owner argued, among other matters, that the organizations lacked standing to bring the suit. On appeal, the United States Court of Appeals for the Fourth Circuit (1) assumed, without deciding, that the organization had initially had standing; (2) concluded that the case had become moot, because the only remedy currently available to the organizations — civil penalties payable to the government — would not have redressed any injury that the organizations had suffered; (3) vacated the district court's order; and (4) remanded the case with instructions to dismiss (149 F3d 303). The owner subsequently asserted that after the court of appeals issued its decision, but before the United States Supreme Court granted *certiorari* in 1999, the entire facility was closed, dismantled, and put up for sale; and that all discharges from the facility permanently ceased.

On *certiorari*, the Supreme Court reversed the court of appeals' judgment and remanded the case for further proceedings. In an opinion by Ginsburg, J., joined by Rehnquist, Ch. J., and Stevens, O'Connor, Kennedy, Souter, and Breyer, JJ., it was held that (1) the organizations had demonstrated sufficient alleged injury in fact, by means of the members' affidavits and testimony, to establish the organizations' standing to bring the suit, even if there was no proof of harm to the environment from the facility owner's alleged violations; (2) the organizations had **standing** to seek civil penalties, as such penalties (a) were for alleged violations that were ongoing at the time of the complaint and that could continue into the future if undeterred, and (b) carried with them a deterrent effect that made it likely that the penalties would redress the organizations' injuries by abating current violations and preventing future ones; (3) the organizations' failure to appeal the district court's denial of injunctive relief did not moot the civil penalties claim on appeal; (4) the civil penalties claim had possibly become moot when the owner came into compliance with the permit or closed the facility, but only if one or the other of these events made it absolutely clear that the permit violations could not reasonably be expected to recur; and (5) the effect of the owner's compliance and the closure of the facility on the prospect of future violations was a disputed factual matter that remained open for consideration on remand.

Concurring opinions — Stevens, J., concurring, expressed the view that the claim for civil penalties would not have been moot even if it had been absolutely clear that (1) the owner of the facility had gone out of business and posed no threat of future permit violations, or (2) the owner's violations could not reasonably have been expected to recur because of the owner's achievement of substantial compliance with the permit requirements after the organizations filed their complaint but before the district court entered judgment. Kennedy, J., concurring, expressed the view that questions concerning the permissibility — in view of the responsibilities committed to the executive by Article II of the Constitution — of exactions of public fines by private litigants and the delegation of executive power that might be inferred from such authorization were best reserved for a later case.

Dissenting opinions — Scalia, J., joined by Thomas, J., dissenting, expressed the view that (1) injury in fact (*for standing*) ought not to have been found on the basis of the affidavits presented by the organizations, becaue those affidavits were vague and were undermined by the district court's express finding that the facility owner's discharges caused no demonstrable harm to the environment; (2) it was a violation of traditional principles of federal standing to hold that a civil penalty, payable to the public, remedied a threatened private harm and sufficed to sustain a private suit; and (3) while the Supreme Court might have been correct in stating that the parallel between standing and mootness was imperfect, this did not change the underlying principle that the requisite personal interest that must exist at the commencement of the litigation must continue throughout the existence of the litigation.

5.8.11   Section 402 — NPDES Permits — Navigable Waters Pollution — Discharge by a POTW — Upstream Users' Liability — Sections 303–304(a)

Navigable Waters Defined — In *Driscoll v. Adams,* 181 F.3d 1285 (11th Cir. 1999) and *Community Assn. for Restoration of the Environment v. Henry Bosma*

Dairy, 65 F. Supp. 2d 1129 (U.S. Dist. Ct., E.D. Wash. 1999), the courts defined navigable waters of the United States under the CWA:

> The Clean Water Act (CWA) defines "navigable waters" as "waters of the United States, including the territorial seas." 33 U.S.C. § 1362(7). This broad definition "makes it clear that the term 'navigable' as used in the Act is of limited import" and that with the CWA Congress chose to regulate waters that would not be deemed navigable under the classical understanding of that term.... Consequently, courts have acknowledged that ditches and canals, as well as streams and creeks, can be "waters of the United States" under § 1362(7). Likewise, there is no reason to suspect that Congress intended to exclude from "waters of the United States" tributaries that flow only intermittently. (Eidson, 108 F.3d at 1341-42, holding that a man-made drainage ditch was a navigable water under the CWA.) (*Driscoll* at 1291.)

Waters of the United States or waters of the U.S. means:

(a) All waters which are currently used, were used in the past, or may be susceptible to use in interstate or foreign commerce, including all waters which are subject to the ebb and flow of the tide;
(b) All interstate waters, including interstate "wetlands";
(c) All other waters such as intrastate lakes, rivers, streams (including intermittent streams), mudflats, sandflats, "wetlands," sloughs, prairie potholes, wet meadows, playa, or natural ponds the use, degradation, or destruction of which would affect or could affect interstate or foreign commerce including any such waters:
  (1) Which are or could be used by interstate or foreign travelers for recreational or other purposes;
  (2) From which fish or shell fish are or could be taken and sold in interstate or foreign commerce; or
  (3) Which are used or could be used for industrial purposes by industries in interstate commerce;
(d) All impoundments of waters otherwise defined as waters of the United States under this definition;
(e) Tributaries of waters identified in paragraphs (a) through (d) of this definition;
(f) The territorial sea; and
(g) "Wetlands" adjacent to waters (other than waters that are themselves wetlands) identified in paragraphs (a) through (f) of this definition (*Community Assn.,* p. 1142).

In *Oklahoma v. EPA,* 908 F.2d 595 (10th Cir. 1990), the EPA had issued an NPDES permit to a new sewage treatment plant in Fayetteville, Arkansas, for wastewater discharges into a river flowing across the state line into Oklahoma. Oklahoma challenged the permit on grounds that the discharges would violate Oklahoma's water quality standards.

The court found that there was substantial evidence that the river in Oklahoma suffers from degraded water quality, caused partly by the types of pollutants that would be discharged by the Fayetteville facility, and that the pollution from Fayetteville would reach Oklahoma waters and would contribute to the deterioration of the river. The court concluded the CWA requires upstream dischargers to comply with federally approved water quality standards of all affected downstream states, and therefore, a new source that lowered downstream quality could not be permitted. The court reversed the permit issued by EPA to the Arkansas facility.

However, 2 years later, that decision in *Oklahoma v. EPA* by the Tenth Circuit on appeal by Arkansas was reversed by the U.S. Supreme Court in *Arkansas v. Oklahoma*, 112 S. Ct. 1046 (1992). The Supreme Court held that, although EPA regulations required ensuring that downstream waters would not be degraded, there was nothing in the water quality standards preventing the issuance of an NPDES permit to an upstream user where the downstream user's waters were already in violation of the water quality standards. There would be a violation of the permit issued only if EPA found the discharge by Arkansas caused a detectable change in Oklahoma's water quality.

In *City of Albuquerque v. Browner*, 97 F.3d 415 (10th Cir. 1996), Albuquerque's water quality was meeting EPA's national discharger standards but the city objected when the EPA imposed higher quality standards on the city's discharged water because of a protest by a downstream Indian tribe. The downstream tribe had developed tribal water quality standards that were more stringent than the national criteria. The court upheld EPA's approval of water quality standards developed by the Indian tribe. The court held that the downstream tribal standards can be more stringent than the national criteria established by the EPA and the EPA can impose those higher tribal standards on dischargers that are upstream of the tribal lands (Mansfield, 1996, p. 251).

In *Umatilla Waterquality Protective Association v. Smith Frozen Foods, Inc.*, 962 F. Supp. 1312 (D. Or. 1997), "the district court held that discharges into groundwater are not subject to NPDES permitting requirements, even when the groundwater at issue is hydrologically connected to navigable surface water. The question was certified for interlocutory appeal in the Ninth Circuit.... Also, the court held that the discharge of residual pollutants purposefully collected in an unlined brine pond can constitute a discharge from a point source. The fact that the collected pollutants migrate through dirt, with the help of rain water and gravity before they reach a navigable water does not change the status of the brine pond as a point source" (Mansfield, 1997, p. 311).

In *U.S. v. Eidson*, 108 F.3d 1336 (11th Cir. 1997), the Eleventh Circuit held that a storm drainage ditch was "a navigable water, reasoning that Congress intended to regulate discharges into all waters that may eventually lead to waters affecting interstate commerce" (Mansfield, 1997, p. 311). Accordingly, is a dripping water faucet in a private home a navigable stream of the United States?

5.8.12 Section 402 — NPDES and Storm Water Permits

In *Friends of Santa Fe County v. Lac Minerals, Inc.*, 892 F. Supp. 1333 (D.N.M. 1995), the District Court of New Mexico held that an arroyo was not a navigable water within the meaning of the CWA where the plaintiffs (Friends of Santa Fe) failed to show that the arroyo either affected interstate commerce or was connected to an interstate watercourse. While acknowledging that hydrologically connected groundwaters are regulated under § 301 of the CWA, the court found that the plaintiffs' evidence failed to suggest that such a connection occurred "at least at some time in the past" and that it was "reasonably likely" to happen again (Mansfield, 1995, p. 270).

In *Beartooth Alliance v. Crown Butte Mines*, 904 F. Supp. 1168 (D. Mont. 1995), the district court interpreted the CWA point source definition to include mine adits, holding the defendant owners liable for the discharge of acid mine drainage without an NPDES permit. The court rejected the contention of defendant Crown Butte that its pending permit application constituted compliance with §§ 301(a) and 402.

In contrast, the district court in *Friends of Santa Fe County* (p. 1359; see above), held that seepages of *Leslie* subsurface waters carrying traces of acid mine drainage were nonpoint sources beyond the CWA's permitting requirements. The court further found that the migration of residual contamination from the former mine operations did not constitute the discharge of a pollutant within the meaning of § 502(12)(A) caused by the current mine owners (Mansfield, 1995, p. 270). (Compare the court holding in *Thompson & Phillips*, discussed in Section 5.8.10.)

In *Driscoll v. Adams*, 181 F.3d 1285 (11th Cir. 1999), the plaintiffs brought suit against Adams for violation of the CWA by discharge of allegedly polluted storm water containing mud, silt, sand, and other materials into a stream and thence flowing into ponds on the plaintiffs' properties. The flow of materials from Adams' property occurred when Adams cut and graded roads, installed storm pipes, and cut and removed timber while developing his property for residential development.

The plaintiffs filed their lawsuit in December 1996 against Adams for violations of the CWA, 33 U.S.C. §§ 1251–1376 (1994), pursuant to its citizen suit provision, 33 U.S.C. § 1365. They included in their complaint pendent **state law claims for nuisance, trespass, and negligence**. Driscoll filed a motion for summary judgment, and Adams filed a motion to dismiss, which the district court treated as a cross motion for summary judgment. The court denied the plaintiffs' motion and granted Adams a dismissal, stating that the requirement of an "NPDES permit was an impossible condition ... and there were no approved federal standards for how much sand, silt and mud could be in the released water." After disposing of the federal law claim, the court declined to retain supplemental jurisdiction over the state law claims and dismissed them without reaching the merits. The plaintiffs appealed.

The court of appeals rejected the district court's findings and reversed both the award of summary judgment to Adams and the denial of summary judgment to plaintiffs on the CWA claim. The case was remanded for further proceedings consistent with its opinion. Dismissal of the state law claims was vacated.

5.8.13 Section 402 — Storm Water — NPDES Permits — Sewer Overflows

In *Northwest Environmental Advocates v. City of Portland*, 56 F.3d 979 (9th Cir. 1995), the Ninth Circuit Court affirmed the district trial court's holding that discharges from 54 combined sewer overflows (CSO) outfalls were covered by the city's 1984 NPDES permit. However, the circuit court reversed in part the district court's holding that violations of water quality standards were actionable only if they were incorporated into an NPDES permit by effluent limitations (Mansfield, 1995, p. 270).

5.8.14 Section 404 — Dredge and Fill Permits

In *Leslie Salt Co. v. U.S.*, 896 F.2d 354 (9th Cir. 1990), the Ninth Circuit Court of Appeals "upheld the U.S. Army Corps of Engineers' assertion of jurisdiction over pits dug on dry land on property that had been mined for salt production by the Leslie Salt Company. In a case of stretching federal authority beyond reasonable limits, the Circuit court had held that the presence of migratory birds on the property created a sufficient connection to 'interstate commerce' to allow the Corps regulation under the CWA" (Mansfield, 1995, p. 360). The U.S. Supreme Court declined to hear the case on *certiorari* (review) over a strong dissent by a Justice.

> The Justice stated, "Other than the occasional presence of migratory birds, there was no showing that the petitioner's land use would have any effect on interstate commerce, much less a substantial effect. This case raises serious and important constitutional questions about the limits of federal land use regulation in the name of the Clean Water Act that provides a compelling reason to grant certiorari in this case" (Mansfield, 1995, p. 271).

In *American Mining Congress v. U.S. Army Corps of Engineers*, 951 F. Supp 267 (D.D.C. 1997), the American Mining Congress (AMC), forerunner of the National Mining Association and others, sued the U.S. Army Corps of Engineers challenging the *Tulloch* rule that incidental fallback accompanying dredging is a discharge into the waters of the United States within the CWA and a violation of the permitting provision for discharge of dredge fill material. The Corps of Engineers was joined by the National Wildlife Federation and other environmental groups.

Incidental fallback is the incidental soil movement from excavation, such as the soil that is disturbed when dirt is shoveled, or the back-spill that comes off a bucket and falls back into the same place from which it was removed. It should be noted that *side casting*, which involves placing removed soil alongside a ditch, and sloppy disposal practices involving significant discharges into waters have always been subject to § 404.

Prior to the agencies' redefining "discharge of dredged material," the interpretation had been that the term did not include *de minimus*, incidental soil movement occurring during normal dredging operations. The Corps' 1986 regulations stated that § 404 clearly directs the Corps to regulate the discharge of dredged material, not the dredging itself. Dredging operations cannot be performed without some fallback. Referring to case law, the court stated that "the Act does not authorize the agencies to regulate incidental fallback. In *Salt Pond Assocs. v. U.S. Army Corps of Engineers*, 815 F. Supp. 766 (D. Del. 1993), the court held that land clearing and excavating activities were outside the reach of § 404."

The district court held that the *Tulloch* rule exceeded the scope of governmental agency statutory authority, was invalid, and was no longer to be applied or enforced by the Army Corps of Engineers or the EPA.

5.8.15 Section 404 — Wetlands — Injunctions — Statute of Limitations

As background, in *U.S. v. Telluride Co.*, 884 F. Supp. 404 (D. Colo. 1995), the United States filed a civil action on October 15, 1993 against Telco in the U.S. District Court for the District of Colorado under § 309 of the CWA, 33 U.S.C. § 1319. The government sought civil monetary penalties and injunctive relief for Telco's illegal filling of approximately 45 acres of wetlands between 1981 and 1989, in violation of 33 U.S.C. § 1311(a). In its request for injunctive relief, the government sought to enjoin Telco from discharging additional material, and to require Telco to restore damaged wetlands to their prior condition or create new wetlands to replace those that could not be restored.

Telco subsequently filed a motion for partial summary judgment on all the government's claims for violations that occurred before October 15, 1988, contending these claims were barred by the 5-year statute of limitations in 28 U.S.C. § 2462. Section 2462 states in relevant part: "except as otherwise provided by Act of Congress, an action, suit or proceeding for the enforcement of any civil fine, penalty, or forfeiture, pecuniary or otherwise, shall not be entertained unless commenced within five years from the date when the claim first accrued." The government conceded § 2462 applied to its claim for civil penalties, but argued the statute did not bar its claims for injunctive relief. The district court disagreed, applying the concurrent remedy rule to hold § 2462 barred the government's claims for injunctive relief. The court interpreted the concurrent remedy rule as providing when legal and equitable relief are available concurrently, and a statute of limitations bars the concurrent legal remedy, the court must withhold the equitable relief. Consequently, because § 2462 barred the government's claims for legal relief, civil monetary penalties, the court held § 2462 also barred its claim for injunctive relief. On May 2, 1995, the court granted Telco's motion for partial summary judgment, dismissing all the government's claims for relief for wetlands illegally filled prior to October 15, 1988. The government appealed the district court judgment, claiming § 2462 does not apply to its claims for injunctive relief, and the district court erred in applying the concurrent remedy rule to bar those claims.

On appeal in *U.S. v. Telluride Co.*, 146 F.3d 1241 (10th Cir. 1998), the issues were whether the 5-year statute of limitations provided in 28 U.S.C. § 2462 applied to the government's claims for injunctive relief, where § 2462 by its terms applied only to the "enforcement of any civil fine, penalty, or forfeiture," and whether the district court erred in applying the concurrent remedy rule to bar those claims.

The Tenth Circuit Court of Appeals reversed the district court's judgment holding that restorative injunctions are not subject to the federal 5-year statute of limitations for civil action set forth in § 2462. The concurrent remedy rule was found to be inapplicable to the government in its enforcement capacity when injunctive, equitable relief is sought. (*Note:* For more on statute of limitations for CWA citizen suits, see *Georges River Tidewater Assn v. Warren Sanitary District,* 2000 U.S. Dist. Lexis [D. Me.] in Section 5.8.17.)

5.8.16 Section 405 — Sewer Sludge

In *Welch v. Rappahannock County (Va.) Board of Supervisors*, 888 F. Supp. 753 (W.D. Va. 1995), the prohibition of application of sewer sludge to land applications was found to be reserved to local authorities and not preempted by § 405 of the CWA. State and local authorities may adopt more stringent requirements than those of the CWA and they do not violate the commerce clause of the U.S. Constitution or conflict with the CWA (Mansfield, 1995, p. 272).

In *Sierra Club v. EPA*, 992 F.2d 337 (D.C. Cir. 1993), "environmental groups petitioned for review of EPA's regulation controlling the disposal of toxic substances and solid waste at landfills." The court decided that the EPA had adequately explained its decision to dispense with numeric limits for toxic substances co-dispersed with sewage sludge in municipal solid waste landfills. It also found that EPA could not measure effects of chemical interactions between pollutants and sludge or between pollutants and solid waste, and that the EPA had insufficient data about the chemical composition of debris in typical MSW landfills, and, therefore, could not establish scientifically defensible numeric limits (Mansfield, 1993, p. 256).

5.8.17 Section 505 — Citizens' Suits — Statute of Limitations

CWA § 505 authorizes citizens to bring private actions against persons who are alleged to be in violation of certain provisions of the act and against the EPA if it has failed to perform a nondiscretionary duty under the act.

However, in *Georges River Tidewater Assn. v. Warren Sanitary District,* 2000 U.S. Dist. Lexis (D. Me.), in a citizen suit filed under the provisions of the CWA, the plaintiffs sought injunctive relief and civil penalties for defendant's alleged 349 violations since the commencement of its treatment facility. Because the CWA does not prescribe a statute of limitations, the defendant argued that many of the alleged violations occurred more than 5 years before the action was filed, and, thus, all prior-to-5-year violations were barred by the 5-year statute of limitations under the federal code. The plaintiffs argued that the State of Maine's 6-year statute of limitations for civil actions should be applied.

The court agreed with the defendant that for purposes of citizen suits to enforce the CWA, state statutes of limitations are unsatisfactory because of nonuniformity among the states. The court held that the 5-year federal statute of limitations applied to citizen suits brought under the CWA. (*Note:* For more on statute of limitations under CWA citizen suits, see *U.S. v. Telluride Co.*, 146 F.3d 1241 [10th Cir. 1998], cited earlier in Section 5.8.15.)

5.8.18 Additional Recent CWA Cases for Reference and Study

1. *American Wildlands v. Browner*, 94 F. Supp. 2d 1150 (D. Colo. 2000). Re: § 303, Water Quality Standards.
2. *Natural Resources Defense Council v. Fox*, 93 F. Supp. 2d 531 (S.D. N.Y. 2000). Re: § 303(d), TMDLs.
3. *U.S. v. Deaton*, 209 F.3d 331 (4th Cir. 2000). Re: § 404, Discharge into Wetlands.

4. *Palm Beach Isles Associates v. U.S.*, 208 F.3d 1374, as modified by 231 F.3d 1354 (Fed. Cir. 2000). Re: § 404, Wetlands; Regulatory Takings.
5. *U.S. v. Alcoa, Inc.*, 98 F. Supp. 2d 1031 (N.D. Inc. 2000). Re: §202, Enforcement; Injunctive Relief.

CHAPTER 6

Other Important and Miscellaneous Environmental Statutes in a "Nutshell"

6.1 TOXIC SUBSTANCES CONTROL

One of the earlier environmental problems to be treated was regulation of toxic pesticides by Congress's enactment of the Federal Insecticide, Fungicide and Rodenticide Act (FIFRA), which was amended in 1972 by the Federal Environmental Pesticide Control Act, and again in 1975, 1978, and 1988. Prior to this period of acute environmental awareness of the necessity for toxic control, only scattered and special controls had been enacted. The Environmental Protection Agency (EPA) estimated that prior to 1976, approximately 90% of all hazardous wastes were improperly disposed of. As a result, the Toxic Substances Contol Act 1976 (TSCA) was enacted as a comprehensive, gap-filling measure to control toxic substances. Amendments were made to TSCA in 1986 and 1988 to deal with two toxic health hazards, viz., asbestos in public building structures and radon.

Various other acts were amended or passed to implement toxic pollution control, e.g., the Safe Drinking Water Act (SDWA) was strengthened in 1986 with controls setting maximum levels for toxic contaminants in water for public water systems, particularly where toxic contaminants polluted underground water supplies; the CWA was amended by the Water Quality Act of 1987 to deal with toxic hot spots needing more stringent controls than those provided by CWA. The CAA was amended in 1990 with the addition of § 112. The EPA's prescribed risk management criteria for hazardous air pollutants, however, which had been denied in *NRDC v. U.S. EPA*, 824 F.2d 1146 (D.C. Cir. 1987), the vinyl chloride case, were endorsed by Congress in the new § 112.

An important step in the management of hazardous wastes took place when Congress amended the Solid Waste Disposal Act (SWDA) in 1976 by passing the Response Conservation and Recovery Act (RCRA). RCRA was intended to provide a so-called "cradle to-the-grave" plan for hazardous wastes.

6.2 UPDATES ON LITIGATION OF SOLID WASTE DISPOSAL — 1999–2000

In *Rensselaer v. Duncan (Deputy Comm. for N.Y. Natural Resources)*, 698 N.Y.S. 2d 113 (App. Div. 1999), one of the parties, 4C's Development Corporation, had previously held an unused but expired permit to operate a landfill in 1988. When a timely renewal application was not filed, the permit expired.

On reapplication for issue of the permit in November 1993, respondent 4C's Development Corporation applied to the Department of Environmental Conservation (DEC) for permits to resume the construction and operation of a 12.4-acre landfill in the Town of East Greenbush, Rensselaer County, for construction and demolition debris. A legislative hearing was convened on January 16, 1996 to receive comments on the proposed project and the draft environmental impact statement. On January 17, 1996 the administrative law judge (ALJ) commenced an issues conference to consider applications for party status, narrow or resolve disputed issues of fact, and determine whether such disputed issues should be adjudicated in a hearing. Two of the four issues identified for the adjudicatory hearing were settled by the parties, that is, the potential impact from dust and storm water management. Left to be decided were the potential adverse impacts from hydrogen sulfide and determination of whether a public need for such landfill existed should such environmental impacts not be adequately mitigated or avoided. After receiving **testimony from numerous expert witnesses** pertaining to these remaining issues, the ALJ rendered a hearing report that recommended the draft permit be granted.

The Deputy Commissioner for Natural Resources for DEC adopted the findings of the ALJ, effective February 1, 1998, and directed that a permit be issued. He found that the landfill would not generate significant amounts of hydrogen sulfide and that no further discussion of the social and economic benefits would be needed because the environmental impacts associated therewith had been mitigated or settled. The City of Rensselaer challenged Deputy Commissioner Duncan's decision seeking an annulment of the landfill permit grant. In its review, the court noted,

> The Deputy Commissioner's determination that the proposed landfill, as controlled by the draft DEC permit, is designed not to produce significant amounts of hydrogen sulfide is supported by a rational basis. Not only does the permit prohibit the disposal of pulverized construction and demolition debris which undisputably contributes to hydrogen sulfide production, the proposed landfill is also designed with a leachate collection system which collects and removes excess moisture in addition to preventing groundwater from seeping into the waste mass. Notably, petitioners' expert agreed that an operating leachate collection system would limit the potential for hydrogen sulfide production.

> Moreover, the record reflects that the design of this facility was amended to facilitate the conversion of its passive vent system into an active gas collection system if conditions made it necessary. With the further incorporation of a prohibition on construction of a new cell until the previously constructed cell is properly closed and capped, we find no basis upon which to conclude that the determination is arbitrary or capricious, represents an abuse of discretion or is predicated upon an error of law

To the extent that conflicting expert opinions, with respect to this issue, were propounded, the Deputy Commissioner had the discretion to choose the experts presented by 4C's over those of petitioners.

Addressing the failure to have required 4C's to conduct a cumulative noise impact analysis for the proposed landfill, both the ALJ and the Deputy Commissioner permitted a deviation since a cumulative noise impact analysis had already been conducted for the neighboring mining operation which took into account 4C's prior operations. Such document was specifically described, its findings summarized and was properly incorporated by reference into the draft environmental impact. Moreover, we note the inclusion of 4C's noise study in the draft environmental impact statement, which study took into account the ambient noise level associated with truck traffic, including traffic traveling to and from the mine. With the parties' further voluntary agreement to place a limit on truck traffic, and with petitioners failing to make an offer of proof with respect to additional impacts of noise from truck traffic, we find the Deputy Commissioner to have taken the requisite 'hard look' at such issue, making a "reasoned elaboration of the basis for [his] determination" (*Rensselaer v. Duncan*, p. 116).

The court found the Deputy Commissioner's interpretation was entitled to deference and declined further review of additional contentions that the court held without merit. The appeal was dismissed, and the permit determination was confirmed.

In *Chittenden Solid Waste District v. Hinesburg Sand & Gravel Co., Inc.,* 169 Vt. 153, 730 A.2d 614 (1999), Hinesburg Sand & Gravel Co. (HS&G) appealed a judgment of the Chittenden Superior Court condemning its land in Williston for use as a landfill. HS&G claims that the trial court erred by (1) not deciding whether a publicly owned landfill was needed in Chittenden County; (2) considering and imposing a condition on Chittenden Solid Waste District (CSWD) to make sand available to HS&G after condemnation (the stockpile plan); (3) excluding evidence that CSWD acted in bad faith by seeking to condemn the property; and (4) making clearly erroneous findings of fact. The Supreme Court of Vermont affirmed the appeal for HS&G.

In March 1987, CSWD was organized as a union municipal district. Its statutory purpose was to provide for efficient, economical and environmentally sound management of solid waste produced by the member municipalities. Included in the approval of CSWD by the legislature was a grant of authority "to exercise the power of eminent domain within the District." After a lengthy process to determine appropriate landfill sites within Chittenden County, CSWD selected land owned by HS&G as the most suitable site. HS&G is a family-owned business, processing sand and sand aggregate. The selected site for the landfill has significant reserves of Redmond sand, a unique sand utilized by HS&G in its processing of specifically formulated sand blends.

Adopting a resolution, Chittenden SWD started condemnation proceedings of the HS&G site for development of a long-term regional landfill. The trial court concluded that necessity required the taking of the land, but in its judgment order it made the taking subject to the condition that CSWD, at its cost and expense, must make the Redmond sand available to HS&G, its successors or assigns, for 30 years, or until all the available sand was retrieved, whichever occurred first. CSWD had the option of making the sand available directly to HS&G or storing it at the site or on nearby land. CSWD was obligated to preserve or enhance the value of the sand during the excavation process and to protect the sand from any significant contamination by litter or landfill leachate.

In its analysis, the court noted that over 100,000 tons of solid waste are generated by Chittenden County residents each year and, absent a landfill, the current plan of indefinitely sending the waste elsewhere is not feasible. Further, the court noted that CSWD was responsible for managing solid waste under Vermont's Solid Waste Plan. Statewide strategies for the management of solid waste under this plan include (1) prioritizing the successful siting of new lined land disposal facilities, (2) obtaining adequate landfill capacity, and (3) opening a new lined landfill in each region by 1991. The finding of necessity was affirmed by the Supreme Court.

In *Concerned Taxpayers of Brunswick County (Va.). et al. v. Department of Environmental Quality, and Aegis Waste Solutions, Inc.*, 31 Va. App. 788, 525 S.E.2d 628 (2000), appellants were an unincorporated organization of Brunswick County taxpayers and property owners and eight individuals who own property adjacent to, or within a short distance of, a solid waste landfill owned and operated by AEGIS Waste Solutions, Inc. (AEGIS).

Concerned Taxpayers of Brunswick County appealed the Brunswick County Circuit Court decision finding that the Department of Environmental Quality (DEQ) and the Director of DEQ (Director) complied with the requirements of Code § 10.1-1408.1(B)(1) in issuing a permit and permit amendments to AEGIS authorizing construction and operation of a solid waste landfill in Brunswick County. Appellants argued that three parcels of land encompassed by the permit and the permit amendments were not certified as complying with all local ordinances as required by the code.

Before the court could dispense with the appellants' arguments on the three parcels of land, it had to rule on appellee AEGIS' challenge of appellants' standing to appeal. With regard to Brunswick's standing, the court stated that the code "permits all unincorporated associations to sue and be sued under the name by which they are commonly known.... The words 'unincorporated association' ... denote a voluntary group of persons joined together by mutual consent for the purpose of promoting some stated objective [*Yonce v. Miners Memorial Hospital Assn.*, 161 F. Supp. 178, 186 (W.D. Va. 1958)]. We find that Concerned Taxpayers of Brunswick County satisfies the Yonce definition of an unincorporated association, and, therefore, qualifies as a 'person' pursuant to the definition set forth in Code § 10.1-1400. Members of the association who sued individually clearly are 'persons' as defined by the Act" (*Concerned Taxpayers*, p. 794). The court noted, "It is apparent from the record that appellants participated in the submittal of written comments in the public comment process." (*Id.*)

"The Act requires appellants to meet the requirements for standing under Article III of the United States Constitution. In *Lujan v. Defenders of Wildlife, et al.*, 504 U.S. 555, 560-61, 119 L. Ed. 2d 351, 112 S. Ct. 2130 (1992), the United States Supreme Court set forth the three requirements for Article III standing:

First, the plaintiff must have suffered an "injury in fact" — an invasion of a legally protected interest which is (a) concrete and particularized ... and (b) "actual or imminent, not 'conjectural' or 'hypothetical,'" ... Second, there must be a causal connection between the injury and the conduct complained of — the injury has to be "fairly ... traceable to the challenged action of the defendant, and not ... the result [of] the independent action of some third party not before the court." ... Third, it must be "likely," as opposed to merely "speculative," that the injury will be "redressed by a favorable decision." (*Id.*)

The court of appeals was satisfied that the appellants qualified for standing and held that Concerned Taxpayers and its members who brought claims individually, have standing to bring their claims pursuant to Code § 10.1-1457(B) and Article III of the United States Constitution. Having established standing for the appellants, the court turned to the issue of the three land parcels that were not certified by the local government.

In October 1993, as part of the permit application process, AEGIS requested certification from Brunswick County that the proposed facility complied with all local ordinances. On October 22, 1993, the Planning Director of Brunswick County issued a certification that the "proposed location and operation of the facility is consistent with all ordinances."

On December 6, 1993, AEGIS submitted Part A of the permit application. The Part A application included the Near Vicinity Map which identified the proposed site boundaries of the solid waste management facility. The Near Vicinity Map submitted by AEGIS with the Part A application included three parcels that were marked by the letter A on the map. One of the notes on the map stated that the parcels designated by the letter A were under negotiation for inclusion in the site. DEQ approved the Part A application on March 25, 1994.

AEGIS submitted the Part B application on June 20, 1994. The Part B application contained a different map, titled "Proposed Site Features." The Proposed Site Features Map included the three parcels within AEGIS' property boundary that were marked by the letter A on the Near Vicinity Map.

DEQ published a draft permit and held a public hearing on March 6, 1995. On April 17, 1995, DEQ issued the permit to AEGIS. The permit stated that the "total site property consists of approximately 854 acres." The approved Part A application acreage was 822 acres. DEQ granted the first amendment to the permit on December 10, 1997, which allowed a change in classification from industrial disposal to sanitary landfill, a liner design change for the existing landfill, and acceptance by the facility of regulated asbestos-containing material. The maps submitted by AEGIS for this amendment fully incorporated the three parcels as part of the property and facility boundary.

AEGIS submitted an application for a second permit and included a second local government certification, dated October 9, 1997. The second certification contained no clarifying language as to the three parcels. DEQ granted the second permit amendment on May 4, 1998, allowing expansion of the sanitary landfill area by 141 acres.

Appellants contended: (1) DEQ lacked authority to consider a solid waste facility permit application complete or to issue the permit when the application contained land parcels that were not certified by the local government pursuant to Code § 10.1-1408.1(B)(1) and (2) DEQ lacked authority to consider and issue amendments to the solid waste facility permit because it contained land parcels that were not certified by the local governing body pursuant to Code § 10.1-1408.1(B)(1).

The court of appeals agreed with appellants and reversed the decision of the circuit court, stating "For these reasons, we hold that appellants have satisfied the standing requirement. Additionally, we hold that DEQ and the Director improperly issued the permit and permit amendments that authorized the landfill facility operated by AEGIS because three parcels which were included in the permit and permit amendments were not certified by the local government. Reversed and final judgment."

6.2.1   Additional Recent Solid Waste Disposal Cases for Reference and Study

1. *Gemstar Corp. v. Department of Environmental Protection*, 726 A.2d 1120 (Pa. Commw., 1999). Re: permit issues for waste tire facility.
2. *Waste Management Holdings v. Gilmore*, 57 F. Supp. 2d 536 (E.D. Va. 2000). Re: intrastate vs. interstate disposal of solid waste.
3. *Cadiz Land Co., Inc. v. Rail Cycle, et al.*, 99 Rptr. 2d 378 (Cal. App. 4th, 2000). Re: challenge of an inadequate EIR and approval of proposed large landfill in Mojave Desert supplied by rail.

6.3 THE RESOURCE CONSERVATION AND RECOVERY ACT (RCRA) — 1976, AND AS AMENDED — 1984, 1986

The RCRA, originally enacted in 1976, was an amendment to the Solid Waste Disposal Act (SWDA). Its creation was spurred by the Love Canal hazardous waste catastrophe in New York. RCRA essentially focuses on the current management of hazardous wastes produced from manufacturing process generators, and the transportation, treatment, storage, and disposal facilities of these wastes. Title C of RCRA's 1984 and 1986 amendments provides for hazardous waste controls. The RCRA amendments shift past hazardous waste management procedure from land disposal to treatment. Under promulgation powers of Title C, the EPA has specified standards under § 3002 for hazardous waste reporting, record keeping, labeling, and proper waste containers. This section also requires hazardous wastes to be processed on its generation site or at a processing facility before it is transported. Section 3003 dictates standards for transporters. Section 3004 deals with storage and disposal of the hazardous wastes. Section 3004 requires that new, and expansions of, landfills and earth burial of hazardous wastes, must employ double liners and leachate collection systems with groundwater monitoring apparatus, or an equally effective practice, to prevent migration of hazardous wastes into groundwater or surface water sources. Regulations for underground storage tanks (USTs) were adopted in the 1984 Hazardous and Solid Waste Amendments (HSWA) to control buried tanks containing petroleum products, solvents, pesticides, and other hazardous liquids. Section 3005 requires permitting for handling of hazardous waste and gives EPA broad powers of inspection, enforcement, and penalties for violations under subsequent §§ 3007 and 3008. An important addition to RCRA's powers of enforcement is § 7002, which provides for citizen suits against parties contributing to hazardous waste pollution.

RCRA has some areas of overlapping jurisdiction with the Clean Air Act (CAA), Clean Water Act (CWA), SWDA, TSCA, Comprehensive Environmental Response, Compensation and Liability Act (CERCLA), et al. Some waste issues may fall under control of RCRA and another act. However, an RCRA provision (42 U.S.C., § 6905[b]) charges EPA to "integrate the law of other regulatory statutes for the purpose of administration and enforcement." The following section on RCRA litigation wlll elaborate on some of the major problems and issues in dealing with application of the act.

6.3.1 Update on RCRA Final Rules Issued by EPA

1. May 26, 1998 — Phase IV Land Disposal Restrictions. Treatment standards for soil containing mineral processing wastes and 13 toxic metals listed under the Bevill amendment. See 63 *Federal Register* 28,556 (1998).
2. November 30, 1998 — New requirements for treated, stored, or disposed cleanup RCRA hazardous remediation wastes. See 63 *Federal Register* 65,874 (1998).
3. January 21, 1999 — Final rule revising standards for hazardous waste treatment for generators, storage and disposal facilities; clarification and technical amendments governing organic air emission standards for tanks and containers, and surface impoundments. See 64 *Federal Register* 3382 (1999).
4. September 30, 1999 — Final rule revising standards for hazardous waste incinerators, hazardous waste-burning cement and lightweight aggregate kilns under the CAA and RCRA. See 64 *Federal Register* 52,828 (1999).
5. January 19, 2000 — Final rule, "Comprehensive Guidelines for Procurement of Products Containing Recovered Materials"; and amendments for 18 new items that can be made with recovered materials. See 65 *Federal Register* 3070 (2000).
6. March 8, 2000 — New regulations promulgated allowing large volume generators of sludges from treated electroplating wastewaters, classified as F006 sludges, which increase the accumulation period of F006 sludge without a hazardous waste storage permit provided certain conditions are met for metal recovery. See "180-Day Accumulation Time under RCRA for Waste Water Treament Sludges from the Metal Finishing Industry," 65 *Federal Register* 12,378 (2000).

6.3.2 Update on Litigated RCRA Cases

At issue in *Aurora National Bank, et al. v. Tri-Star Marketing, Inc., et al.,* 990 F. Supp. 1020 (N.D. Ill. 1998), was assigning liability under the RCRA for remediating contamination on plaintiffs' property. In 1959, plaintiff property owners leased the property to North States Oil, a now defunct company not a named defendant in this lawsuit. North States Oil operated a gas station on the property from 1959 until September 30, 1981. In 1959, North States Oil installed three steel gasoline USTs on the property that were in existence until 1989. There was also a kerosene UST in place on the property during the tenancy of North States Oil. The court attempted to establish liability for soil contamination resulting from leakage of the USTs.

The court held that petroleum contamination of soil and goundwater from a gasoline service station's USTs is solid waste under RCRA. It also found that the lessor of the land on which the service station was located was not to be classified as a contributor.

In *Resource Investments, Inc. v. U.S. Army Corps of Engineers*, 151 F.3d 1162 (9th Cir. 1998), the case concerned a landfill site in the State of Washington and raised the question whether § 404 of CWA authorizes the United States Army Corps of Engineers (Corps) to require a landowner to obtain a dredge and fill permit from the Corps before constructing a municipal solid waste landfill on a wetlands site.

The court of appeals held that the construction of a municipal solid waste landfill on a wetlands site is regulated by the Environmental Protection Agency (EPA) or states with solid waste permit programs approved by the EPA under the RCRA and not by the Corps under § 404 of the CWA. Accordingly, the court reversed the district court order upholding the Corps' decision to deny a permit and remanded with instructions to vacate that decision.

U.S. v. Fiorillo (and Art Krueger), 183 F.3d 1136 (9th Cir. 1999), concerned criminal violations of RCRA, convictions, and penalties. Frank Fiorillo and Art Krueger appealed their convictions for wire fraud, violations of RCRA, and Fiorillo for receiving explosives without a permit.

Case facts: "Frank Fiorillo was the president and CEO of West Coast Industries, Inc. ('West Coast'). The company's primary business was the storage of a number of products at a warehouse located in Sacramento, California. Fiorillo, who had provided warehouse services to Diversey Corp. in the past, submitted a proposal for the disposal of the products to Farris on behalf of West Coast and SafeWaste Corp. ('SafeWaste'), Art Krueger's company. Farris agreed to the proposal and the parties entered into a contract on February 24, 1993, for the disposal of 10,000 gallons of Diversey's products, Slurry and Eclipse. Under the contract, Diversey agreed to pay 50% of the contract costs when the products were transported to Fiorillo's warehouse and the remaining 50% upon submission of compliance documentation.

"Diversey Corp. ('Diversey') is a company engaged in the manufacture and sale of industrial cleaning products. In 1992, Diversey discovered that two of its products, Slurry and Eclipse, would leak out of their containers in warm or humid weather. The two products are industrial-strength cleansers used in institutional settings and both are highly caustic.* After determining that the products were unsaleable, Diversey authorized its corporate distribution manager, Adrian Farris, to dispose of 30,000 gallons of the products. (*Fiorillo* at 1142.)

"Diversey periodically received compliance documentation from Fiorillo and Krueger in the form of certificates of disposal, which were signed by Krueger. Ultimately, Diversey paid Krueger and Fiorillo $254,000 for the disposal of 30,000

---

* Materials with a pH level greater than 12.5 are classified as hazardous under federal regulations. See 40 C.F.R. § 261.22(a)(1). One of the Government's expert witnesses testified that Slurry and Eclipse had pH levels of 13 to 14. Each one-point difference in pH represents a tenfold increase in the materials' corrosiveness.

gallons of the hazardous products. In reality, Fiorillo and Krueger only properly disposed of two of the eleven truckloads of Slurry and Eclipse by sending it to a facility in Nevada, which met the requirements set out in RCRA. The rest of the Slurry and Eclipse was stored at Fiorillo's warehouse in Sacramento in a cold room that Krueger leased from Fiorillo.

"In August 1993, Rick Knighton, a former West Coast employee, informed David DeMello, a Sacramento County Fire Department official, that West Coast was storing Class A explosives at its warehouse.* DeMello, who had conducted earlier fire inspections of the warehouse, and another fire inspector, Robert Billett, went to the warehouse where they informed the receptionist that they were there to conduct an inspection. DeMello's and Billett's testimony conflicts over what happened next.

"According to DeMello, the receptionist phoned someone who authorized the inspectors to enter the warehouse. Billett did not recall the receptionist getting permission to let them in. Rather, he remembered that she simply allowed them to proceed with the inspection. Regardless, before the men discovered any explosives, they were met by Fiorillo. DeMello testified that Fiorillo was cordial and polite when he greeted the two men. Fiorillo agreed to accompany the inspectors during the inspection. DeMello and Billett then discovered the Class A explosives, consisting of approximately 17,000 artillery shells, taking up about one-third of the warehouse.** DeMello also discovered hazardous material, which covered an additional one-third of the warehouse, leaking from its containers about six feet from the explosives.

"Over the course of the next few days, members of the fire department returned to the warehouse to ensure that proper cleanup was occurring and that no further violations were happening. About eight days after DeMello's discovery, the fire captain, Ed Vasques, received an anonymous tip that additional hazardous materials were being stored in a room that the fire inspectors had not discovered. DeMello, Vasques, and other officials conducted a re-inspection of the warehouse and discovered an unmarked door that was hidden behind several pallets of food and beverages.

"Peter Bishop, an independent contractor hired by West Coast to assist in the cleanup, entered Fiorillo's office to get keys to the room. An investigator from the Sacramento County environmental office overheard Fiorillo say that there was nothing in the room, that he had done everything they wanted and that he had had enough. Nevertheless, Bishop came back out with the keys. A door outside the warehouse led into the cold room as did a door inside the warehouse. The keys did not work on the outside door, and when Bishop went to unlock the inside door, it was apparently unlocked. At this point, the county officials discovered the Slurry and Eclipse, which Krueger and Fiorillo had told Diversey was destroyed." (Id., 1143).

The warrantless search of the warehouse was sustained by the court because it had been sufficiently consented to. Fiorillo was charged with 12 counts of wire fraud (four of the counts were dismissed by the government prior to trial), two counts of violating provisions of RCRA, and two counts of receiving Class A explosives

---

* When Knighton was a West Coast employee, he had contacted DeMello to inquire whether West Coast met the requirements to store Class A explosives in the Sacramento warehouse. DeMello informed Knighton that West Coast would not be approved to store Class A explosives at its warehouse.
** DeMello later learned that West Coast had previously stored a similar shipment of Class A explosives, which had been shipped out in January 1993.

without a permit. Krueger was charged with all of the same counts except those relating to the explosives. A jury found both men guilty of all the counts against them. (*Id.,* 1143)

In *Assn. of Battery Recyclers, Inc. et al. v. U.S. EPA*, 208 F. 3d 1047 (D.C. Cir. 2000), the Court of Appeals for the District of Columbia consolidated petitions for judicial review of EPA's regulations promulgated on May 26, 1998, under the RCRA of 1976, Public Law 94-580, 90 Stat. 2795. The regulations, known collectively as the Land Disposal Restrictions Phase IV Rule, deal with residual or secondary materials generated in mining and mineral processing operations and the EPA's classification of these materials as solid waste; with the treatment standards for a specific category of hazardous waste; and with the EPA test for determining whether certain wastes are hazardous.

The court's case analysis was to decide three issues, viz., the first was to decide whether EPA properly defined solid waste. The court unanimously decided that it did not. The second part decided unanimously that EPA's treatment standards for a particular category of hazardous waste are lawful. The third part decided that the EPA test for determining toxicity is valid for certain wastes, but not for others.

The court proceeded to define solid waste as follows:

Two petitioners — the National Mining Association and the American Iron and Steel Institute — and an intervenor — the Chemical Manufacturers Association — challenge the portion of EPA's Phase IV Rule defining a "solid waste" in terms of how materials "generated and reclaimed within the primary mineral processing industry" are stored. 40 C.F.R. § 261.2(e)(iii). The question is of substantial importance to these petitioners because, together, they represent most of the nation's producers of coal, metals, and industrial and agricultural minerals; two thirds of the nation's steel production; and more than ninety percent of the nation's productive capacity of basic industrial chemicals.

RCRA defines "solid waste" as "any garbage, refuse, sludge from a waste treatment plant, water supply treatment plant, or air pollution control facility and other discarded material...." [42 U.S.C. § 6903(27)]. Solid wastes are "considered hazardous if they possess one of four characteristics (ignitability, corrosivity, reactivity, and toxicity), or if EPA lists them as hazardous following a rulemaking." Disposal of hazardous waste is forbidden unless the waste is treated to reduce its hazardous constituents or stored in a manner ensuring that the hazardous constituents will not migrate from the disposal unit.

To understand the contentions of the parties, it will be helpful to outline the current solid waste classification system (most of which predates the Phase IV Rule and is not being challenged). EPA's general regulation defining "solid waste" begins by repeating a portion of the statutory definition: "a solid waste is any discarded material." 40 C.F.R. § 261.2(a)(1). It then defines "discarded material" to mean "any material which is Abandoned ... or Recycled, as explained in paragraph (c) of this section...." [*Id.*§ 261.2(a)(2)]. Paragraph (c) identifies four situations in which "recycled" materials will be considered "solid waste": when the materials are "used in a manner constituting disposal"; when the materials are "burned for energy recovery";

when the materials are "reclaimed"; and when the materials are "accumulated speculatively." [40 C.F.R. § 261.2(c)(1)–(4)].

The Phase IV Rule revised only the reclamation provision. Before the revision, EPA classified reclaimed spent materials and scrap metal as solid waste. See 40 C.F.R. § 261.2(c)(3) and tbl.1 (1996). Reclaimed sludges and by-products were classified as solid waste only if they had been specifically listed in 40 C.F.R. pt. 261 as a hazardous waste following an EPA rulemaking. See 40 C.F.R. § 261.2(c)(3) and [*1051] tbl.1 (1996). Reclaimed sludges and by-products exhibiting a characteristic of hazardous waste, but not specifically listed as hazardous wastes, were not classified as solid waste. See *id.* This classification system applied without regard to the industry that produced the materials.

The Phase IV Rule purported to take materials reclaimed by the mineral processing industry outside this framework and to subject these secondary materials to a new test for determining whether they constituted "solid waste." See 40 C.F.R. § 261.2(c)(3) and tbl.1. We say "purported" because it is not clear to us that EPA accomplished its objective. The relevant part of the new recycling-reclamation provision reads:

Materials [listed in a table] are not solid wastes when reclaimed (except as provided under 40 CFR 261.4(a)(17)).* The new § 261.4(a)(17) gave a so-called "conditional exclusion": if the provision's criteria were met, reclaimed mineral processing secondary materials would not be classified as solid waste. We have trouble making sense of these two provisions. The first provision (§ 261.2(c)(3)) broadly describes what is not a solid waste, unless it complies with the other provision. But the other provision, § 261.4(a)(17) — is an exclusion, and the consequence of not complying with the provision is, of course, loss of exclusion. In other words, read together, the provisions seem to say that something is not a solid waste unless it is not excluded from being a solid waste. But petitioners make nothing of the point and we shall therefore assume that if secondary material of this sort — derived from mineral processing — does not meet the conditions specified in § 261.4(a)(17), EPA will consider the material "solid waste" potentially subject to full RCRA Subtitle C regulation.

As to the conditions set forth in § 261.4(a)(17), **EPA's dividing line between "waste" and nonwaste is the manner of storage** [emphasis added]. If the mineral processor stores secondary material destined for recycling in tanks, containers, buildings, or on properly maintained pads, the materials are not considered "solid waste." See *id.* § 261.4(a)(17)(iii), (iv). Given our assumption (and that of the parties), if by-products and sludges exhibiting a characteristic of hazardous waste are not stored in such a manner prior to being recycled, they may be regulated as hazardous "waste."

How long the materials are stored is of no consequence according to the regulation. See Fed. Reg. 28,556, 28,582–83 (1998). They could be placed on the ground for only a few minutes before being put back into the production process, yet they would still be subject to RCRA if not stored in accord with § 261.4(a)(17). Petitioners say this rule extends EPA's authority far beyond the statute. They ask how secondary

---

* The final rule published in the *Federal Register* incorrectly cited § 261.4(a)(15). See 63 Fed. Reg. 28,556, 28,636 (1998). EPA later corrected its mistake. See 64 Fed. Reg. 25,408, 25,408 (1999).

material held for recycling in production could possibly qualify as "waste" when the statute defines "waste" as "discarded materials"? [42 U.S.C. § 6903(27)].

In unraveling the issues and terms, the court decided that EPA's rule classifying a solid waste based on the method of storage was beyond EPA's authority under the statute. Under RCRA, materials must be disposed of, thrown away, or abandoned before EPA may consider them waste. Therefore, RCRA does not apply to materials that are intended for reuse in an ongoing industrial process, including recycling by the industry.

In its findings, the court rejected the *Battery Recyclers'* challenge to the Land Disposal Restrictions Phase IV treatment standards. The court also found that in EPA's procedure for determining the toxicity of mineral processing wastes, the agency had proven effective its use of Toxicity Characteristics Leaching Procedure (TCLP). However, the court found that EPA had not justified its application of TCLP criteria to waste from manufactured gas plant (MGP) waste and voided that part of Phase IV as applied to MGP waste.

6.3.3   Additional Recent RCRA Cases for Reference and Study

1. *Harmon Industries v. Browner*, 19 F. Supp. 2d 988 (W.D. Mo. 1998). Re: EPA overfiling on a state-authorized program to impose penalties for continuing RCRA violations after a consent decree had been reached in a state court.
2. *Columbia Falls Aluminum Co. v. EPA,* 139 F.3d 914 (D.C. Cir. 1998). Re: Court voiding some of the EPA land disposal restrictions for aluminum smelters spent "potliner" wastes (classified as K088 wastes).
3. *Lefebvre v. Central Maine Power Co.*, 7 F. Supp. 2d 64 (D. Me. 1998). Re: Timing of RCRA citizen suits is not governed by a statute of limitations, but by the imminent and substantial endangerment requirement; also, see *Mutual Life Insurance Co. of N.Y. v. Mobil Corp.*, WL 160820 (N.D. N.Y. [March 31, 1998]), an RCRA citizens suit not subject to statute of limitations.
4. *Moly Corp., Inc. v. U.S. EPA*, 197 F.3d 543 (D.C. Cir. 1999). Re: EPA technical background document on Bevill amendments for 49 different mineral commodities was advisory only and not subject to litigated review.
5. *Assn. of Battery Recyclers, Inc. et al v. U.S. EPA*, 208 F. 3d 1047 (D.C. Cir. 2000) (study of full case recommended).
6. *Christie-Spencer Corp. v. Hausman Realty Co.*, 118 F. Supp. 2d 408 (S.D. N.Y. 2000). Re: Injunction against voluntary cleanup; failure of RCRA citizen suit for lack of proof of imminent and substantial endangerment.

6.4 COMPREHENSIVE ENVIRONMENTAL RESPONSE COMPENSATION AND LIABILITY ACT — 1980 (SUPERFUND); SUPERFUND AMENDMENT AND REAUTHORIZATION ACT (SARA) — 1986

This environmental act is best known by its acronym, CERCLA, or the Superfund law. Its principal purpose is the cleaning up of contaminated and hazardous sites in the United States. EPA has prepared a list, called the National Priorities

List (NPL), of known nationwide, dangerous, contaminated and hazardous waste sites requiring cleanup procedures by potentially responsible parties (PRPs). The main sections of CERCLA are:

1. Hazardous and substantially dangerous site listing for treatment (the NPL)
2. Authorization and empowerment for emergency responses, removal and remedial actions
3. Authorization of a Hazardous Substances Trust Fund to underwrite the cost of removal and remedial actions
4. Establishing liability and responsibility of PRPs for cleanup and restitution costs of removal and remedial actions

For a review of early CERCLA litigation on establishing PRPs and responsibility for cleanup costs, the following three cases are suggested: (1) *State of New York v. Shore Reality Corp.*, 759 F.2d 1032 (2d Cir. 1985); (2) *U.S. v. Northeastern Pharmaceutical & Chemical Co., Inc.*, 810 F.2d 726 (8th Cir. 1986); and *U.S. v. Monsanto Co.*, 858 F.2d 160 (4th Cir. 1988).

It should be noted that Chapter 7 on water pollution by abandoned mine sites, acid mine drainage and mined land reclamation that follows is an extensive study dealing with rulings on four CERCLA hazardous sites involving their litigated cleanup and fixing of PRPs.

6.4.1   Update on Litigated CERCLA Cases

In *U.S. v. Dico, Inc.*, 136 F.3d 572 (8th Cir. 1998), the case arose under CERCLA where the EPA has broad "authority to direct clean-up operations prior to a final judicial determination of the rights and liabilities of the parties affected." (See *Solid State Circuits, Inc.v. U.S. EPA*, 812 F.2d 383, 387 [8th Cir. 1987].) In the 1986 SARA amendments, Congress provided procedures by which a party who pays for cleanup pursuant to an order from the EPA, but does not believe it is liable, may petition the president to recover its response costs from the Hazardous Substance Response Trust Fund (Superfund) established by CERCLA, and may bring the same claim in federal court if it receives an adverse ruling from the president. This was the basis for Dico's counterclaim. Also, CERCLA provides that the EPA may bring an action in district court to recover from responsible parties the removal and remediation costs the government has incurred in association with the cleanup of a hazardous waste site. This was the basis of EPA's claim.

The *Dico* case has its roots in the discovery in the mid-1970s of contamination in the Des Moines, Iowa, public water supply. EPA tests determined that the Des Moines Water Works (DMWW) was contaminated by trichloroethylene (TCE) and other substances designated as hazardous by the EPA. The EPA identified the land area determined to be the source of the contamination (the Des Moines TCE site) and in 1983 this site was placed on the EPA's NPL. Dico's property was included in the site. Over the years, Dico and its corporate predecessors had used TCE in industrial degreasing operations and other activities, and other businesses in the vicinity apparently also had used the compound on their properties. In 1986, the

EPA ordered Dico as a PRP to capture and treat the contaminated groundwater in Operable Unit 1 (OU-1), one of four operable units that the EPA defined for purposes of cleaning up the site and the only one at issue here. The remedial system Dico constructed began operating in December 1987.

The following summer, in July 1988, Dico sought costs the company had incurred and would continue to incur from the Superfund reimbursement with respect to its remediation efforts at the site. The EPA denied the petition, holding that the 1986 SARA amendment that permitted such reimbursement did not apply retroactively to EPA's cleanup order issued to Dico before the effective date of the amendment. The Environmental Appeals Board (EAB) sustained the EPA after an administrative hearing. Dico then brought suit in the district court, but the court granted summary judgment for the EPA, deferring to the agency's interpretation of the statute. Dico appealed. The appeals court reversed and remanded to the district court with instructions to remand for further proceedings.

On April 21, 1995, with Dico's administrative claim for reimbursement pending before the EAB, the United States filed this action in the district court seeking recovery from Dico alone of costs that the EPA had incurred in connection with cleanup of the groundwater at the Des Moines site. Dico filed a counterclaim for reimbursement of its costs, and also moved the EAB to stay the administrative proceedings on the ground that the claim pending before the EAB was the same as Dico's counterclaim filed in federal court. The EAB granted the motion.

On September 13, 1996, the district court granted the EPA's motion to dismiss Dico's counterclaim for failure to exhaust administrative remedies. On April 1, 1997, the court granted summary judgment to the EPA on its claim for response costs, including indirect and oversight costs, in the amount of $4,378,110.66. (See *U.S. v. Dico, Inc.*, 979 F. Supp. 1255 [S.D. Iowa 1997].) On April 25, 1997, the EPA moved the EAB to deny Dico's administrative action for reimbursement without a hearing, arguing that it was barred by *res judicata*. Counsel for Dico represented to this court at oral argument that the motion had been granted. Dico appealed the orders of the district court.

Dico appealed the grant of summary judgment to the U.S. on the government claim to recover from Dico response costs incurred in association with the environmental cleanup of groundwater determined by EPA to be contaminated. Dico also appealed the order dismissing its counterclaim, in which the company sought reimbursement of amounts it expended cleaning up the site.

The court of appeals found that (1) Dico had not exhausted its administrative remedies, on its counterclaim, and (2) EPA had failed to establish a causal *nexus* (connection) with adequate evidence to support its claim of a release of TCE by Dico onto the soil of the defendant's facility causing the groundwater contamination. The judgment entered for the United States was vacated and the money judgment was set aside. The case was remanded to the district court for further consistent proceedings.

The issue in *A & W Smelter and Refiners, Inc. v. Clinton*, 146 F.3d 1107 (9th Cir. 1998), was whether: (1) the piles of low-grade gold–silver ore, containing small amounts of lead (in its natural mineral form), stored on the plaintiff's property were properly considered waste; (2) EPA's characterization of the materials as hazardous substances and an imminent threat for contaminating the ground was supportable; and (3) the government's order to remove them as an imminent threat or endangerment was valid.

The short, full decision of the court of appeals is reproduced as an enlightening study of EPA's enforcement measures and a court analysis.

A & W SMELTER AND REFINERS, INC,
a California corporation, Plaintiff-Appellant,
v.
WILLIAM J. CLINTON, in his official capacity as President of the United States;
CAROL M. BROWNER, in her official capacity as Administrator of the U.S.
Environmental Protection Agency; and ENVIRONMENTAL PROTECTION
AGENCY, Defendants-Appellees.

No. 97-15596

UNITED STATES COURT OF APPEALS FOR THE NINTH CIRCUIT
146 F.3d 1107; 1998 U.S. App. LEXIS 13462; 46 ERC (BNA) 1918; 98 Cal. Daily Op.
Service 4864; 98 Daily Journal DAR 6868; 28 ELR 21341
February 10, 1998, Argued, Submitted, San Francisco, California
June 24, 1998, Filed

DISPOSITION: AFFIRMED in part, REVERSED in part and REMANDED.
Each party shall bear its own costs.

JUDGES: Before: Alfred T. Goodwin, Alex Kozinski and David R. Thompson, Circuit Judges. Opinion by Judge Kozinski, Circuit Judge.

The Environmental Protection Agency ordered plaintiff to dispose of ore containing gold and silver because it also contained small quantities of lead. We must resolve various questions of statutory etymology on the way to deciding who must pay for this disposal.*

The Ore in Motion and at Rest

"Ore Pile # 2" (O2P) at A & W Smelter's processing facility in the Mojave Desert contained ore that had been piling up since the Sixties. The ore included small amounts of silver and gold, not quite enough to justify smelting by A & W's methods. Unhappily for A & W, the ore also contained some naturally occurring lead, with some slag mixed in as well. Slag is a waste product of smelting.

Under pressure from state and federal authorities, A & W decided to move O2P. To that end, it contracted with Roelof Mining Company to process the ore in Baja, Mexico. A & W packed the ore into drums and started shipping it, but several trucks were stopped at the border and their contents labeled hazardous because of the lead. Several months later, Mexico returned these trucks to the United States. The EPA ordered A & W to take this ore back within three days, but A & W was unable to arrange a pickup on short notice.

---

* A & W does not seek reimbursement for the value of the ore, as indeed it could not. Any such claim would have to be brought in the Court of Federal Claims under the Tucker Act. See 28 U.S.C. §§ 1346(a)(2), 1491.

EPA then declared the ore abandoned, brought it to a storage facility and issued Order 93-06 directing A & W to dispose of the ore in an approved landfill.

After seeing some of its trucks impounded, A & W diverted other trucks. Six truckloads wound up in Nevada. A & W claims this was a temporary resting place as the ore waited to be processed at the nearby Durga Mine facility. The EPA found this ore as well and issued Order 93-03 directing A & W to ship it to a hazardous waste landfill.*

A & W complied with both disposal orders and then filed a complaint seeking reimbursement of its compliance costs. See 42 U.S.C. § 9606(b)(2). The EPA and its co-defendants moved for summary judgment, which the district court granted. See A & W Smelter & Refiners, Inc. v. Clinton, 962 F. Supp. 1232 (N.D. Cal. 1997).**

The Statutory Framework

The Comprehensive Environmental Response, Compensation, and Liability Act regulates cleanup and disposal of hazardous substances. CERCLA gives the EPA several mechanisms for enforcing the Act against violators. Under one of these, 42 U.S.C. § 9606(a), "when the President determines that there may be an imminent and substantial endangerment to the public health or welfare or the environment because of an actual or threatened [**5] release of a hazardous substance from a facility," he may issue orders, as the EPA did here in his name. Those who violate such orders face fines of up to $25,000 a day. See 42 U.S.C. § 9606(b)(1).

Those who pay for the cleanup, but believe they should not have, may petition for reimbursement [*1110] of reasonable costs incurred. See 42 U.S.C. § 9606(b)(2)(A). If the EPA refuses, they may sue in district court, see 42 U.S.C. § 9606(b)(2)(B), as A & W did here. A & W is entitled to reimbursement if it was not liable for response costs under section 9607(a). See 42 U.S.C. § 9606(b)(2)(C). Even if otherwise liable, it may be reimbursed if the order was arbitrary and capricious. See 42 U.S.C. § 9606(b)(2)(D). A & W claims reimbursement on both of these grounds.

The EPA claims A & W is responsible for the cleanup costs pursuant to sections 9607(a)(3) and (4), which hold liable "any person who by contract, agreement, or otherwise arranged for disposal or treatment ... of hazardous substances owned or possessed by such person ... from which there is a release, or a threatened release which causes the incurrence of response costs, of a hazardous substance...." We must ponder the meaning of several of these statutory words and phrases: "hazardous substance," "release," "disposal," "treatment," "waste" and "imminent and substantial endangerment."***

* A & W claims the EPA orders do not establish the proper delegation of authority. Both orders start by explaining the chain of delegation. Executive Order No. 12,580, 52 Fed. Reg. 2923 (1987), delegates authority from the President to the Administrator of the EPA. Delegation No. 14-14-B further delegates authority from the Administrator to Regional Administrators. This delegation says the Regional Administrators must consult with the Assistant Administrator for Solid Waste and Emergency Response, but a clarification dated April 4, 1990, provides that the "consultation occurs during the decision-making process and is therefore implicit in the delegation process...." Regional delegation 1290.41 in turn delegates authority from the Regional Administrator to the Director, Toxics and Waste Management Division. The orders here were issued by the Director, Hazardous Waste Management Division. A & W makes much of the difference in title. The EPA replies that the division's title has changed, but does not point to any evidence that this is the same division with a new name. On remand, the district court shall verify whether the EPA's assertion is correct.

** A & W claims the district court improperly made credibility findings in granting summary judgment, but it is mistaken. It also claims that the court improperly considered inadmissible evidence, but we see no abuse of discretion.

"Hazardous substance"

EPA labeled O2P a hazardous substance because it contained lead. All agree that at some level of concentration lead is hazardous. **But is it hazardous at this level?** CERCLA defines hazardous substance as "any substance designated pursuant to" several other statutes, or to EPA regulations promulgated under CERCLA. See 42 U.S.C. § 9601(14). The EPA points to regulations promulgated under CERCLA at 40 C.F.R. § 302.4, where lead is listed, along with various lead compounds.* The EPA also points to Clean Water Act regulations listing lead. See 33 U.S.C. § 1321(b)(4); 40 C.F.R. § 401.15.

A & W asks us to read a minimum level requirement into the statute and regulations. It argues that trace levels of hazardous substances are present just about everywhere. **Read as the EPA suggests, CERCLA seems to give the agency** *carte blanche* **to hold liable anyone who disposes of just about anything.** Drop an old nickel that actually contains nickel — A CERCLA violation? Throw out an old lemon? It's full of citric acid, another hazardous substance.

It's not surprising that an agency would urge an interpretation which gives it such broad discretion. Perhaps more surprising is that CERCLA leaves us little choice but to agree. Section 9601(14) refers simply to "any substance" designated under one of the various regulations, and the regulations in turn give no minimum levels. The table in 40 C.F.R. § 302.4 does list reportable quantities, but this refers to notification requirements under 42 U.S.C. §§ 9602 & 9603. Under these sections, anyone who owns a facility which stores hazardous substances and releases a quantity of a substance above the reportable level must notify the EPA. Nothing in the law suggests that quantities of a hazardous substance below its reportable level render it no longer hazardous. The Second, Third and Fifth Circuits have faced this very question and all agree that CERCLA's definition of hazardous substance has no minimum level requirement. See *U.S. v. Alcan Aluminum Corp.*, 990 F.2d 711, 720 (2d Cir. 1993) (Alcan II); *U.S. v. Alcan Aluminum Corp.*, 964 F.2d 252, 260-63 (3d Cir. 1992) (Alcan I); *B.F. Goodrich Co. v. Murtha*, 958 F.2d 1192, 1199-1201 (2d Cir. 1992); *Amoco Oil Co. v. Borden, Inc.*, 889 F.2d 664, 669 (5th Cir. 1989). We see no basis for parting company.

The Fifth Circuit has imposed a minimum level requirement through the back door. It did so by focusing on language in section 9607(a)(4) which defines a liable person as one responsible for a release "which causes the incurrence of response costs." 42 U.S.C. § 9607(a)(4) (emphasis added); see *Amoco*, 889 F.2d at 669. The Fifth Circuit held that the release causes the EPA's response only if it poses a serious enough threat to justify the response; otherwise, the response is caused by the agency's overzealousness. See *id.* at 671. In our view, this reads much too much into the word "causes" in section 9607(a)(4). [*1111] Where a party is responsible for a particular release,** that party is a cause of any response to that release, although on occasion EPA overzealousness may be another cause as well. The Fifth Circuit rather explicitly adopted this strained definition of the causation requirement as a way of getting around the lack of a minimum level requirement.

*** A & W cites *Lara v. Secretary of the Interior*, 820 F.2d 1535 (9th Cir. 1987), for the proposition that precious metal mining is subject to different rules. *Lara* concerns mining discovery claims. It has nothing to do with CERCLA or this case.

* A & W also argues that mineral lead is not a hazardous substance under any of the relevant regulations. Mineral lead is still lead.

** Causation becomes trickier where multiple parties have contributed to the pollution at one site. See *Alcan II*, 990 F.2d at 722 (collecting cases).

See id. at 670. We sympathize, but we believe it is not our function to read into the statute a limitation that Congress did not put there.* Other circuits agree. See *Alcan II*, 990 F.2d at 721 (collecting cases).

"Release"

To issue an order under section 9606(a), there must be "an actual or threatened release of a hazardous substance." Release is defined as "any spilling, leaking, pumping, pouring, emitting, emptying, discharging, injecting, escaping, leaching, dumping, or disposing into the environment (including the abandonment or discarding of barrels, containers, and other closed receptacles containing any hazardous substance or pollutant or contaminant)...." (whew) 42 U.S.C. § 9601(22). For the shipment left near Durga Mine, there was unquestionably a release: Wind was blowing particles from the pile; A & W does not dispute this.

There is a **serious** question, though, as to the shipment the government sent to the Appropriate Technologies facility. As to this shipment, there was no release into the environment (in the ordinary sense): The lead-bearing ore stayed safely within the shipping drums. The government, however, claims that A & W abandoned the ore, which is one form of release under the statute. The EPA's abandonment theory rests on A & W's failure to move the shipment within the time allowed (3 days) by the EPA. A & W claims that it intended to retrieve the material, but as a small company it couldn't respond within the three days the EPA allowed. The question is whether the agency can deem the shipment abandoned (and therefore released) for purposes of CERCLA because A & W failed to move it within the time allotted by the EPA.

We are unaware of case law defining abandonment under section 9601(22).** Nor is this a term of art specific to CERCLA. Rather, it's a term with a rich common law tradition. Property is abandoned when the owner intends to divest himself of all interest in it. See 1 Am. Jur. 2d, Abandoned, Lost, and Unclaimed Property §§ 11, 13 (1994). A & W argues that it did not intend to abandon this property, that it intended to retrieve it but just didn't have the means to do so within the 3-day time allotted by the EPA. The EPA order does not seem to have considered A & W's intent; rather, the EPA deemed the property abandoned just because the agency said it was abandoned. Had the EPA published regulations defining abandonment quite differently from the common law, those regulations would be due considerable deference. See *Chevron U.S.A. Inc. v. Natural Resources Defense Council, Inc.*, 467 U.S. 837, 844, 81 L. Ed. 2d 694, 104 S. Ct. 2778 (1984). Published regulations would also have given A & W notice as to what constitutes abandonment. Even a well-established agency practice of defining abandonment so as to cover

---

* In a situation where it is truly unreasonable for the EPA to respond to a release, the party held liable could claim that the EPA's action was arbitrary and capricious. See 42 U.S.C. § 9606(b)(2)(D). Although A & W has claimed that the EPA's action was arbitrary and capricious, its claim is based on totally different grounds. See *infra* at 1998 U.S. App. LEXIS 13462 at *18.

** A & W points to customs regulations defining abandonment, giving time ranges from thirty days to five years. See 19 C.F.R. § 127.12. Depending on the facts, these regulations may well be relevant. If the trucks carrying the ore were in the Custom Service's custody, A & W might be justified in relying on these regulations to define intent to abandon, although as the text explains that might not be so in an emergency. See *infra* at 1998 U.S. App. LEXIS 13462 at *14. The record is somewhat unclear as to the Custom Service's involvement. Three trucks were impounded by Mexican customs while four trucks detained by U.S. Customs were held at the L & Z Trucking yard in Chula Vista. The trucks from Mexico were returned to the U.S., and the U.S. Government sent them to the Appropriate Technologies storage facility. The EPA seems to have coordinated the movement of the trucks, but it is not clear whether the Customs Service also played a role, and if so what that role was.

situations like this would be due deference. [*1112] However, the EPA presented no evidence of such a practice.

Ad hoc agency action such as here is also entitled to some deference, but not all deference is created equal. How much deference an agency decision is due depends in part on such factors as how much deliberation went into reaching it and whether the decision fits with a policy the agency has consistently followed. See *Atchison, Topeka & Santa Fe Ry. v. Pena*, 44 F.3d 437, 442 (7th Cir. 1994), aff'd sub nom. *Brotherhood of Locomotive Eng'rs v. Atchison, Topeka & Santa Fe R.R.*, 516 U.S. 152, 116 S. Ct. 595, 133 L. Ed. 2d 535 (1996); *Barnett v. Weinberger*, 260 U.S. App. D.C. 304, 818 F.2d 953 (D.C. Cir. 1987). Here, one EPA administrator, faced with an allegedly imminent threat, wrote an administrative order ignoring entirely A & W's intent to abandon. A & W had no opportunity to respond before the order was issued. In a single sentence, that order simply stated that the ore was abandoned and that this was a release under CERCLA. See Administrative Order 93-06 at 7-8. There's no sign the agency considered the possible objections to this position, and no recognition of the fact that so defined the term becomes something very different from what Congress likely had in mind when it used it. We see no reason to defer to an implausible interpretation of the statute under these circumstances.*

The matter might well be different in an emergency. When hazardous waste presents a serious, immediate threat and the waste's owner is not present to take control, it may be reasonable for the EPA to declare the waste abandoned and take appropriate steps to dispose of it safely. This, however, would require a substantial showing on the EPA's part that the trucks had to be moved quickly to avoid danger to health or the environment. The EPA is free to make such a showing on remand.

"Disposal," "Treatment" and "Waste"

A & W is a liable party under section 9607(a)(3) only if it arranged for "disposal or treatment" of hazardous substances. Disposal, as defined in the Solid Waste Disposal Act, see 42 U.S.C § 9601(29), is "the discharge [and so on] of any solid waste or hazardous waste...." 42 U.S.C. § 6903(3) (emphasis added). "Treatment" is defined as "any method, technique, or process ... designed to change the ... character or composition of any hazardous waste so as to neutralize such waste or so as to render such waste nonhazardous, safer for transport, amenable for recovery, amenable for storage, or reduced in volume." 42 U.S.C. § 6903(34) (emphasis added). Thus, A & W disposed of or treated the ore only if it was waste. See *3500 Stevens Creek Assocs. v. Barclays Bank of Cal.*, 915 F.2d 1355, 1362 (9th Cir. 1990). If the ore was a useful product, then it was not waste and not subject to CERCLA. This has been called the useful product defense. See *California v. Summer Del Caribe, Inc.*, 821 F. Supp. 574, 581 (N.D. Cal. 1993). The district court held that the ore here wasn't a useful product because it couldn't be used in its present state: "In order to extract what A & W alleges to be valuable minerals (gold and silver), the material would have to be subjected to a heap leaching

---

* We note another aspect of the definition of abandonment that the EPA's order ignores. Abandonment is a form of release under section 9601(22), which says release means any spilling, leaking and so on "into the environment (including the abandonment or discarding of barrels, containers, and other closed receptacles....)" Any release, including abandonment, must thus be "into the environment." Had A & W left the drums in the middle of the Mojave, with no intent to retrieve them, that would be an abandonment because, over time, there could be a release into the environment. But here the ore was in the hands of the U.S. and Mexican authorities, who had confiscated it. We are hard-pressed to imagine how this could be a release into the environment.

process, which would still leave a by-product requiring disposal." 962 F. Supp. at 1238. But raw materials, by definition, cannot be used in their current state. They must be refined — a process which separates the useful portion from the waste. The hard cases occur when a company uses the by-products of its main manufacturing process, or sells them for use by others — i.e., when it uses waste products as the raw material for further processing. See, e.g., *Cadillac Fairview/California, Inc. v. United States*, 41 F.3d 562, 566 (9th Cir. 1994); *Louisiana-Pacific* [*1113] *Corp. v. Asarco Inc.*, 24 F.3d 1565, 1574–75 (9th Cir.1994). Here, though, the ore wasn't a by-product: It hadn't been processed even once. Smelting unprocessed ore was A & W's business. See *Louisiana-Pacific*, 24 F.3d at 1575 (n.6) (drawing distinction between "products that were produced as the producers' principal business products [and] by-products that the producers had to get rid of.") That the smelting would leave by-products after processing fails to establish that the ore was a by-product or waste.

The slag mixed in with the ore complicates the analysis. Slag is indeed a waste by-product; had O2P been only lead-bearing slag, it would have been hazardous waste. But how do we deal with the mix of useful product and waste? Adding waste to an otherwise useful product may make it all waste (e.g. adding a small quantity of lead to a carton of milk) or it may only decrease its usefulness (e.g. adding a small quantity of lead to a bar of gold). It all depends on the materials and their relative quantity. This may look like a repeat of the minimum level requirement question — does a little slag make the whole pile waste? But, the question here isn't whether this material was a hazardous substance — it was. The question is whether the whole pile was waste or a useful product.

If the ore was mixed with enough slag so that it was no longer usable for A & W's principal business, then it was waste. This can be determined by looking both to A & W's actions and to commercial reality. See *Louisiana-Pacific Corp.*, 24 F.3d at 1575. The contract with Roelof for processing suggests the material was a useful product, although A & W's attempt to dispose of the pile in a municipal landfill before contracting with Roelof points the other way. O2P's market value, if any, would also be quite relevant. This is a factual question on which the record is not particularly clear. The EPA orders don't speak to this point. On remand, the district court shall consider this issue.

"Imminent and substantial endangerment"

The EPA may issue an order under section 9606(a) only if it determines there is "an imminent and substantial endangerment to the public health or welfare or the environment." CERCLA directs the EPA to establish guidelines for using this power. 42 U.S.C. § 9606(c). Although the EPA has done so, see 47 Fed. Reg. 20664 (1982), A & W argues the guidelines do so little guiding that the EPA's actions are arbitrary and capricious, and hence it should be reimbursed. See 42 U.S.C. § 9606(b)(2)(D).

Given the minimal rationality required to withstand arbitrary and capricious review, A & W's argument fails. However, the EPA's guidelines are skimpy, and courts must therefore give meaning to "imminent and substantial." In particular, substantial implies that the release must present a more than minimal threat to health, welfare or the environment. However, since A & W has not argued that the release was not a substantial endangerment to health or the environment, we need not resolve that issue here. It remains a question that could be raised on remand.

Conclusion

CERCLA gives the EPA broad powers to respond to the improper disposal of hazardous substances. However, those powers aren't quite as broad as the EPA or the district court thought. It is not clear that the ore was waste, nor that all of it was released. The EPA may yet prevail on these issues, but so far summary judgment is premature.

AFFIRMED in part, REVERSED in part and REMANDED. Each party shall bear its own costs.

Browning-Ferris Industries of Illinois Inc. v. Ter Maat, 195 F.3d 953 (7th Cir. 1999), concerned a court determination for individual and personal liability for environmental response costs. The case is very instructional in CERCLA cleanup legal liability analysis and, consequently for the enlightenment of the student, extensively quotes Chief Judge Posner's opinion.

Browning-Ferris and several other companies brought a suit for contribution under CERCLA (the Superfund statute). The suit is against Richard Ter Maat and two corporations of which he was the president and principal shareholder; vis., M.I.G. Investments, Inc. and AAA Disposal Systems, Inc.

In 1971, the owners of a landfill had leased it to a predecessor of Browning-Ferris, which operated it until the fall of 1975. Between then and 1988 it was operated by M.I.G. and AAA. In June of that year, after AAA was sold and Ter Maat moved to Florida, M.I.G. abandoned the landfill without covering it properly. For tax reasons, M.I.G. had been operated with very little capital, and it lacked funds for a proper cover. Two years after the abandonment, the EPA placed the site on the NPL, the list of the toxic waste sites that the Superfund statute requires be cleaned up; and shortly afterward Browning-Ferris and the other plaintiffs, which shared responsibility for some of the pollution at the site, agreed to clean it up.

Section 113(f)(1) of the Superfund law authorizes any person who incurs costs in cleaning up a toxic waste site to "seek contribution from any other person who is liable or potentially liable under section 9607(a) of this title.... In resolving contribution claims, the court may allocate response costs among liable parties using such equitable factors as the court determines are appropriate." A part of the statutory provision includes in the set of potentially liable persons anyone who owned or operated a landfill when a hazardous substance was deposited in it, and this set was conceded to include both M.I.G. and AAA. The district judge held that Ter Maat was not himself a potentially liable person because he had done nothing that would subject him to liability on a "piercing the corporate veil" theory for the actions of the two corporations. So far as corporate liability for cleanup costs was concerned, the judge ruled that of the 55% of those costs that he deemed allocable to transporters and operators (the other 45% he allocated to the owners of the landfill and the generators of the toxic wastes dumped in it), 40% was the responsibility of Browning-Ferris and the other 60% the responsibility of M.I.G. and AAA. As between those two, the judge allocated responsibility equally, holding that, although the two corporations had operated the landfill jointly, the statute required him to allocate liability severally rather than jointly.

Browning-Ferris argued that the district court allocated too much of the liability to them for the pollution at the site relative to M.I.G.'s and AAA's operations.

The court's opinion stated:

Two issues are relatively simple ... one is whether an individual can shield himself from liability for operating a hazardous-waste facility merely by being an officer or shareholder of a corporation that also operates the facility. The answer is no. The principle of limited liability shields a shareholder from liability for the debts (including debts arising from tortious conduct) of the corporation in which he owns shares (with the exception discussed later for "veil piercing" situations), but not for his personal debts, including debts arising from torts that he commits himself. In other words, the status of being a shareholder does not immunize a person for liability for his, as distinct from the corporation's, acts. (citations omitted) There is no liability shield at all for an officer. If he commits an act that is outside the scope of his official duties, his employer may not be liable; but he is, whether or not the act was within that scope. (citations omitted) Which is not to say, however, that the officer is automatically liable for the acts of the corporation; there is no doctrine of "superiors' liability," comparable to the doctrine of respondeat superior, that is, the employer's strict liability for torts of the employee committed within the scope of his employment.

So if Ter Maat operated the landfill personally, rather than merely directing the business of the corporations of which he was the president and which either formally, or jointly with him (as well as with each other), operated it, he is personally liable. (citations omitted) The line between a personal act and an act that is purely an act of the corporation (or of some other employee) and so not imputed to the president or to other corporate officers is sometimes a fine one, but often it is clear on which side of the line a particular act falls. If an individual is hit by a negligently operated train, the railroad is liable in tort to him but the president of the railroad is not. Or rather, not usually; had the president been driving the train when it hit the plaintiff, or had been sitting beside the driver and ordered him to exceed the speed limit, he would be jointly liable with the railroad. If Ter Maat did not merely direct the general operations of M.I.G. and AAA, or specific operations unrelated to pollution, (citations omitted), but supervised the day-to-day operations of the landfill, for example, negotiating waste-dumping contracts with the owners of the wastes, or directing where the wastes were to be dumped, or designing, or directing measures for preventing toxic substances in the wastes from leaching into the ground and, thence, into the groundwater, then he would be deemed the operator, jointly with his companies, of the site itself. (citations omitted) Unfortunately the district court did not consider this possibility, although urged to do so by the plaintiffs, and so a remand is necessary to determine Ter Maat's status. (*Browning-Ferris Industries*, 956.)

In traditional common law, when two or more persons inflict an indivisible injury each is fully liable for the injury. That is, the (injured) plaintiff can, if he wants, sue one of the tortfeasors for the entire damages and let the other go, and the one who is sued has no remedy against the one who got off scot-free. CERCLA modifies the traditional common law rule (as many other statutes do and as many state courts have done by modifying the common law) by allowing one liable party to sue another for contribution. It does not follow that if, as in this case, contribution is sought from more than one party, the defendants cannot be held jointly liable. It is up to the district judge, guided only by equitable considerations — a broad and loose standard, to decide, and it is easy to imagine cases, of which this may be one, where such considerations weigh heavily in favor of joint liability.

The next issue is whether the district judge allocated too large a share (40 percent) of responsibility for the cost of the clean up to Browning-Ferris relative to the defendants, who had operated the landfill for a lot longer time and had dumped a much larger quantity of wastes in it. The judge allocated as large a share as he did to Browning-Ferris because he found that it had operated the landfill poorly and had dumped particularly toxic wastes from a nearby Chrysler plant in violation of its operating permit, and the liquid character of the wastes had hastened their absorption into groundwater. Browning-Ferris argues both that these findings are erroneous and that, in any event, there is no evidence that the wastes from the Chrysler plant increased the cost of cleaning up the site and anyway the amount dumped in the landfill was not as great as the district judge found. From evidence that a considerable portion of the Chrysler wastes were dumped elsewhere, Browning-Ferris argues that defendants' **expert** had exaggerated the amount deposited in the landfill. Browning-Ferris may be correct on all these factual points, but we cannot say that the district court committed any clear errors in finding as it did, and that of course is our criterion.

The trickier question, which returns us to the issue of the district court's equitable discretion in allocating liability among polluters, is whether the court must find a causal relation between a party's pollution and the actual cost of cleaning up the site. To answer this question, we have to distinguish between a necessary condition (or "but-for cause") and a sufficient condition. If event A is a necessary condition of event B, this means that, without A, B will not occur. If A is a sufficient condition of B, this means that, if A occurs, B will occur. If A is that the murder weapon was loaded and B is the murder, then A is a necessary condition. If A is shooting a person through the heart and B is the death of the shooting victim, then A is a sufficient condition of B but not a necessary condition, because a wound to another part of the victim's body might have been fatal as well.

This distinction may sometimes be important in the pollution context. It is easy to imagine a case in which, had X not polluted a site, no clean-up costs would have been incurred; X's pollution would be a necessary condition of those costs and it would be natural to think that he should pay at least a part of them. But suppose that even if X had not polluted the site, it would have to be cleaned up — and at the same cost — because of the amount of pollution by Y. (That would be a case, perhaps rare, in which the clean-up costs were sensitive neither to the amount of pollution nor to any synergistic interaction between the different pollutants.) Then X's pollution would not be a necessary condition of the clean up, or of any of the costs incurred in the clean up. But that should not necessarily let X off the hook. For, suppose that if X had not polluted the site at all, there still would have been enough pollution from Y to require a clean up, but that if Y had not polluted the site X's pollution would have been sufficient to require the clean up. In that case, the conduct of X and the conduct of Y would each be a sufficient but not a necessary condition of the clean up, and it would be entirely arbitrary to let either (or, even worse, both) off the hook on this basis. So far as appears, this is such a case; Browning-Ferris's pollution was serious enough (if indeed it dumped a large quantity of Chrysler's particularly toxic wastes) to require that the site be cleaned up, but the other pollution at the site was also enough. If Browning-Ferris's conduct was thus a sufficient, though not a necessary condition of the clean up, it is not inequitable to make it contribute substantially to the cost. (*Id.*, 958)

We do not suggest that this is one of the presumably rare cases in which the total costs of clean up are unaffected by the number of polluters or the specific amounts or types of pollution contributed by each. Browning-Ferris's pollution was not, so far as it appears, so serious all by itself as to have required the incurring of all the clean-up costs that were incurred. Even so, no principle of law, logic, or common sense required the court to allocate those total costs among the polluters on the basis of the volume of wastes alone. Not only do wastes differ in their toxicity, harm to the environment, and costs of cleaning up, and so relative volume is not a reliable guide to the marginal costs imposed by each polluter; but polluters differ in the blameworthiness of the decisions or omissions that led to the pollution, and blameworthiness is relevant to an equitable allocation of joint costs. (Presumably it would not entitle the judge to make one polluter pay for separable costs wholly imposed by other polluters.) The district judge did not abuse his discretion in deciding that all these factors warranted making Browning-Ferris bear more than its proportional volumetric share of the pollution.

It remains to consider two issues of derivative liability. If Ter Maat is deemed on remand to be personally liable as an operator of the landfill, and especially, if the judge allocates 100 percent of the joint liability of M.I.G. and AAA to AAA, the solvent one of the pair (there is no suggestion that the change of ownership gets AAA off the liability hook), then Browning-Ferris and the other plaintiffs will be able to collect the amount of contribution to which they are entitled from the defendants. But these are big "ifs" (how big is for the district judge to decide, in the first instance, on remand). The judge may find that Ter Maat was not an operator and that AAA should not bear the entirety of its joint liability with M.I.G. In that event, it will become important whether Ter Maat and AAA are derivatively liable for the conduct of M.I.G., and as the issue is fully briefed we shall resolve it now and hope to head off a further appeal.

The argument is that the corporate veil should be pierced, and Ter Maat, as shareholder of M.I.G. and AAA, and AAA as an affiliate of M.I.G. (both dominated by Ter Maat, and AAA in addition a 30 percent shareholder of M.I.G.), should be held liable for M.I.G.'s debt to the plaintiffs. Although there is a split of authority on whether federal or state law governs the piercing of the corporate veil in CERCLA cases (the Supreme Court expressly left the question open in *United States v. Bestfoods*, *supra*, 524 U.S. at 63, fn. 9), that is not a problem here because the parties have assumed that state law (Illinois state law, specifically) governs and so any contention that federal law should instead govern has been waived.

The general rule, of course, in Illinois as elsewhere, is that a shareholder *qua* shareholder, and a parent, subsidiary, or other affiliate, *qua* affiliate, is not liable for a corporation's debts. (citations omitted) That is the principle of limited liability and it serves the important social purpose of encouraging investment by individuals who are risk averse and, therefore, will not invest (or will insist on a much higher return) in an enterprise if by doing so they expose their entire wealth to the hazards of litigation. But in some circumstances, the corporate veil is pierced and the corpora-tion's debtor allowed to collect his debt from the shareholder or affiliate. (*Id.*, 959)

...Analysis is more difficult in the case of an **involuntary creditor, such as the plaintiffs here,** (emphasis added) who wish to be compensated for having, in effect, "lent" the money to clean up the site of the former landfill but lent under compul-sion, having been forced to clean it up by the Superfund law rather than pursuant to a contract with the "debtors," such as M.I.G. (citations omitted) In such a case, there is no issue of protecting reliance induced by misrepresentations by the debtor. The plaintiffs in dumping toxic wastes to the landfill obviously were not relying on M.I.G.'s appearing to have greater assets than it actually had. In these circum-stances we can think of only two arguments for piercing the corporate veil. The first is that the owners may have so far neglected the legal requirements (require-ments not intended solely for the protection of creditors) for operating in the corporate form that they should be taken to have forfeited its protections, forfeiture thus operating to enforce the legal requirements that the state has seen fit to impose on investors who want the benefits of limited liability. Moreover, if the formalities have been flouted, it becomes hard to see how the investors could reasonably have relied on the protections of limited liability; they would have known they were skating on thin ice. In such a case the investment-encouraging function of limited liability is attenuated.

Second, it could be argued that enterprises engaged in potentially hazardous activ-ities should be prevented from externalizing the costs of those activities, by being required to maintain or at least endeavor to maintain a sufficient capital cushion to be answerable in a tort suit should its activities cause harm for which liability would attach, on pain of its shareholders' and affiliates' losing their limited liability should the corporation fail to do this. This argument has not carried the day in any jurisdiction that we are aware of, presumably because of the risks that it would impose on shareholders and because the potential victims of the corporation's hazardous activities can be protected without making inroads into limited liability by requiring enterprises engaged in such activities to post a bond large enough to assure that any judgement against the corporation will be collectible. Courts do, it is true, frequently mention "undercapitalization" as a separate ground from neglect of corporate formalities for piercing the corporate veil. (citations omitted) They do not do so on the basis of unusual risks to potential tort victims or other involuntary creditors, however, though conceivably such concerns are in the background of their thinking.(*Id.*, 760)

There is no evidence that either M.I.G. or AAA failed to comply with the legal requirements for operating in the corporate form, except with regard to keeping minutes of their corporate meetings, not a failure significant enough to warrant forfeiture of limited liability, given that penalties should be proportioned to the gravity of the misconduct being penalized. The plaintiffs further argue, however, that M.I.G. was undercapitalized. The clearest case — here merging, though, with neglect of corporate formalities — for forfeiture of limited liability on this ground is where the corporation has failed to maintain the minimum capitalization required by law. But such cases are few (citations omitted), and do not bear on the "pierceability" of the corporate veil of M.I.G., which is not accused of ever having maintained a lower capitalization than the law required. The fact that its motivation in operating with little capital was to reduce taxes does not argue for piercing the corporate veil, since taking advantage of lawful opportunities for avoiding taxes is not wrongful activity.... Undercapitalization is rarely, if ever, the sole factor in a decision to pierce the corporate veil, and we think is best regarded simply as a factor helpful in identifying a corporation as a pure shell, which M.I.G. was not. (*Id.*, 961)

So on remand, to summarize, there will be no issues of veil piercing to consider but the court will have to decide Ter Maat's personal liability and whether AAA should pay more than 50 percent of the two corporations' liability to the plaintiffs.

The court affirmed the lower court's judgement in part, reversed in part, and remanded the case for judgement in keeping with this Opinion.

Note: In 1999, 52 new sites were added to the NPL.

6.4.2   Additional Recent CERCLA Cases for Reference and Study

U.S. v. Bestfoods, 118 S. Ct. 1876 (1998). Re: guidelines for establishing liability of parent companies for cleanup actions under CERCLA.

U.S. v. Vertac Chemical Corp., 33 F. Supp. 2d 769 (E.D. Ark. 1998). Re: retroactive application of CERCLA affirmed.

Dent v. Beazer Materials and Services, Inc., 156 F.3d 523 (4th Cir. 1998). Re: no liability for property owners where hazardous substances had reached their property by subsurface migration from a Superfund site, and where the quantities were too low to require remediation.

Franklin County Convention Facilities Authority v. American Premier Underwriters, Inc., 61 F. Supp. 2d 740 (S.D. Ohio 1999). In this action, Franklin County Convention Facilities Authority (CFA) sought to hold defendant, American Premier Underwriters, Inc. (APU) liable for costs relating to the cleanup following an alleged release of hazardous substances on property now owned by the CFA. The parties stipulated that APU is the successor of several railroads that, between 1864 and 1973, owned property on which a large wooden box was buried. In 1990, the box was uncovered and split open by the CFA in the course of construction activity. Consequently, a portion of the contents of the box was released into the environment. CFA claimed that APU was liable under CERCLA, as amended for the clean-up costs that it has incurred. In addition to denying liability under the statute, APU has also asserted that imposition of CERCLA liability for preenactment conduct is unconstitutional. Specifically, APU contends that application of CERCLA under such circumstances would violate APU's substantive due process rights under the Fifth Amendment.

In *NutraSweet Co. and Monsanto Co. v. X-L Engineering Co., et al.*, 227 F.3d 776 (7th Cir. 2000), NutraSweet/Monsanto sued X-L Engineering and its president and principal shareholder, Paul Prikos, for improperly disposing of hazardous compounds that contaminated NutraSweet's property. The district court entered partial summary judgment in favor of NutraSweet, finding X-L to be at least partly responsible for the hazardous waste on NutraSweet's property. After a bench trial, the district court found that X-L was, in fact, 100% liable for these wastes and awarded NutraSweet the full amount of its requested damages. On appeal, X-L raised numerous issues concerning the proceedings below. The appeals court affirmed the decision of the district court for NutraSweet in all respects.

Blaisland, Bouck & Lee, Inc. v. City of North Miami, 96 F. Supp. 2d 1375 (S.D. Fla. 2000), concerns a CERCLA dispute between a municipality and an environmental engineering firm. The dispute centered around a municipal landfill owned by the City of North Miami, known as Munisport. Munisport operated from 1974 to 1980. In 1983, the EPA placed Munisport on the NPL of uncontrolled hazardous release sites due to the decomposition of waste into ammonia. The ammonia was leaching into the underlying groundwater and then migrating toward a mangrove preserve. In 1990, the EPA issued its Record of Decision (ROD) pursuant to which it promulgated a comprehensive cleanup plan for the landfill.

By written contract dated July 14, 1992, the city hired BB&L to assist in implementing the cleanup plan for Munisport. BB&L was primarily responsible for conducting hydrogeologic studies of the aquifer and using those studies to design a leachate collection system. This work was to be performed for the fixed price of $1.4 million.

Between July 1992 and June 1995, BB&L performed a variety of work at Munisport, including aquifer pump tests, computer modeling, and surface and groundwater sampling. In addition, BB&L breached a causeway (permitting tidal flow into a mangrove area), realigned a dike, and negotiated with the state concerning the city's landfill closure permit application. One pump test (No. 3) BB&L conducted had particular significance. That test yielded a 3 gallon-per-minute extraction rate. The extraction rate was substantially lower than both BB&L's expectations and the rates from other pump tests. Despite this seemingly anomalistic result, BB&L insisted on using the 3 gallon-per-minute rate as the basis for its design of the leachate collection system. The 3 gallon-per-minute rate was inaccurate and resulted from BB&L's failure to comply with procedures and standards governing environmental engineers.

North Miami brought a CERCLA claim against BB& L as an arranger or operator at the Munisport site. However, the court held that BB&L was not a PRP under CERCLA.

Note: **Review of this full case is recommended for engineering and geoscience students.**

Geraghty and Miller, Inc., v. Conoco Inc. and Condea Vista Chemical Co., 234 F.3d 917 (5th Cir. 2000), concerns a district court summary judgment decision in favor of G&M, an environmental engineering firm, in an environmental cleanup action. Conoco/Vista appealed. The case included a claim under CERCLA as well as state common law claims.

This dispute arose out of an environmental cleanup at the Lake Charles Chemical Complex in Westlake, Louisiana. The complex has been owned and operated by Conoco or Vista since 1961, and in 1968 Conoco began managing ethylene dichloride at the facility. Ethylene dichloride, a feedstock in the production of vinyl chloride monomer, is a *hazardous substance* as CERCLA defines that term. As a result of historic releases and migration, ethylene dichloride contamination occurred in soil from the surface to at least 25 feet down and in shallow groundwater zones.

The Louisiana Department of Environmental Quality required Conoco to investigate and address the ethylene dichloride contamination under state groundwater protection laws and regulations, and federal and state solid waste laws and regulations. It also required Conoco to put in place a groundwater monitoring and assessment program pursuant to RCRA and its corresponding regulations and their Louisiana state counterparts. As is often the case in such cleanups, the process was set to take place in stages. Conoco and G&M entered into a contract on March 12, 1985 under which G&M was to furnish all required services for groundwater quality assessment, phase 2, at the vinyl chloride monomer plant area and wastewater treatment area (the plant) of the complex.

Under the contract, G&M was to assess possible contamination beneath several suspected source areas at the complex. It was to perform that assessment by: (1) preparing design specifications for the installation of groundwater monitor wells and piezometers used to monitor possible groundwater contamination at the complex; (2) installing the monitor wells and piezometers; and (3) sampling the monitor wells following installation. G&M completed the installation of 50 monitor wells on July 23, 1985.

Sometime before May 1988, Conoco/Vista began to suspect potentially serious technical and physical deficiencies in three of the monitor wells G&M had installed. They were concerned that such deficiencies were aggravating the contamination. Conoco/Vista received approval from the Louisiana Department in May 1988 to plug and abandon those wells. Conoco/Vista alleged that they uncovered physical evidence that the three wells were not installed according to the contract specifications, and they sent a series of letters to G&M concerning the deficiencies. In December 1989, Conoco/Vista plugged and abandoned a fourth well, and they alleged that this well also was not installed according to specifications.

These experiences caused Conoco/Vista to question the soundness of the remaining wells and other parts of the groundwater monitoring system. Conoco/Vista and G&M discussed who was responsible for the costs associated with these allegedly defective wells, but they were unable to resolve the issue. On August 27, 1990, the parties entered into the Groundwater Wells Interim Agreement. The interim agreement called for the parties to agree on criteria for determining whether a given monitor well was suspect or not suspect of being improperly installed, and to agree on criteria for determining whether a given well was properly or not properly installed. Once the criteria were in place, the parties would apply them to each of the wells to determine which needed to be removed and who would bear the costs. That never occurred, however, because the parties never agreed on the criteria.

The parties entered into the interim agreement to allocate responsibility between them for the costs of plugging and abandoning additional wells that they were to agree on as being suspect. G&M maintained that there was no other purpose for entering into the deal, but Conoco/Vista insisted that the interim agreement gave the parties time to investigate the integrity of the wells while not allowing the statute of limitations to run on any defect claims that remained unresolved. Conoco/Vista asserted that G&M received, as consideration for the deal, a release from approximately $250,000 in monitor well plugging and abandonment costs (*G&M v. Conoco*, p. 921).

Conoco/Vista retained other environmental consulting firms to continue the groundwater assessment program and on the recommendation of one such consultant, Conoco/Vista plugged and abandoned the remaining G&M-installed wells in 1993 and replaced them (*Id.*, 922).

The issues on appeal include (1) **Conoco/Vista's state common law claims were barred by Texas statutes of limitations;** (2) G&M was not liable for contribution because it was not a covered person under CERCLA as an operator, arranger, or transporter of hazardous materials and amenable to a contribution claim; (3) the 6-year limitation period of 42 U.S.C. § 9613(g)(2) (1994) barred Conoco/Vista's CERCLA claim for contribution; and (4) the district court entered summary judgment without giving Conoco/Vista notice and an opportunity to respond.

In answering the issues, the court reversed the district court's judgment and order with respect to its conclusions that Conoco/Vista's CERCLA counterclaim is barred by the statute of limitations; that G&M cannot be found liable in a CERCLA contribution claim as an operator or an arranger; and that Conoco/Vista's state common law claims for breach of contract and fraud are time-barred. We affirm the district court's judgment and order insofar as it dismisses the CERCLA claim seeking to hold G&M liable as a transporter and Conoco/Vista's counterclaim for breach of warranty and negligence. The suit was remanded to the district court for further proceedings consistent with its opinion.

Note: **Review of this full case is recommended for engineering and geoscience students.**

CHAPTER 7

Water Pollution by Abandoned Mine Sites; Acid Mine Drainage; Mined Land Reclamation

7.1 INTRODUCTION

Surface and groundwater polluting discharges originate from two source types and are classified as: (1) identifiable point sources, and (2) nonpoint sources (NPS). The latter source produces diffused wastes into waters by land runoff, washout from the atmosphere, or other means.

Examples of point sources for polluting waters are municipal sewer systems, storm water runoff, industrial wastes, and animal wastes from commercial feedlots. Examples of NPS are natural drainage sediments, agricultural chemicals, mine drainage, and spills of hazardous substances. NPS pollution does not result from a discharge at a single, specific location, but results from land runoff, precipitation, snowmelt, atmospheric deposition, drainage, or seepage.

In this chapter, the subject matter of concern is the nonpoint water pollution source of natural acid mine drainage (AMD) from abandoned and active mine sites. Natural drainage from abandoned and active mines pours harmful acids, minerals, metals, and sediments into streams and lakes and is referred to as AMD. Moisture seepage from mine waste rock dumps and mill tailings sediments also pass AMD into the groundwaters.

With the advent of active and militant environmentalism, the passage of the National Environmental Policy Act (NEPA) proved the nation's determination to clean up the environment, particularly focusing on obtaining clean water in the navigable waters of the United States. In analyzing the numerous contributing NPS, abandoned mine sites were found to be hazardous sources of water pollution. The modern statutory remedy at law is found in the Comprehensive Environmental Response, Compensation and Liability Act (CERCLA, a/k/a Superfund) and the Superfund Amendments and Reauthorization Act of 1986 (SARA).

The number of hardrock abandoned mines sites in the United States that contribute to this NPS has been a focal point of environmental argument and contention for two decades in attempting to establish the blame on the mining industry that thousands of abandoned mine sites were hazardous sources of water pollution degrading the quality of streams and rivers. Reported counts and estimates of unreclaimed, abandoned, hardrock mine sites are unsubstantiated and unreliable. The counts range from a fantastic 550,000 or 560,000 abandoned mine sites on public and private lands as estimated by the environmentalist group, the Mineral Policy Center (MPC) of Washington, D.C., to 2500 hardrock mine sites on public lands as reported by the General Accounting Office (GAO), the investigative arm of Congress.

The number of abandoned mines in the United States remains a mystery, according to the GAO. The GAO said, "The problem is that the sites are defined and counted differently in each survey and that the results from different groups cannot be compared ... At the same time, estimated clean-up costs also run along a broad gamut, ranging from the former U.S. Bureau of Mines' estimates between $4 B and $35.3 B to the Mineral Policy Center's projections between $33 B and $72 B."

The GAO's report noted that the types of hazards posed by abandoned mines tend to vary along with the count (*Eng. Min. J.*, p. 16-II).

A more realistic comment on the MPC's fantastic figures was made in an analytical note appearing in the *Mining Journal (London)*.

Of the 550,000 estimated abandoned sites, only 50 are on the Superfund hazardous waste list with an estimated clean-up cost of $12.5 million to $17.5 million. A further 500 sites with groundwater contamination could cost $2.5 million to $7.5 million whilst another 14,000 sites with surface water contamination would cost between $14 million and $43 million. The report identifies some 230,000 sites where the landscape has been disturbed and 195,000 where little, if any, remedial measures would be required (*Min. J.*, 1993, p. 51).

A later, analytical and cogent comment of note made in 1999 in the book *Surface Mining Law and Reclamation by Landfilling* is appropriate here concerning the MPC's alarming abandoned mine site figures.

Clarification is required for the initially sounding and alarming "cleaning up of 550,000 abandoned mine sites" complained of by the Mineral Policy Center (MPC) in the preceding article. As further reading of the article clarifies, the alarming figure can be immediately reduced by some 325,000 sites leaving a far lesser number to be concerned over. According to MPC's own published figures, about 195,000 sites are "reclaimed and/or benign" and require "little, if any, remedial measures." Also, MPC states 231,900 sites have only landscape disturbance, thus needing only cosmetics to beautify the land. (Then why does MPC include these large numbers of benign sites? to inflame the public?) Out of the 557,650 sites complained of by MPC, the truth of the matter is that the number of sites to be environmentally concerned about should be reduced by 76%, leaving only 24%, or 131,750 sites of any potential environmental harm. Of the total, it should be noted and emphasised that only 50 of the sites (0.0089%) are on the Superfund clean-up list because of environmentally hazardous conditions (Aston, 1999).

Selected for this study are four abandoned hardrock mine sites discharging AMD, two of which are on the National Priority List (NPL) and still being treated by CERCLA/Superfund, viz., Iron Mountain and Summitville. The third site, Idarado, came under the CERCLA statute, but is distinguishable in that it was treated as a state CERCLA project instead of a federal one. The fourth site, the Penn Mine, is also a variation of a hardrock site polluting national waters with hazardous AMD. The Penn Mine site varies from the others in that the site was taken over and title acquired to the contaminated property by a state agency. Under CERCLA an exception was thought to have been created by virtue of its operation by a sovereign. However, the assumed exception to CERCLA liability was found not to apply as the sovereign continued to contribute to the release of the hazardous substances from the facility and, therefore, liable.

7.2 STATUTORY TREATMENT OF WATER POLLUTION BY ACID MINE DRAINAGE

7.2.1 Governmental Remedies — CERCLA, Superfund, SARA, and Clean Water Act

Abandoned mine sites emitting AMD are treated under CERCLA (a/k/a the Superfund law), which was enacted in 1980 and revised by SARA. Its principal purpose is the cleanup of all leaking hazardous waste disposal sites, mineral or other. It is the primary statutory tool for the enforcement of cleaning up abandoned mine sites contributing AMD contamination to the streams and rivers of the United States and for fixing the liability and costs of the cleanup.

CERCLA was structured to accomplish four goals, viz., (1) to collect data to locate and characterize hazardous waste sites, and to analyze that data to enable federal and state governments to determine priorities for responsive actions; (2) it authorizes under § 104, the president, by means of empowered executive agencies, to establish a hazard ranking system for responsive actions employed in the removal or remedial action necessary for treating the releases of hazardous substances into the environment, and to remedy or remove the hazardous sites by actions consistent with the National Contingency Plan (NCP) in § 105; (3) it created a Hazardous Substances Trust Fund to finance the removal of the hazardous material or remedial actions necessary for the sites on the NPL; and (4) it authorizes the finding of potentially responsible parties (PRPs) who are responsible not only for releasing hazardous substances but also for fixing the liability of the costs of the cleanup of the hazardous site and/or restitutional costs. Section 106 authorizes the president to issue orders directing the PRPs to take immediate responsive action, or alternatively for injunctive relief by the attorney general where actual or threatened releases pose "imminent or substantial endangerment" to the public's health or to the environment.

Further detailed, statutory features of CERCLA will be noted in the litigated case reviews of the four hazardous abandoned mine sites selected for this study.

A fifth AMD, non-CERCLA case review follows, *Johansen, et al. v. Combustion Engineering Inc.*, 170 F.3d 1320 (11th Cir. 1999) (see Chapter 6, Section 6.4). The *Johansen, et al.* claim for damages to their lands from AMD was made under the common law and, therefore, is not a CERCLA case.

7.3 EXAMPLES OF AMD SOURCE POINT ABANDONED HARDROCK MINE AND MILL SITES LITIGATED UNDER CERCLA, SARA, AND CWA

The following four hazardous, abandoned hardrock mines sites, two of which have been placed on the NPL and selected for this study, are

1. The NPL Iron Mountain Mine Site, Redding, California
2. The NPL Summitville Mine, Summitville Consolidated Mining Company and Galactic Resources, Inc., Summitville, Rio Grande County, Colorado
3. The Idarado Mine, San Miguel, Ouray, and San Juan Counties, Colorado
4. The Penn Zinc and Copper Mine, Calaveras County, California

7.3.1   NPL Iron Mountain Mine Site, Redding, California

7.3.1.1   Introduction — General Description

The Iron Mountain mine site is located 9 miles northwest of Redding, Shasta County, California. (See Figure 7.1, the location map of the Iron Mountain abandoned mine site.) The Iron Mountain mines are situated on the southeastern border of the Klamath Mountains and are actually a group of old mines in Iron Mountain.

> Gold, silver, copper, zinc, iron and pyrite were mined at various times during a one-hundred-year period beginning in the early 1860s and ending with the termination of open-pit mining in 1962. Iron Mountain was the largest producer of copper in California, and now produces some of the most acidic waters in the world. Approximately 730 tons of dissolved copper, zinc, and cadmium drain into Spring Creek every year, ultimately entering the Sacramento River below Shasta Dam and above Keswick Dam. At the confluence point of Spring Creek with the Sacramento River, the acid waters are neutralized upon mixing, the metals are precipitated, and large sediment piles have formed in Keswick Reservoir (Prokovich, 1965; Banta and Nordstrom, 1997).

The extensive mining exposed unmined sulfide deposits (of pyrite) that react with rainwater and groundwater to form AMD, a pollutant harmful to fish and drinking water sources. The Iron Mountain AMD also poses a threat to the drinking water source and supply of the residents of the Redding area who obtain their water from the Sacramento River below Keswick Dam.

The Iron Mountain Mine site was listed on the NPL in 1983. After that, the Environmental Protection Agency (EPA) was investigating, designing, and implementing responses to the environmental pollution at the site. From 1983 until 1985, EPA conducted its initial remedial investigation (RI) that characterized the Iron

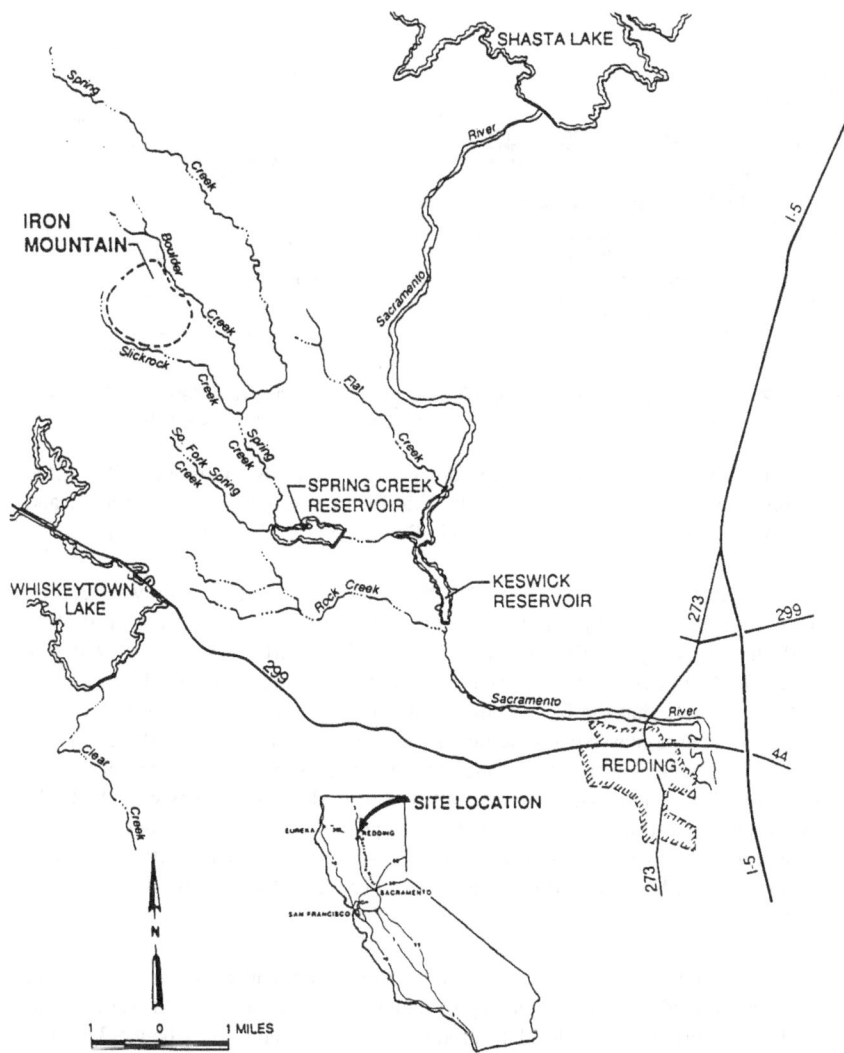

Figure 7.1 Location map of Iron Mountain abandoned mine site.

Mountain site and the pollution found there, and developed its initial feasibility study (FS) that described and evaluated possible cleanup measures.

7.3.1.2 Iron Mountain Mining History

Nordstrom has written a fine mining history of the Iron Mountain site for *The Mining Environmental Handbook*, Volume 1. The following quote from this source is found to be much more edifying to geological engineers and scientists than the scant information supplied for the courts in the CERCLA litigation of the Iron Mountain site (Figure 7.2).

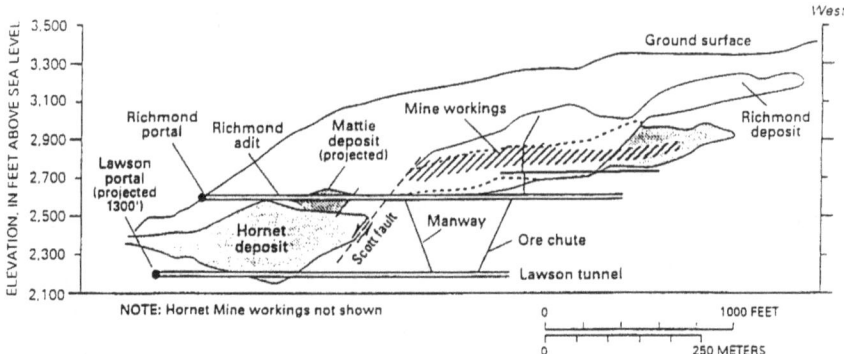

Figure 7.2   Cross section through Iron Mountain mine workings and deposit. (Section Credit
– *Mining Environmental Handbook*, Imperial College Press, U.K.)

A brief history of mining has been compiled from the review by Kett (1947), the
reports of CH2M Hill contracted to the U.S. Environmental Protection Agency and
the various publications of the California Department of Mines and Geology.

Gossan outcropping was discovered in the 1860s and Iron Mountain was secured
as an iron mine, although nothing was mined at that time. It was not until 1879
with the discovery of silver in the gossan by three partners and in 1894 it was
sold to British interests who formed the Mountain Mining Company, Ltd. in 1895.
[It is noted here that the Iron Mountain site at this time consisted of 4,400 acres
— Aston.] Large massive sulfide deposits were discovered beneath the gossan in
1895, and smelters were built at nearby Keswick to process the ore which was
transported on a narrow-gauge railroad from Iron Mountain to Keswick. In 1897,
the property was transferred to Mountain Copper Company, Ltd. of London, which
maintained the operations until 1967 when it was purchased by Stauffer Chemical
Company. Iron Mountain Mines, Inc., took over the property from Stauffer at the
end of 1976.

Copper mining ceased in 1919 due to a decrease in the market price of copper and
only very limited and intermittent copper mining took place until World War II, when
the U.S. Government subsidized the production of copper and zinc. About 5.2 million
tons of sulfide ore have been mined by underground methods from Iron Mountain.
From 1955 to 1962, 9.5 million tons of waste from the top of Iron Mountain were
removed, and 600,000 tons of pyrite were open-pit mined for sulfuric acid production.
More than 2.6 million tons of gossan were mined for gold and silver. Most of the
gossan was mined and processed by cyanide extraction from 1929–1942. Copper
cementation was also used to extract copper from the effluent mine waters. From
1962 to the present it has been the only active process for metal recovery but it has
also served as a remediation measure to decrease the discharge of copper to the
Sacramento River (Banta and Nordstrom, 1997, p. 684).

7.3.1.3 *Earlier Environmental Problems and Remediation Attempts*

The first ore reduction for obtaining copper in the Iron Mountain area was
done in the 1890s by open-air roasting of the ore heaped onto burning log piles,

which was the common method employed in that era for sulfide ores. The cutting of large amounts of timber in the local area to feed the ore roasting obviously was the first large step in totally denuding the area of vegetation. The final step in completing the denuding of the area of all vegetation followed quickly from the emissions of noxious sulfurous gases from the ore reduction roasting piles. The sulfurous gases readily combined with moisture in the air to form sulfurous acid, or, if sufficient moisture, to sulfuric acid, which fell upon the remaining grasses and vegetation thereby completing the denudation. The surface was easily prey to rapid erosion. The toxic emissions from the open air roasting of ore not only resulted in total denudation of the surface for miles around Iron Mountain but also contaminated the soils while eroding the surface. While smoke and emitted gases added metals, such as antimony, arsenic, lead, zinc, and cadmium, to the soil, sedimentation and turbidity rates in the Sacramento River were reportedly increased. By 1895, smelters were clustered at nearby Spring Creek. A flurry of lawsuits were filed against the companies roasting ores by affected private landowners and the U.S. Forest Service claiming damages to their lands from the noxious smoke, gases, and fumes, which by 1907 brought about the closure of all smelters at Spring Creek. Ore was then shipped to Martinez for smelting and refining. (For a similar situation, see the decades of litigation, from 1895 to 1918 over smoke [gaseous] damage from the open air roasters of sulfide ores in the Ducktown Copper Basin of East Tennessee.)

A further, detailed description by Nordstrom of the Iron Mountain Mine site is included in Appendix A, and includes (1) the geology and hydrology of the Iron Mountain site; (2) environmental contaminating conditions at Iron Mountain; and (3) site remediation steps prior to litigated CERCLA proceedings.

7.3.1.4 Environmental Contaminating Conditions at Iron Mountain

Conditions at Iron Mountain are optimal for the maximum production of acid mine waters from pyrite oxidation. The concentration of pyrite is nearly 100% in single large masses excavated by tunnels, manways, and stopes that allow rapid transport of gaseous oxygen advection. The massive sulfides are at or above the water table (phreatic surface) so that moisture and oxygen are always present. It is thought that the airflow is probably aided by thermally driven convective cells due to high heat generated from pyrite oxidation (Banta and Nordstrom, 1997, p. 684).

7.3.1.5 Site Remediation Steps Prior to Litigated CERCLA Proceedings

From 1939 to the present, various studies of the environmental impact of Iron Mountain have been conducted by the California Department of Fish and Game, the U.S. Fish and Wildlife Service, the Regional Water Quality Control Board (RWQCB), the U.S. Geological Survey, the U.S. Bureau of Reclamation, and the EPA. Since 1983, studies have been conducted as part of the Superfund investigations authorized by CERCLA.

Over 20 fish kill events have occurred since 1963 with at least 47,000 trout killed during 1 week in 1967 (Nordstrom and others, 1977). Continued fish kills kept the RWQCB actively pursuing remediation of the site. Following a thesis study by Nordstrom in 1977 at the site, a cleanup and abatement order was issued to the mine owner, Stauffer Chemical Company.

From 1977 to 1989, six orders were issued to reduce toxic metal discharges that were in violation of state law. The orders to cease and desist as well as for emergency treatment measures have been through both the Shasta County and the State of California courts. Stauffer Chemical Company has since become a part of Rhône-Poulenc, which then became liable for the site under CERCLA.

Because Iron Mountain was officially listed on the EPA NPL for Superfund in 1983, the first CERCLA RI/FS was submitted in 1985. The remedial investigation report (EPA, 1985b) identified the five major point sources of pollution discharges through a comprehensive surface water sampling survey. The greatest discharge source was identified as the Richmond portal effluent. EPA (1985b) also documented the occurrence of increased concentrations of copper, zinc, and cadmium from portal effluents following heavy rainstorm events and related this phenomenon to rapid flow of surface water into the mine workings through areas of subsidence. The feasibility study (EPA, 1985b) considered more than a dozen alternative treatment possibilities and estimated the costs and anticipated benefits from each individual alternative as well as several possible combinations.

The EPA had evaluated the modified mine plugging alternative as part of the second RI/FS completed in 1992 (EPA, 1992). The EPA had also considered air sealing, but favored a complete capping treatment as the most cost-effective solution in conjunction with the surface water diversions that have already been initiated.

During the second Remidial Investiation phase (1986–1992) of EPA Superfund activities, the Richmond tunnel and part of the Richmond mine workings were made accessible to underground surveys. On September 11, 1990, water and mineral samples were collected during one of these surveys, which resulted in the discovery of extremely acidic seeps with pH values as low as 3.4 and a total dissolved solids concentration of about 935 grams per liter. These acid iron sulfate waters were precipitating or efflorescing soluble iron sulfate salts, often coating tunnel walls and muck piles (Banta and Nordstrom, 1997, pp. 686–687).

7.3.1.6 CERCLA Litigation Begins — 1991–1997 — Establishing PRPs and Liability for Iron Mountain Cleanup

Over the 6-year period, from the summer of 1991 until October 1997, the plaintiff parties — the U.S. EPA and the State of California — litigated with the principal defendant parties, Iron Mountain Mines, Inc. and T. W. Arman (hereafter, IMMI/Arman), and Rhône-Poulenc Basic Chemicals Company (hereafter, Rhône) over the issues of PRPs and liability for the cleanup costs of the Iron Mountain site. The federal litigation was cited as *United States v. Iron Mountain Mines, Inc., et al.* and *State of California v. Iron Mountain Mines, Inc., et al.*, commencing at 812 F. Supp. 1528 (U.S. Dist. Ct. E.D. Cal., 1992).

In the 6 years, 1991–1997, some ten actions were filed and heard by the U.S. District Court for the Eastern District of California on the Iron Mountain case. This review for engineers and engineering students will touch on only the more important arguments and defenses submitted by the parties to illustrate the extensive powers of CERCLA. The three major trials and decisions occurred in (1) September 21, 1992; (2) March 31,1995; and (3) September 30, 1997. However, various filings by both parties were made in between the three major court decisions, of interest from a legal viewpoint but not from an engineering view and study of CERCLA.

7.3.1.6.1 Litigation and Decision of September 21, 1992

Beginning in June 1991, the United States/EPA filed its complaint against the PRPs. The State of California filed its complaint in late August of 1991. The two cases were consolidated by the court in November 1991. In August 1992, the court heard the arguments (in 812 F. Supp.1528) of the parties, with both sides moving for summary judgment (i.e., either dismissal for the defendants, or judgment for plaintiffs).

The court heard and discussed the following major points in making its analysis:

A. Has the U.S. EPA stated a claim under CERCLA?
  1. Are the releases *at* rather than *from* a facility?
  2. Are mining wastes excluded under CERCLA by the Bevill amendment?
  3. Were the releases permitted under federal law, thus relieving liability?
  4. Does mining constitute disposal under CERCLA?
  5. Is failure of government to mitigate damages an unnecessary response cost?
B. Constitutionally based defenses
  1. Deprivation of procedural due process in remedy selection
  2. Deprivation of substantive due process
  3. Equal protection as to liability and remedy selection
  4. *Ex post facto* law
  5. Taking for public use without just compensation (eminent domain power)
  6. Equitable defenses
  7. Other U.S. Bureau of Reclamation (USBR) defenses
C. Release of a naturally occurring substance
D. Is a mine a covered facility?
E. Who are covered persons?
F. Must plaintiff prove that ARARs were exceeded?
G. Prejudgment interest
H. Preenactment and prelisting response costs
I. Defenses conceded by IMMI/Arman
J. Recoupment and setoff
K. Must defendants have disposed to be liable?
L. Are California's costs response costs?
M. Propriety of joint and severable liability
N. Presence of releases warranting response

Each of the preceding defensive arguments by the defendants is discussed next to illustrate points of CERCLA's legally enforceable parameters.

A. Has the U.S. EPA Stated a Claim Under CERCLA? — Under the Federal Rules of Civil Procedure, this challenge to an adverse party's complaint is a standard challenge, invariably made by the defendant party when answering a complaint.

With the two complaints and cases being joined, on August 11, 1992, the federal district court heard the government's motions to strike (invalidate) the defenses of defendant Rhône and, also to grant defendants' motion for summary judgment for failure to state a claim on which relief can be granted. (Summary judgment means that, if it can be demonstrated by the moving party — here, U.S. EPA — that there is no genuine issue as to any material fact of the opposing party's case, then the moving party shall be entitled to summary judgment as a matter of law and the complaint dismissed.) In summary judgment practice, the moving party always bears the initial responsibility of showing the court the basis for its motion and identifying those portions of the pleadings, depositions, answers to interrogatories, or admissions on file, together with affidavits, if any, which it believes demonstrate the absence of a genuine issue of material fact. If the moving party meets its initial responsibility of showing no material or genuine fact exists as claimed by the opposing party, the burden shifts to the opposing party to establish that a genuine issue of material fact does exist; to tender evidence of fact that a dispute exists, is material, and might affect the outcome of the suit; and to show the evidence is such that a reasonable jury could return a verdict for the nonmoving party.

Here, the defendants, Rhône and IMMI moved for dismissal of the charges for the government's failure to state a claim. Defendant's motion was denied by the court. The court quoted the Eighth Federal Circuit's 1989 statement in a prior case that "most courts have held CERCLA imposes strict liability," relying on CERCLA § 107(a)(1)(A) that holds certain persons liable for "all costs of removal or remedial action incurred by the United States government, or a State," so long as those costs are "not inconsistent with the National Contingency Plan (NCP)." CERCLA § 107(a)(1 and 2) states persons liable include "the owner and operator of a vessel or facility" and "any person who at the time of deposal of any hazardous substance owned or operated any facility at which such hazardous substances were disposed of." CERCLA enumerates three exceptions for liability of hazardous waste disposal, viz., (1) act of God (a natural catastrophe, such as an earthquake), (2) act of war, (3) act or omission of a third party other than an employee or agent of the defendant. Only these, or any combination of same defenses, will establish no liability. The court added that "there is no real question that the plaintiffs have stated formally sufficient claims for relief under CERCLA," i.e., that (1) the site is a facility; (2) defendants are the current owners of the site; (3) defendant Rhône owned or operated the mine during the time in which hazardous substances were disposed of; (4) the substances disposed of are hazardous within the meaning of § 101(14); (5) releases or threatened release of hazardous substances at or from the mine have occurred; (6) the releases or threatened releases have caused the United States to incur response costs; (7) each defendant is a person liable under § 107; and (8) such persons are liable for all removal and remedial costs not inconsistent with the NCP. The court added that, "the pleading exceeds the requirements" to state a claim under CERCLA.

A. 1. Are the Alleged Releases at Rather Than from a Facility? — The question to be answered here was whether a release *at* a facility — meaning a release that never leaves the facility — is a covered (CERCLA) release. The court cited *U.S. v. Mottolo*, 695 F. Supp. 615, 623 (D. N.H. 1988), which held "that a release of hazardous wastes *at* a facility is sufficient to trigger § 107 liability for costs." In *Mottolo*, hazardous chemicals dumped onto a site surface constitute releases or threatened releases within the meaning of CERCLA. *Mottolo* also noted that there is no requirement under CERCLA that offsite pollution occur (*Id.*). The court denied Rhône's defense of the release being confined to their property was not a valid defense.

A. 2. Are Mining Wastes Excluded under CERCLA by the Bevill Amendment?

Amendment? — CERCLA defines *hazardous substance* at length; among the definitions in § 1321(b)(2)(A) of Title 33, it is defined in (C) as: "any hazardous waste having the characteristic identified under or listed pursuant to § 3001 of the Solid Waste Disposal Act [42 U.S.C § 6921] *(but not including any waste the regulation of which under the Solid Waste Disposal Act [42 U.S.C. § 690] has been suspended by Act of Congress)* (emphasis added). The court noted that the "operative language is, 'but not including waste, the regulation of which under the Solid Waste Disposal Act has been suspended by Congress'."

In 1978, EPA had proposed a classification of *special wastes* that it defined as high-volume, low-toxicity wastes, including mining wastes, and that they be subject to fewer regulations. Before the EPA's final regulations took effect, Congress enacted the Solid Waste Disposal Act Amendments of 1980, known as the Bevill amendment, which suspended RCRA regulation of "solid wastes from extraction, beneficiation, and processing of ores and minerals" until at least 6 months after EPA completed a comprehensive study of the adverse effects of these low-hazard, high-volume wastes. In 1980, Congress also enacted CERCLA, incorporating the Bevill amendment mining waste exclusion in CERCLA's definition of hazardous substances. Also, in a 1986 regulation, EPA exempted extraction and beneficiation wastes from RCRA although EPA did not immediately decide which particular wastes would be excluded under the Bevill amendment. In 1988, the District of Columbia Circuit Court held in *Environmental Defense Fund v. U.S. EPA*, 852 F.2d 1309, 1329 (D.D.C. 1988) that Congress intended the Bevill amendment exclusion to encapsulate the special waste concept articulated by EPA in 1978 (see, *supra*) meaning that all high-volume, low-toxicity wastes would be excluded from regulation under subtitle C of RCRA. The final list of excluded mining wastes that qualified was published about September 1989 and is found at 40 CFR 261.4(b)(7) (1990).

The *Iron Mountain* court stated that, "Several courts have considered the question whether mining wastes excluded from RCRA regulation by the Bevill Amendment may be regulated under CERCLA notwithstanding the exclusionary reference in § 101(14)(C). The leading case is *Eagle-Picher Indus. v. U.S EPA*, 759 F.2d 922 (D.C. Cir. 1985).... In this Court's view, (the) *Eagle-Picher* (decision) misreads the statute. First, to give effect to the exclusion in § 101(14)(C), the statute must be read to exclude all Bevill Amendment wastes from CERCLA regulation. It is an 'elementary canon of construction that a statute should be interpreted so as not to render one part inoperative.' [*Mtn. States Tel. & Tel. v. Pueblo of Santa Ana*, 472 U.S. 237, 249 (1985).] Second, placement of the exclusion within subdivision (C) is sensible in that all SWDA wastes are defined in that subdivision. Third, to effectuate the exclusion ... comports with clear legislative history.... In this Court's view, even if mining wastes are covered by CERCLA, certain wastes are excluded from coverage by reference to the Bevill Amendment in 101(14)(C).

"Having decided that Bevill Amendment wastes are excluded from coverage even though they may fall within other parts of section 101(14), the next question is whether the particular waste at issue here — AMD — is among those excluded from coverage by the amendment.... In fact, the Bevill Amendment excludes only 'special' mining wastes defined at 40 C.F.R 261, 4(b)(7). Not all mining wastes are excluded by the regulation.

"As a matter of law, Plaintiffs' motion must be denied, ... it also appears that some mining wastes are excluded, namely those found at 40 C.F.R. 261.4(b)(7). It cannot be decided on the present record that the part of Rhône's arguments pertaining to exclusion of mining wastes is insufficient."

A. 3. Were the Releases Permitted under Federal Law, Thus Relieving Liability? — Rhône's defense here requires a review of liability for releases permitted by federal law. In its analysis, the court stated

Rhône asserts that plaintiff United States' complaint also fails to state a claim to the extent that plaintiff seeks to recover response costs resulting from federally permitted releases; another similar allegation by Rhône which asserts that, because releases that may have occurred when RP's predecessors owned the mine were federally permitted, plaintiff cannot recover under section 107 of CERCLA. Plaintiff United States moves for summary judgement as to both defenses, asserting that they are legally insufficient.

Section 107(j) provides that "recovery ... for response costs ... resulting from a federally permitted release shall be pursuant to existing law in lieu of this section. 42 U.S.C § 9607(j).

One such permitted release may arise under § 402 of the Clean Water Act (CWA), which authorizes issuance of a permit to discharge pollutants under the National Pollutant Discharge Elimination System (NPDES), 33 U.S.C. § 1342. Plaintiff acknowledges that from 1978 to 1983 an NPDES permit governed discharge of metals from two copper cementation plants at Iron Mountain.

"Under CERCLA, costs of responding to a federally permitted release may not be recovered unless it is shown that non-federally permitted releases contributed to the natural injury. [See *In re Acushnet River & New Bedford Harbor*, 722 F. Supp. 893, 897 (D. Mass. 1989).] While *Acushnet* suggests that plaintiffs have the burden to prove that non-permitted releases contributed to the harm, it places on defendants the burden to prove that the injury is divisible, (*id.* at 897), so that the award of response costs may be reduced to reflect the unrecoverable portion attributable to a permitted release. Even where releases may have been permitted, response costs may be recovered for any releases that (1) were not expressly permitted; (2) exceeded the limitations of the permit; or (3) occurred at a time when there was no permit. [*Idaho v. Bunker Hill*, 635 F. Supp. 665, 673–674 (D. Idaho 1986)].

"Plaintiff insists that flows covered by the 1978–1983 permit represent only a portion of AMD flows at the site. Plaintiff submits evidence that 'a large percentage of metals released into the environment at the site come from waste rock piles, tailings piles, seeps, and sediments. These metals-bearing flows are not treated in the copper cementation plants operated at the sites.' (Decl. of R. Sugarek at P 12.) Plaintiff also submits evidence that all flows since 1983 have been unpermitted (Decl. of J. Pedri at PP 6 and 8); that permits were limited to copper loading; and that full compliance with the permits was never achieved, *id.* at pp. 6–7."

The court's analysis continued. "Rhône does not address the question whether nonpermitted releases contributed to the alleged injury; instead, Rhône argues that plaintiff has not met burden to produce evidence that the harm is divisible. Plaintiff replies that, even if the burden of production is properly plaintiff's, that burden is satisfied by its evidence that only some releases were permitted. Plaintiff is correct: there is evidence that some wastes were not treated in the permit-covered cementation plant and that post-1983 releases were not permitted is sufficient to suggest that nonpermitted releases contributed to the harm. More important, that a release may have been federally permitted does not alone prevent recovery of response costs. Recovery is prevented only where defendant proves that the injury is divisible. *Acushnet* places the burden to prove divisibility on defendant RP, who has offered no evidence to establish that the injury alleged by plaintiff may be attributed in measurable proportion to permitted releases. Rhône, thus, fails to sufficiently establish an essential element of its defense: divisibility of the injury. Accordingly, summary judgement is appropriately in plaintiff-U.S.'s favor." Rhône had stated no claim to the extent that it sought to recover for permitted releases.

A. 4. Does Mining Constitute Disposal under CERCLA? — In the fourth part of Rhône's defensive arguments, Rhône and IMMI alleged that plaintiffs failed to state a claim for relief because mining, including the creation of tailings piles, does not constitute disposal within the meaning of CERCLA. Both plaintiffs asserted that discharge of AMD and abandonment of tailings constitute disposal under CERCLA.

CERCLA defines *disposal* by reference to the Safe Water Drinking Act (SWDA) § 1004, codified at 42 U.S.C. § 6903 and 42 U.S.C. § 9601(29). Under SWDA, *disposal* means the discharge, deposit, injection, dumping, spilling, leaking, or placing of any solid waste or hazardous waste into, or on, any land or water so that such solid waste or hazardous waste or any constituent thereof may enter the environment or be emitted into the air or discharged into any waters, including groundwaters.

The court took note of witness testimony in evidence for plaintiff United States that ore extraction at the mine has fractured mineralized zones, allowing formation of AMD that is discharged through the watersheds into Spring Creek Dam and the Sacramento River (Declaration of R. Sugarek at pp. 6–8, 11). Plaintiff also submitted evidence that heavy metals may leach from tailings piles that were created during mining (*Id.*, at p. 12). Plaintiff contended that discharge of AMD and abandonment of tailings constitute disposal under CERCLA.

Opposing summary judgment, Rhône first argued that only disposal of solid or hazardous waste, instead of disposal of hazardous substances, can subject it to CERCLA liability (citing *Stevens Creek Assn. v. Barclays Bank*, 915 F.2d 1355, 1361 [9th Cir. 1990]). The *Stevens Creek* court refused to allow cost recovery for removal of asbestos-laden building materials on grounds that the materials were not solid or hazardous wastes. That decision the *Iron Mountain* court noted, "draws a fine distinction in the statutory scheme." First, it notes that liability under CERCLA depends on occurrence of "a 'release' or 'threatened release' of a hazardous substance..." (*Id.* at 1359; 42 U.S.C. § 9607[a][4]). Second, it notes that a CERCLA defendant must be a person "who at the time of disposal of any hazardous substance owned or operated any facility at which such hazardous substances were disposed of" (*Id.*, 42 U.S.C. § 9607(a)(2)). Third, it notes that the action constituting "disposal of any hazardous substance must involve 'any solid waste or hazardous waste...'" (*Id.* at 1361); 42 U.S.C. § 9601(29) and 42 U.S.C. § 6903. "Thus, on its face, 'disposal' pertains to 'solid waste or hazardous waste,' not to building materials which are neither. (*Id.*) Having established that a covered disposal must involve waste, Rhône narrows its argument, insisting that because CERCLA incorporates the RCRA definition of hazardous waste but not its definition of solid waste, a disposal of solid mining wastes is not a disposal within the meaning of CERCLA" [*Iron Mtn. v. U.S. EPA*, op. cit at p. 1539; Lexis-Nexis version, p. 11]. In fact, the *Stevens Creek* court read CERCLA to reach disposal of either solid or hazardous waste, not hazardous waste alone (915 F.2d at 1361).

The court noted here that Rhône's conclusion that disposal under CERCLA pertains only to hazardous wastes was "a leap not supported by the statute," adding that "Rhône does not otherwise dispute that 'discharge' of AMD from the mine constitutes a 'disposal' under CERCLA. For this reason alone, summary judgement must go to plaintiff on the assertion by Rhône that mining does not constitute disposal."

A. 5. Is Failure of Government to Mitigate Damages an Unnecessary Response Cost? — Plaintiff United States moved for summary judgment on Rhône's defense that alleged that it was not liable for costs that are unnecessary or inconsistent with the NCP. Rhône argued, "if it is liable for any response costs, it is liable only for costs that are necessary and not inconsistent with the NCP." Rhône also argued that plaintiff failed to mitigate its damages.

Under § 107, a person, other than the United States, a state, or an Indian tribe, is liable under CERCLA for "any other necessary costs." Because the statute provides that all costs may be recovered, federal courts have rejected the defense of failure to mitigate damages. The court ruled that CERCLA does not impose a duty upon the Government to mitigate response costs.

Opposing such a rule, Rhône cited *U.S. v. Hardage* (W.D. Okla. 1987), which refused to strike the failure-to-mitigate defense in light of potential delay and misfeasance by the government. In this instant case, Rhône, neither suggested, nor offered evidence of misfeasance or delay by the government; nor did it offer any persuasive reason to maintain an otherwise insufficient defense.

As to recovery of unnecessary costs, governments have generally been allowed to recover *all* response costs, as is permitted by the statute. (See *U.S. v. R.W. Meyer, Inc.*, 889 F.2d 1497, 1504 [6th Cir. 1989] where the government was entitled to recover indirect, or overhead costs, although costs were higher than other entities might have incurred.) Also see *U.S. v. Bell Petroleum Servs., Inc.*, 734 F. Supp. 771, 780–781 (W.D. Tex. 1990) where costs need not be shown to be reasonable and **necessary**; and *U.S. v. Hardage*, 733 F. Supp. 1424, 1432 (W.D. Okla. 1989) where "courts have emphasized that liability extends to *all* response costs" (emphasis added).

For these reasons, the court granted plaintiff United States' motion for summary judgment to the extent that Rhône's defenses were in error in denying liability for unnecessary response costs.

B. Constitutionally Based Defenses — Numerous constitutional issues were raised by Rhône, which alleged that imposition of CERCLA liability violated the procedural and substantive due process, equal protection, taking (eminent domain) and ex post facto clauses of the United States Constitution. Rhône alleged that the remedy selection process (the process of determining responses necessary to releases or potential releases of hazardous substances) violates the due process and equal protection clauses.

Due process of law (1) is defined as a course of legal proceedings according to those rules and principles that have been established in our systems of jurisprudence for the enforcement and protection of private rights; (2) implies the right of the person affected thereby to be before the tribunal that pronounces judgment on the question of life, liberty, or property, in its most comprehensive sense — to be heard, by testimony or otherwise, and to have the right of controverting, by proof, every material fact that bears on the question of right in the matter involved; (3) has the essential elements of notice and opportunity to be heard and to defend in orderly proceeding adapted to the nature of the case, and the guarantee of due process requires that every man have the protection of "his day in court" and benefit of general law; (4) is defined by Daniel Webster to mean, "a law which hears before it condemns, which proceeds on inquiry and renders judgement only after trial" (Black, 1968, p. 590).

B. 1. Deprivation of Procedural Due Process in Remedy Selection — Rhône and IMMI alleged that it was denied procedural due process in the selection of remedies to be used at Iron Mountain Mine, to examine witnesses, and to have the remedy selection issues resolved by an impartial decision maker.

The court noted that the same contentions by Rhône had been considered at the June 1992 hearing on the United States' motion to limit review of the remedy selection to the administrative record. In that record, the court had already stated that "it is satisfied that the informal remedy selection hearing process in which the administrative record is made comports with due process.... The court also rejected Rhône's argument regarding partiality of the remedy-selection decision-maker." Therefore, the court found that it had already ruled on Rhône's deprivation of procedural due process challenge.

B. 2. Deprivation of Substantive Due Process — Substantive law is defined as that part of law that creates, defines, and regulates rights, as opposed to adjective or remedial law, which proscribes method of enforcing the rights or obtaining redress for their invasion (Black, 1968, p. 1598; *Maurizi v. Western Coal & Mining Co.*, 321 Mo. 378, 11 S.W. 2d 268, 272 [Mo. 1928]).

Rhône argued in its defense that CERCLA deprived it of substantive due process in that the statute may lead to imposition of substantial monetary liability on a corporate entity that neither was responsible for, created, nor profited from releases from the Iron Mountain superfund site. Rhône further argued that its due process rights were violated in that its immediate predecessor, Stauffer Chemical Company, "had no connection with the mining activities at Iron Mountain and in no way profited from those activities." The court responded that "the burden is on the one complaining of a due process violation to establish that the legislature has acted in an arbitrary and irrational way" or that the legislative act is unconstitutional. "Rhône makes no showing. But more important, Rhône's liability is not predicated on Rhône's or Stauffer's status as the miner of land; rather, liability is premised on Rhône's status as successor to the liabilities of the former miner, Mountain Copper, Ltd."

Rhône failed to overcome the presumption of constitutionality of CERCLA; its substantive due process defenses must be rejected.

B. 3. Equal Protection as to Liability and Remedy Selection — In pleading the equal protection defense, Rhône argued its two components (as it did with the due process defenses, *supra*, i.e., (1) due process remedial law, and (2) due process substantive law. Here, Rhône pled remedial law violated its equal protection as did substantive law, asserting that imposition of CERCLA liability also violates the equal protection clause.

In analyzing Rhône's pleading under equal protection, the court countered, "Because Rhône does not assert membership in a 'suspect' class, an equal protection review is done under the rational basis test. Rhône asserts that statutory joint and several liability under CERCLA encourages selective prosecution of those who are best able to reimburse governments for response costs, i.e., who are wealthiest."

Because Rhône advanced no evidence or established no legal argument in support of its "deep pockets" theory for selective prosecution under CERCLA, the court found the Rhône defensive argument irrational. In counterevidence, the court cited *U.S. v. Conservation Chemical Co.*, 619 F. Supp. 162, 214 (8th Cir., W.D. Mo. 1985) and *U.S. v. Kramer,* 757 F. Supp. at 242 (3d Cir., N.J. 1991), which held that "CERCLA's imposition of joint and several ... does not raise a constitutional equal protection issue and, even if it did, would not violate the constitutional standard requiring equal protection of the laws."

The court held that Rhône gave no evidence that its equal protection rights had been violated, and accordingly, that defense was denied and plaintiffs' motion to strike was granted.

B. 4. Ex Post Facto Law — An *ex post facto* law is defined as "any law that makes an action, done before the passing of the law, and which was innocent when done, criminal, and punishes such action." Also, a law which, assuming to regulate civil rights and remedies only, in effect imposes a penalty which, when done was lawful; and, any statute that changes the prescribed punishment which may be imposed for a crime theretofore committed is *ex post facto* only if it prescribes or permits imposition of a greater sentence (Black, 1968, p. 662).

As a further constitutional defense, Rhône alleged that CERCLA constitutes an *ex post facto* law in violation of Article 1, Sections 9(3) and 10(1) of the United States Constitution. The *Iron Mountain* court declared that the *ex post facto* issue had been rejected by all (federal) courts entertaining such claims that CERCLA is an *ex post facto* law (citing *U.S. v. Kramer*, 757 F. Supp. at 431 [3d Cir., D. N.J. 1991]), that it is not a penal law in nature. The court cited *Louis Vuitton SA v. Spencer Handbags Corp.*, 765 F.2d 966 (2d Cir. 1985), which noted that the *ex post facto* "generally applies to criminal statutes [but] may also be applied [to civil law] where the civil disabilities [are] disguise(d) [as] criminal penalties" (*Id.* at 972). However, Rhône argued that the *ex post facto* clause had been applied analogically in civil cases.

For its decision, the court stated, "But Rhône gives no reason for (an) analogical application (to civil cases, viz. CERCLA) of the *ex post facto* clause to this case.... Here it can hardly be said that CERCLA's liability scheme, which limits liability to response costs incurred by the government, provides for imposition of criminal penalties."

For the preceding reasons, the court rejected Rhône's *ex post facto* challenge and granted plaintiffs on that part assailng CERCLA as an *ex post facto* law.

B. 5. Taking for Public Use without Just Compensation (Eminent Domain Power) — The final constitutional challenge pled by Rhône was that CERCLA violates the "taking clause" (the power of eminent domain) under the Fifth Amendment, which states that private property shall not be taken for public use without payment of just compensation.

It should be noted here that the federal courts have held that to constitute a "taking" of private property under the Fifth Amendment, the actual confiscation of private property does not have to occur. It has been held by the courts that the denial by the government of all economic use and value of a private property is sufficient to constitute a "taking."

The court rejected Rhône's defense, quoting several prior decisions in *U.S. v. Northeastern Pharmacetical & Chemical Co. (NEPACCO)*, 810 F.2d 726, 734 (8th Cir. 1986): "government's cleanup does not deprive property owner of any property interest"; and *Conservation Chemical, supra*, 619 F. Supp. 215–217: "What defendants have referred to as a 'taking' is, in reality, nothing more than an attempt to transform a substantive due process challenge of an economic regulation into a confiscation of defendant's property rights." The court rejected Rhône's challenge as lacking support or cases to support its conclusion.

B. 6. Equitable Defenses — Equitable defenses refer to jurisprudence rooted in equity (sometimes called natural justice) which is a separate body of law or "jurisprudence, differing in its [English] origin, theory, and methods from the common law." It is a body of rules or maxims existing by the side of the old civil law, founded

on distinct principles claiming to supersede the civil law in virtue inherent in its principles (Black, 1968, p. 634).

Courts of equity, called Chancery Courts, still exist in England and the United States. Many courts in the United States, including the federal system, practice equity along with the common law. In its broadest sense, equity denotes the spirit of fairness, justice, honesty, morality, and man's innate sense of right and fairplay, i.e., observance of the golden rule of doing to all others as we desire them to do to us (Black, 1968, p. 634).

Rhône put forth the equitable defenses of "estoppel" and "unclean hands," that is, the government is estopped to recover expenses by former actions by the U.S. Bureau of Reclamation (USBR), and coming to court with "unclean hands" because of some prior improper act.

The court noted the plaintiff-government's argument that "equitable defenses are precluded by CERCLA's limitation in § 107(a) to defenses which specifies that only three defenses are listed in § 107(b), viz., (1) an act of God; (2) an act of war; and (3) an act or omission of a third party other than an employee or agent of the defendant (*Id.*). Because equitable defenses, such as "waiver, estoppel, and unclean hands are not listed as § 107(b) defenses, they are not available to defeat CERCLA."

In response to the government's argument, the court stated, "Courts are divided on the question of whether equitable defenses are available in § 107 cases. A number of cases have allowed equitable defenses reasoning that because the government in a cost recovery case seeks the equitable remedy of restitution, defendants should not be barred from raising equitable defenses. (See *Conservation Chemical*, *supra*, 619 F. Supp. at 206.)

Other decisions have held that CERCLA liability under § 107(a) is "subject only to the defenses set forth at § 107(b), is a valid restriction." (See *Kramer*, *supra*, 757 F. Supp. at 425–428, striking defenses of unclean hands, estoppel, waiver, laches, and *in pari delicto* [in equal fault; parties at equal fault or culpability].)

In its opinion, the court found the cases precluding equitable defenses more persuasive. Additionally, the court held that "the defenses of unclean hands, waiver, and estoppel may not be asserted against sovereigns who act to protect the public welfare" (quoting *Delaware Canal Co. v. U.S.*, 250 U.S 123, 125 (1919), et al.). Judgment was given for plaintiff U.S.

B. 7. Other U.S. Bureau of Reclamation (USBR) Defenses — Rhône alleged a defense that the government's dam-building activities caused the injury and thereby assumed the risk of the injury for which response recovery costs are now sought. IMMI/Arman also claimed that USBR caused the plaintiff's injuries.

CERCLA liability is "joint and several," and the government too may be held liable under § 120(a) which states: "… each department, agency, and instrumentality of the United States is subject to CERCLA in the same manner, and to the same extent, both procedurally and substantively, as any nongovernmental entity, including liability…." Plaintiff argued that there was no error or irrationality in finding defendants Rhône and IMMI/Arman jointly and severally liable for damage that may have been done by USBR since § 113 permits the defendants to seek contribution from other responsible parties during or after the § 107(a) action at bar. Plaintiff continued its argument that, consequently, no constitutional error or harm is done by holding the defendants liable for damage that may have been done by USBR where Rhône's own liability for injury is established. If Rhône is found here to be liable under § 107, then it is liable for all the damage even though it may not have caused the damage.

In its analysis, the court noted that all defendants had already pled the third party defense, that is, "(3) an act or omission of a third party." "To the extent that a release, if caused by USBR, was not caused solely by USBR, defendants are protected by § 113(f) which permits equitable contribution." The court ruled that the government motion for summary judgment on this defense was denied on grounds that "such a holding would be premature until the question of USBR's liability for damages is determined."

C. Release of a Naturally Occurring Substance

C. Release of a Naturally Occurring Substance — Section 104(a)(3)(A) of CERCLA states, "The President shall not provide for a removal or remedial action … in response to a release or threat of release … of a naturally occurring substance in its unaltered form, or altered solely through naturally occurring processes or phenomena, from a location where it is naturally found" [42 U.S.C. § 9604(a)(3)].

Thus, Rhône argued that it is not liable for releases of naturally occurring substances, and that EPA is not authorized to respond to such naturally occurring releases. The government countered that this was not the case at Iron Mountain; that the AMD flowing from the mine consists of naturally occurring substances, but the AMD flow did not occur naturally, but instead was created by mining (Declaration of R. Sugarek at pp. 6–10*).

The court stated that Rhône was mistaken; the statute permits response to release of any natural substance released in altered form, or to releases of natural substances in altered or unaltered, natural form. The court also noted that although Rhône presented evidence that AMD was discharged from the property before mining occurred, it presented no evidence to support the proposition that the former, naturally occurring AMD drainage equaled or exceeded the present AMD flow from mining; defendants' evidence did not show that natural flows of AMD, if any, occurred in measurable amounts.

The court ruled against Rhône's argument for this defense saying that no reasonable jury could find in favor of the defendants.

D. Is a Mine a Covered Facility?

D. Is a Mine a Covered Facility? — Rhône and IMMI/Arman alleged that a mine is not a covered facility under CERCLA arguing that "because mines were extensively regulated when CERCLA was enacted, it was no accident that Congress did not specifically identify 'mines' in the statute (§ 101(9)(A)."

The court responded that defendants' argument is not supported by law. "While it is correct that Congress did not specifically identify mines in the provision, … (it) did not specifically identify factories, plants, impoundments, laboratories, dumps,

---

* As an aside comment here, relative to Chapter 8 about geoscientific and engineering expert witnessing and admissible evidence, is that the defendants complained that plaintiffs' expert witness, R. Sugarek, was an incompetent witness; still, defendants failed to move the court to strike his testimony as inadmissible.

quarries, ... *etc.*, ... where a hazardous substance has been deposited.... There is no real question that the broad language would reach the parts and the whole of Iron Mountain."

The court denied Rhône's defense and found for the government on this issue.

Other defenses, E through N — With regard to the remaining defenses of the defendants, E through N as listed earlier in Section 7.3.1.6, there is little value in further discussion that would contribute to an engineering understanding.

Court Decision — The court held that mining waste was excluded from the coverage of CERCLA under the Bevill amendment. The claims against the defendants Rhône and IMMI/Arman were dismissed on that basis.

Reconsideration Hearing and Decision of January 20, 1993 — In answer to a plaintiff-government's motion for reconsideration of that part of the court's decision of September 21, 1992, concerning the Bevill amendment excluded wastes, the court concluded that some, but not all, of the mining wastes were excluded from the CERCLA definition of hazardous substances. Excluded substances were those listed by the EPA as Bevill amendment wastes. On January 20, 1993, the district court rendered the same conclusion, that is, "the 1980 legislative history dictates exclusion of Bevill Amendment wastes from regulation within the meaning of § 101(14)." Plaintiffs' motion for summary judgment was again denied. By orders of May 11 and August 23, 1993, the Federal District Court for the Eastern District of California dismissed the EPA's response charges against the defendants Rhône and IMMI/Arman.

7.3.1.6.2 Litigation and Decision of March 31, 1995

On review of the California District Court decision in *U.S. EPA v. Iron Mountain Mines, Inc. Arman and Rhône Poulenc*, 881 F. Supp. 1432, U.S. Dist. N. Cal. 1995 by the Ninth Circuit Court of Appeals, the U.S. EPA complaints were reinstated by the appeals court holding that releases of substances in the Bevill amendment may create liability under CERCLA.

Defendants Rhône and IMMI/Arman subsequently filed third party claims against EPA and California. In their subsequent complaints, defendants assumed new approaches to involve past acts of the government as contributors to the AMD flow from Iron Mountain. Defendants' new claims charged the U.S and California governments with (i) ownership and operation of dams and power plant; (ii) operator activities of Iron Mountain Mines during and after World War II; and (iii) recoupment counterclaims. Plaintiff-government countered with (iv) sovereign immunity under § 702 of the Flood Control Act; and (v) government immunity for regulatory or remedial activities.

Only highlights of the 1995 litigation and arguments are reviewed to acquaint engineers with their effectiveness on CERCLA enforcement.

(i) Government Ownership and Operation of Dams and Power Plant — The AMD from Iron Mountain Mine past mining activities enters two creeks, in turn flowing into Spring Creek. Spring Creek terminates in Spring Creek reservoir behind the Spring Creek Dam. This dam was constructed in 1963 by the USBR. Its power

plant was also constructed by USBR about the same time just downstream of the dam. Spring Creek Dam releases "AMD-tainted" water into the Keswick Reservoir at a point on the Sacramento River between the upstream Shasta Dam and the downstream Keswick Dam.

Rhône argued that the United States is liable for response costs associated with the release of AMD from Iron Mountain because of its construction of the Spring Creek, Keswick and Shasta Dams, and the Spring Creek power plant. Defendants reasoned that had not the government dammed the Sacramento River and Spring Creek, the natural flow of the watershed would have diluted the AMD and there would have been no response costs [*U.S. v. Iron Mountain Mines, Inc.*, 881 F. Supp. 1432, 1435 (1995)]. Rhône alleged "failure upon failure to control the pollution problems stemming from construction of the Shasta Dam in the 1940s." Further, when Keswick Dam was built in 1950 for the purpose of regulating flow from Shasta Dam, it caught and impounded hazardous sediments from Spring Creek. In 1963, the Spring Creek Dam was built to prevent debris from clogging the Spring Creek power plant and to remedy the AMD problem created by the Shasta and Keswick Dams (*Id.* at 1435) According to Rhône, the Spring Creek Dam is composed partly of sediments containing hazardous substances, and the USBR disposed of hazardous sediments elsewhere in the facility during its construction (*Id.* at 1435). In the language of CERCLA, Rhône contends that the government is liable as an owner or operator of the dams behind which AMD is concentrated, or, alternatively, the government is an "arranger" because Spring Creek was built partly for the reason to dispose of the AMD (*Id.* at 1443–1446). Defendants also charged USBR with mismanagement of the dams and power plant by failing to coordinate water releases from the dams to minimize the metals content in the river, once lowering the water in the Keswick reservoir to an abnormally low level and flushing a surge of sediments containing hazardous sediments downstream, and failing to keep the louvers of the Spring Creek dam in proper condition (*Id.* at 1435).

(ii) Operator Activities of Iron Mountain Mines during and after World War II —
As an alternative claim for charging the government with response liability, defendants claim that the United States was involved in the Iron Mountain mine operations during and after World War II, thereby making it liable as an operator.

Defendants alleged that when World War II started, there had been no copper mining at Iron Mountain since 1930. Mining activities since then had been limited to surface gold mining and intermittent mining for pyrite. By directive, the federal government prohibited gold mining as nonessential to the war effort. The government also established the Premium Price Plan as an incentive for the production of copper and zinc. Prices were set to encourage maximum production (*Id.* at 1435, 1436). The operator at that time, Mountain Copper was directed to cease gold mining, and it subsequently contracted with the government to purchase its entire copper and zinc production. The government controlled the marketing and pricing of the ore produced. Under the contract, a new ore body was opened at the mine and a new mill was built. Production of pyrite ore was increased. If Mountain Copper had not increased pyrite production, the government could have seized the mine to assure production demanded. Additional involvement with the mine by the government included the labor force and shipments of the metal ores (*Id.*).

Defendants also claimed that the government involvement as an operator also included mine inspections in 1943 to eliminate hazards thereby interrupting production, and installed measures against sabotage. The government hired workers for the mine, granted deferments to mine workers, and furloughed active military personnel to work at the mine. The government surfaced and improved Mountain Copper's road to Keswick to facilitate shipments from the mine, and refused to permit the Southern Pacific Railroad to abandon the rail line from Iron Mountain mines to Redding (*Id.*).

After the war, when the mining and milling of copper and zinc ore ceased, the government continued its involvement in Iron Mountain by paying for exploration for new ore and for equipment and operations and participating in the methods to search for new ore (*Id.*).

Based on those two allegations, defendants claimed liability for response costs under CERCLA § 107(a) were attributable to the government and contribution by the government under § 113(f) and common law was due. Rhône also noted that the United States is the owner of at least four parcels of land in the Iron Mountain Mine area, implying that the government is an owner of land producing hazardous substances.

In its analysis of the defendants' allegations of operator liability, the court responded, saying, "Even assuming the historical accuracy of these assertions, the complaint does not state a cause of action for 'operator' liability under CERCLA." The court granted that the United States played the role of a very interested consumer during the war, but Rhône's allegation failed to show that the government was involved in any day-to-day hands on management of the mine, or that the United States controlled the cause of contamination and the mining equipment, or made any decisions concerning the disposal of the mining waste. "Purchasing a product and encouraging its production are not the same as controlling the cause of contamination."

Defendants' allegations for postwar operation of the mine also fell short of establishing operator liability. Mining and milling of copper and zinc ore had ceased with the war. Participation by the government in methods of searching for new ore did not result in any production or release of hazardous substances from the mine.

For the forgoing reasons, defendants' claims that the government was liable as an operator were denied for failure to state a claim on which relief could be granted (*Id.* at 1451).

Concerning California as an Operator — The defendants asserted ten counterclaims against the State of California, viz., (1 and 2) an owner and operator of facilities, respectively, the dams and the streambeds from which hazardous substances have been released; (3) the state is an "arranger" for disposal under CERCLA; (4) a CERCLA contribution claim; (5) a nuisance; (6) the creation of a dangerous condition on public property; (7) a breach of public trust; (8) a breach of a mandatory duty; (9) an equitable indemnification; and (10) a recoupment counterclaim incorporating all other claims. The U.S. and state moved the court to dismiss the defendants' counterclaims.

(iii) Recoupment Counterclaims — Under a legal theory of recoupment, a defendant in an action brought by the government may assert any counterclaim arising from the same transaction or occurrence as the government's action, even though the counterclaim otherwise would be barred by immunity or the statute of limitations were it brought as a separate action. The claim in recoupment may not exceed the government's recovery but may only offset it. Defendants had recast their CERCLA and state law counterclaims as additional recoupment counterclaims to seek recovery up to the limit. The *Iron Mountain* court declined to extend the recoupment doctrine to CERCLA for the defendants.

A summary of the court's 1995 decisions in *U.S. v. Iron Mountain Mines* follows:

1. Defendants' claims against the United States for its activities at the mine during and after World War II are dismissed for failure to state a claim under CERCLA. Defendants may seek leave to amend to allege facts sufficient to state a claim under CERCLA based on these activities if there are other facts not included in the current counterclaims that would support such a claim.
2. The common law and recoupment claims against the United States are dismissed because they are barred by sovereign immunity.
3. The motion by the United States to dismiss Rhône's and IMMI/Arman's claims under CERCLA is denied. The court holds that the United States is not immune under the Flood Control Act and that CERCLA does not extend remedial or regulatory immunity to the United States.
4. Rhône's CERCLA claim against California for "arranger" liability is dismissed without prejudice for failure to state a claim.
5. California's motion to dismiss Rhône's CERCLA claim for operator liability is denied because the facts alleged are sufficient to state a cause of action under CERCLA, and because the "remedial" and "regulatory" exception advanced as a basis for dismissal is unsupported by the language of CERCLA or its legislative history.
6. California's motion to dismiss Rhône's CERCLA claim for owner liability is denied, because the sovereign function exception will not protect it if it is liable on Rhône's operator claim.
7. All Rhône's state law and recoupment claims against California are dismissed because they are barred by California's Eleventh Amendment immunity, which California did not waive by bringing this CERCLA cost recovery action.

(iv) Sovereign Immunity Under Section 702 of the Flood Control Act — Section 702(c) of the Flood Control Act of 1928 provides in relevant part that "no liability of any kind shall attach or rest upon the United States for any damage from or by floods, or flood waters at any place." The United States argued that Rhône's claims based on construction and operation of the three dams and power plant were barred by that provision *(Id.* at 1438). Two major questions to this plaintiffs' defense had to be resolved by the court, viz., (1) whether the waters at the Shasta, Keswick, and Spring Creek dams and the Spring Creek power plant are flood waters within the meaning of § 702; and (2) whether CERCLA supersedes the § 702 immunity.

In the court's analysis of immunity under the Flood Control Act, it found that the United States is entitled to sovereign immunity. It was clear from § 702 that governments enjoy immunity from any and all claims of injuries or damages associated with all waters contained in or carried through a federal flood control project for the purposes of, or related to, flood control...." The U.S. Supreme Court had held that unless the injury results from an act "totally unrelated" to flood control or a flood control project, sovereign immunity prevails.

However, in searching CERCLA, the court determined that its finding of sovereign immunity under § 702(c) of the Flood Control Act was nullified under CERCLA. Section 120(a)(1) of CERCLA is the liability section which establishes that government entities are included for liability "notwithstanding any other provision or rule of law, and subject only to the defenses set forth in sub§ (b) of this section," which provides limited defenses for releases caused by an Act of God, war, or third party in certain narrow circumstances. The court found that there was no immunity for the acts of the United States as charged by defendants in connection with the three dams and the Spring Creek power plant (*Id.* at 1441).

(v) Government Immunity for Regulatory or Remedial Activities — After an extensive search and investigation of statutory law and case law, the court held that "there is no 'regulatory' or 'remedial' exception to CERCLA liability. Where a governmental entity's regulatory or remedial activities, of whatever nature, bring the entity within the definition of the terms 'owner, operator, arranger, or transporter,' as those terms are applied to private parties, the government will be liable. Further, when a government agency undertakes to remediate a site within the terms of § 107(d), it will face liability under the standards set forth in that section. The court denied the United States' and the State of California's motions to dismiss Rhône's claims as barred by immunity for government acts undertaken in a regulatory or remedial capacity.

7.3.1.6.3 Litigation and Decision of September 30, 1997

U.S. v. Iron Mountain Mines, Inc., et al., State of California v. Iron Mountain Mines, Inc., et al., and Related Cross-, Counter-, and Third Party Claims, 987 F. Supp. 1233 (E.D. Cal. 1997). As noted earlier in Section 7.3.1.6, "various filings by both parties were made in between the three major court decisions." After the court's decision of March 31, 1995, various legal proceedings were also filed by both parties introducing legal arguments important in law, but not for this engineering study of environmental law. However, as a simplified explanation for the addition of the terms in the case citation, such as "Related Cross-, Counter-, and Third Party Claims," defendants filed **impleader** claims, or third party claims under Rule 14 of the Federal Rules of Procedure. The purpose of Rule 14 is to provide suitable legal machinery whereby the rights of all parties may be settled in one proceeding. (**Impleader** is a procedure whereby the defendant is permitted to bring into the lawsuit a third person who is, or may be, liable for all or part of plaintiff's claim against the defendant.)

Since defendants had already brought counterclaims against the United States and California in the litigation decided March 31, 1995, alleging the government as culpable contributors and PRPs in the capacity of owners, operators, and arrangers to the contamination processes of hazardous substances with which the defendants were charged, the defendants were seeking assurance of contribution and indemnification for reimbursement should the plaintiffs/government be found guilty. Such assurance of contribution by third parties may be secured through the impleader process. Thus, in the continuation of the Iron Mountain litigation, cross-claims and counterclaims by both plaintiffs and defendants were included by the court, to be settled in this one court action instead of necessitating separate suits to be filed afterward to recover any awarded contributions.

In *Iron Mountain*, 1997, it was indisputably found that Iron Mountain Copper Company Ltd. and its subsidiaries which operated the Iron Mountain Mines from 1894 to 1968 was the "responsible person" as defined by CERCLA. However, Mountain Copper had been dissolved in 1968, nearly 30 years prior, when Stauffer Chemical Company acquired Mountain Copper's assets, including the Iron Mountain Mines.

The question to be settled in this 1997 action was whether Rhône-Poulenc succeeded to Mountain Copper's liability hinged on whether Stauffer became the corporate successor to Mountain Copper. Stauffer's possible successorship to Mountain Copper was the necessary determination to be made at this time.

Under federal law, successor liability generally does not arise when one corporation purchases the assets of a second corporation, unless one of the following exceptions applies:

1. The purchasing corporation expressly or impliedly agrees to assume the liability.
2. The purchasing corporation amounts to a *de facto* (in fact, in deed, actually) consolidation or merger.
3. The purchasing corporation is merely a continuation of the selling corporation.
4. The transaction was fraudulently entered into to escape liability.

In the court's analysis of the history of transactions and acquisitions of Iron Mountain by, first, Stauffer and then, second, by Rhône-Poulenc, the court found that: "Given the unambiguous language of the assignments in which Stauffer assumed 'all the liabilities' of Mountain Copper and its subsidiaries, and the circumstances surrounding (i.e., the water pollution investigations to that date) the signing of the assignments, Stauffer expressly assumed the environmental liability (hence CERCLA, although not yet enacted) of Mountain Copper; that Stauffer is the corporate successor of Mountain Copper, and that Rhône-Poulenc, standing in the shoes of Stauffer, is now the corporate successor of Mountain Copper" (*Iron Mountain Mines*, 987 F. Supp 1233, 1243–1244).

The next step in motion hearings of the Iron Mountain litigation continued on September 30, 1997. The court heard defendants' motion on two issues as to whether (1) EPA was prohibited by § 104(a)(3)(A) of CERCLA from responding to releases of **naturally occurring** metals (emphasis added), and (2) EPA bore the burden of proving it was not responding to such releases. Rhône-Poulenc alleged that some of the remedial actions ordered by EPA would affect naturally occurring substances, which incurred unnecessary response costs.

To date, EPA had issued three records of decision (ROD) concerning Iron Mountain. An ROD is the vehicle by which EPA selects remedies to be implemented at a particular site. EPA issued ROD 1 in October 1986. ROD 1 selected the construction of a cap over a portion of Iron Mountain Mine to reduce the release of heavy metals from areas disturbed by mining. It also selected a series of stream diversions to divert clean water from upper Slickrock and upper Spring Creeks around the area affected by AMD; and conducted an evaluation of the feasibility of filling the underground mine workings with low-density concrete. ROD 2, issued in 1992, selected the construction of a treatment plant to neutralize the discharges coming from the underground mines through the Richmond and Lawson Portals. ROD 2 also selected the capping of seven mining waste piles that were eroding and discharging metals into Boulder Creek. Finally, ROD 3, issued in 1993, selected the construction of a treatment plant to neutralize the discharges coming from underground mines through the old No. 8 mine.

The court commented, "Thus, the United States would appear correct in stating that 'all of the RODs issued to date have specifically targeted contamination from the mine workings and mining waste piles.' ... [It] can be stated with assurance because each of the three RODs is narrowly focused on a particular source of contamination clearly generated by mining activity at Iron Mountain Mine.... It may well be that the remedies selected in RODs 1, 2, and 3 will have some effect on naturally occurring metals, whether such metals make up a minute but measurable portion of the flows or a more substantial percentage. Even so, § 104(a)(3)(A) is not implicated merely because a response to mining activity will also have the side benefit of catching naturally occurring hazardous substances."

Because defendant did not adequately demonstrate with evidence that RODs 1, 2, and 3 were in any measure in response to only naturally occurring substances, its motion was denied.

The final holding of the September 30, 1997 court's several decisions was in response to the government's motion to limit repetitious review for defendants of certain EPA response actions, specifically concerning RODs 1 and 2, that had already been heard by the court. After another hearing of the defendants' objections to EPA's ROD 1 and 2 response actions, the court granted the government's request for limitation of review.

Remaining Actions for Court's Decisions of *Iron Mountain Mines* — The final two holdings in *Iron Mountain Mines* were handed down by the court on October 28, 1997. In 987 F. Supp. 1277, the court reviewed Rhône's counterclaim that the United States was liable under CERCLA as an operator of the Iron Mountain Mines for the period during and immediately after World War II. This argument had already been substantially covered in the litigation of March 31, 1995 (see Section 7.3.1.6.2).

(ii) Operator Activities of Iron Mountain Mines during and after World War II (supra) — Reconsideration and review had been granted on the basis of new or additional evidence. However, the court found no new arguments that supported Rhône's pleading that the United States was an operator of the mine at any time during or after World War II. The court denied defendants' counterclaim and granted summary judgment to the United States on this counterclaim motion.

Again, on the final rehearing of Rhône's counterclaim that the governments were owners, operators of facilities, and arrangers under CERCLA and were liable for response costs (see Section 7.3.1.6.2[i] on the government ownership and operation of dams and power plant), the court rejected Rhône's newly improvised arguments of the same claims. The court granted the United States' motion for partial summary judgment.

Comments on the Iron Mountain Mines CERCLA Litigation — The defense attorneys in the *Iron Mountain Mines* litigation put forth an extremely strong, well prepared and valiantly fought case for the defendants, using almost every legal ploy, defense, and tactic conceivable, but to no avail. Under CERCLA's primary requirement for liability for the costs of a responsive action that the property owner and/or operator are the potentially responsible persons, there was no way for Rhône-Poulenc and IMMI/Arman to avoid losing the action.

The seemingly most erroneous legal argument in the *Iron Mountain Mines* litigation is the federal government's finding that CERCLA is not an *ex post facto* law prohibited by the U.S. Constitution. CERCLA has very strong aspects about it that are of an *ex post facto* law nature. Under CERCLA, the party held responsible is now being punished for the former acts of others, and perhaps as well as their former own. Under CERCLA the owners, past or present, are made to correct those former acts, or to remedy a resulting, or inherited, situation, that when performed, was not illegal or wrong. Instead, the acts were an accepted, legitimate practice in the industry and nation at the time done. Now, at a later date, those same legal acts have been made illegal by statute and made punishable by huge penalties of money. That description exactly fits the definition of an *ex post facto* law. *Ex post facto* laws are specifically prohibited by the U.S. Constitution. At the time the mining occurred, society was glad to receive the profits and benefits of the mining operations that extracted the minerals from the earth, all without blame for the accepted process and practice. With environmental concerns of decades later, society absolved itself and placed the blame on the mining industry, when society itself was as responsible and guilty as the miners themselves for the alleged contamination by their acts of condoning the exploitation at the time, and also for accepting the benefits. A fairer way to distribute the costs of remedial environmental responses is for society to share in the expenses of cleaning up the environment, not to single out and punish the mining industry alone, or a present property owner who may be totally innocent of the formerly acceptable act.

The court in *Iron Mountain Mines* argued against CERCLA being an *ex post facto* law saying that *ex post facto* laws generally refer to criminal actions, although admitting that a few exceptions have been made in civil actions. Since the invention, enforcement, and application of jail sentences by the federal government for environmental crimes, environmental statutes no longer are confined to civil actions, but may constitute criminal actions. Although the CERCLA statute does not call for imprisonment sentencing for violators, or those found "guilty" as operators or owners of a polluting mining property, there is no certain guarantee that a court-decreed violator could not be sentenced to imprisonment if that person, or persons, refused to pay responsive costs levied on them.

7.3.2 NPL Summitville Mine, Summitville Consolidated Mining Company and Galactic Resources, Inc., Summitville, Rio Grande County, Colorado

7.3.2.1 Background Information of Summitville Mine

The Summitville mining district is located at the headwaters of the Rio Grande River on the northeast flank of South Mountain in the San Juan Mountain Range of south central Colorado near the New Mexico border. Mining in the district began with gold placer discovery in 1870 and lode discovery in 1872. Major mining activities continued until 1887 with mining declining until 1894. Interest in mining was sporadic until another active mining period from 1926 to 1942. Thereafter, with the depleted high-grade ores, the district was quiet until 1984 when three mineral property owners, Aztec Minerals Corporation, Gray Eagle Mining Corporation, and South Mountain Minerals Corporation leased their properties, on which the Summitville Mine is located, to Galactic Resources, Ltd., of Vancouver, Canada. The Canadian company formed the wholly owned U.S. subsidiaries, Galactic Resources Inc., and Summitville Consolidated Mining Company (hereafter, all three are referred to as Galactic).

In early 1984, Galactic drilled the properties. Following the mineral evaluation, it applied and received a state permit granted by the Mined Land Reclamation Division (MLRD) (now renamed the Division of Mines and Geology [DMG]), to conduct a limited impact test pit for heap leaching. By the summer of 1984, the pilot project tests were said to be successful. In August 1984, Galactic applied for the first large cyanide heap leach reclamation permit in Colorado. No public opposition was recorded by the MLRD. In fact, strong community support of the project was voiced in anticipation of a positive, economic impact on the area. The permit for a full-scale, open-pit and heap leach operation was approved in October 1984. Construction of the mine and heap leach pad and liner started in August 1985.

The heap leach pad liner was composed of an impermeable plastic membrane underlaid by a sand layer. The sand layer incorporated a leak detection system and was underlaid in part by a geotextile membrane that rested on top of a clay liner. A French drain system, consisting of crushed rock, collected groundwater that flowed under the liner. The purpose of the liner in addition to permitting the collection of the gold bearing cyanide solution for processing was to prevent the solution from entering into the surrounding environment.

Galactic encountered considerable difficulties during the winter construction at the 3,500-m/11,500-ft elevation. Snow avalanche damage resulted in the liner being ripped and torn. Although Galactic made repairs to the liner prior to loading the heap leach pad with ore, leaks were detected between the upper and lower liners within a week after heap leaching operations began in June 1986. It was soon discovered that the lower liner was also leaking and that the cyanide solution was entering the French drain system. As a result of the cyanide leaks, Galactic proposed to the MLRB that it be allowed to pump water from the leak detection system and the French drain back onto the heap leach pad. This additional water, however, exacerbated a problem, which later became apparent, concerning the amount of

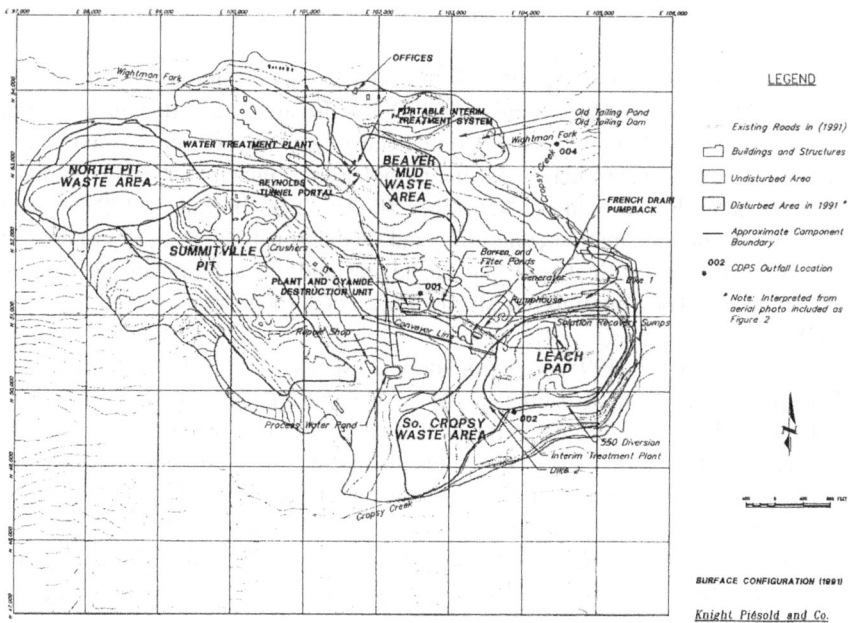

Figure 7.3   Map of Summitville mine site surface workings.

water that could be contained by the heap leach pad without processing. The mine was originally conceived as a zero discharge facility anticipating no discharge of water from the overall project (*Aztec Mining Corp., et al v. Colorado*, 940 P.2d 1025, et seq., Colo. App., Dist. 4, 1996). (See Figure 7.3.)

The no-discharge design became a major factor in the Summitville mishap. The state should have closely inspected the engineering design assumptions before allowing operations to commence. The assumptions by Galactic indicated a negative water balance that would require additional process water to operate the plant. "As now known, this water balance was in material error and underwent six major revisions from 1984 to 1992. The actual water balance required year-round discharges" (Jones, 19).

Commencing in June 1987, the Summitville mine experienced a number of system failures that resulted in cyanide-contaminated water being discharged into a nearby creek and into settling ponds on the site. To handle the increased volume of water in the heap leach pad, Galactic entered into negotiations in 1988 with the Water Quality Control Division (WQCD) to obtain a point source discharge permit. A permit was subsequently approved by the WQCD in May 1989.

In 1989, Galactic also obtained approval from the DMG and the MLRB to construct a process water treatment plant. The plant, however, could not sufficiently treat the water to the standards required by the WQCD. Galactic then sought, and obtained, permission from the MLRB to allow for land application of the water. The land application system called for Galactic to spray contaminated liquid on surrounding lands where it would evaporate and percolate at a controlled rate into the ground. However, because of problems associated with this process, contaminated water flowed into the Wightman Fork and Cropsy Creek, tributaries of the Alamosa River.

In 1991, Galactic, the MLRB, and the WQCD entered into a settlement agreement to address Galactic's continuing problems at the site. The settlement agreement, which was amended in 1992, provided that Galactic would submit a comprehensive plan to remedy its water quality discharge violations. Galactic was also required to submit a final closure plan by November 30, 1992. In the fall of 1992, Galactic performed some reclamation contouring and seeding on 144 acres (58 ha). Galactic also installed a water treatment plant to kill cyanide in solution in the heap and related ponds. The Summitville site facilities disturbed area covered 631 acres (1558.6 ha).

On December 4, 1992, approximately 2 weeks after Galactic submitted its cleanup plan in which it estimated the cost of cleanup at $20 million, Galactic declared bankruptcy and gave notice that it would no longer fund its operations at the site after December 15. Immediately thereafter, the Colorado Department of Health sought the assistance of the EPA to undertake an emergency response action at the site if one became necessary.

On December 15, 1992, the day Galactic abandoned the site, the state obtained a temporary restraining order (TRO) against Galactic to require its continued operation of the water treatment facilities. Less than 2 weeks later, on December 28, the state obtained a preliminary injunction extending the TRO.

On December 16, 1992, the EPA entered the site pursuant to its authority under CERCLA, began operating the water treatment facilities, and sought to stabilize the site conditions. The state and the EPA subsequently entered into several response action contracts (Cropsy Phase I and Phase II agreements) to remedy the environmental damage at the site. The agreements sought to eliminate the need for long-term treatment of hazardous substances releases from the Cropsy waste rock pile, reduce the subsurface infiltration of water through the mine pit, and improve the quality of the effluent from the Reynolds Adit. Pursuant to the agreements, the state agreed to fund a portion of the response costs using the financial warranties forfeited by Galactic. As a result of problems associated with the Summitville site, the EPA, on May 31, 1994, listed Summitville on its NPL (*Aztec Mining Corp., Id.,* 1027–1029).

7.3.2.2 Summitville Litigation Begins

Aztec Mining Corporation, Gray Eagle Mining Corporation and South Mountain Minerals Corporation v. Colorado, 940 P.2d 1025 (Colo. App. Dist. 4, 1996). Since Galactic Resources, as the mine operator, became judgment proof by its filing of bankruptcy, the property owners were left holding the proverbial bag and became the only available PRPs on which to pin the response cost liability.

In December 1994, the Summitville property lessors to Galactic brought an action against the State of Colorado, and several officers of the state in their respective official capacities, viz., Colorado Governor Romer, the Colorado Department of Public Health and Environment (CDPHE), and the Colorado Department of Natural Resources (CDNR), for damages under four separate claims. The claims were that (1) defendant CDPHE, acting through the WQCD, was negligent in failing to inform plaintiffs of the potential risks and liabilities associated with environmental contamination based on Galactic's activities at the Summitville Mine; (2) defendant DNR, acting through the MLRB and the MLRD/DMG, was negligent in failing to adequately regulate the mining operation, including evaluating and granting the mining permit; (3) all defendants, by assisting the EPA in moving the Cropsy waste rock pile and in plugging the historic tunnels and shafts at the Summitville site, trespassed on plaintiffs' property resulting in a taking of and damage to both the surface and mineral estates without just compensation in violation of Colorado Constitution, Article II, § 15; and (4) all defendants, by engaging in the action described in preceding claim 3 have denied plaintiffs due process of law in violation of Colorado Constitution Article II, § 25.

The defendants' moved the court for dismissal of all plaintiffs' charges. Defendants asserted that plaintiffs' first and second claims for relief based on negligence were barred by the Colorado Governmental Immunity Act (GIA), because they were based in tort and because there was no express waiver of immunity in the GIA for such claims. Regarding plaintiffs' third claim for relief, asserting trespass and a taking, should be dismissed because: (1) a common law claim for trespass could not be joined with a claim for inverse condemnation (**inverse condemnation** is a cause of action against a government agency to recover the value of property taken by the agency though no formal exercise of the power of eminent domain has been undertaken) (Black, 1970, p. 740); (2) a common law claim for trespass, because it lies in tort, is barred by the GIA; (3) a claim for inverse condemnation could not be maintained because none of the state agencies had the power of eminent domain; and (4) no trespass or violation of Colorado Constitution, Article II, § 15, had occurred because defendants were acting pursuant to their police powers to abate a public nuisance. Further, as to plaintiffs' fourth claim for relief, asserting a violation of due process, defendants asserted that it should be dismissed because, in the absence of a statute creating liability, Colorado Constitution, Article II, § 25 neither created a substantive right, nor authorized a direct cause of action for damages. Defendants, later, additionally asserted that the third and fourth claims should be dismissed for failure to join the EPA as an indispensable party (*Aztec Mining Corp., Id.*).

7.3.2.3 The Court's Analysis and Rulings

Sovereign immunity — The court ruled that plaintiffs' first two claims for relief, which the plaintiffs characterized as negligence claims, were, in fact, torts. In addition, the court determined that plaintiffs' third claim for relief also sounded in tort, and like plaintiffs' negligence claims, was subject to the immunity (GIA) statute. As to plaintiffs' fourth claim for relief, the court ruled that it was meritless because the due process provisions of the Colorado Constitution did not create substantive rights where none otherwise existed and that the GIA did not confer any substantive rights on plaintiffs. The trial court found that none of the waiver exceptions in the GIA were applicable to any of plaintiffs' claims for relief. The court, therefore, concluded that it lacked subject matter jurisdiction under the

GIA as to all four of plaintiffs' claims for relief. In response to plaintiffs' argument that the GIA did not apply when the public entity undertakes a duty and breaches that duty, the court noted that "a governmental entity will not be deemed to have assumed a duty by adopting a regulation or policy to protect the public's health or safety or by enforcing or failing to enforce such a policy or regulation." The court rejected plaintiffs' argument that defendants had assumed a duty toward them. It thus concluded that because plaintiffs' first three claims for relief were based on such a duty, those claims were barred by the GIA (*Aztec Mining Corp., Id.,* 1029–1030).

Trespass claim — The trial court also concluded that even if plaintiffs' third claim for relief was not barred by the GIA, plaintiffs had failed to establish that DMG had trespassed on their property by exceeding its statutory authority in entering into the agreements with the EPA. The court also determined that DMG was acting within its statutory authority to reclaim the mine when it entered the property after Galactic abandoned the site.

Takings claim — The trial court concluded that defendants did not state a claim for relief and ruled that plaintiffs failed to allege that the state had taken or damaged its property for a public use, which it concluded was an essential element for such a claim. It also determined that because the state was acting pursuant to its police power to regulate private property for the benefit of the public pursuant to its statutory authority to abate a public nuisance, no claim could be asserted.

Due process claim — The court found that plaintiffs' fourth claim for relief failed because, in the absence of a statutory right creating liability, the statute neither created a substantive right, nor authorized a direct cause of action for damages.

EPA as an indispensable party — Alternatively, as to plaintiffs' third and fourth claims for relief, the trial court determined that, because these claims were based on the state's participation in the agreements with the EPA, the EPA was an indispensable party under the state's Rule 19 of Civil Procedure. The court found that plaintiffs were attempting to challenge the actions EPA had undertaken pursuant to its authority under § 104 of CERCLA. The court also determined that without the EPA's presence in the suit, the EPA's interest in the site might be prejudiced and defendants might be subject to double, multiple, or inconsistent obligations. Because jurisdiction under CERCLA rested exclusively in the federal court, the court determined that the EPA could not be joined as a party and that plaintiffs' third and fourth claims must be dismissed (*Aztec Mining Corp., Id.,* 1030–1031).

Taking of private property — A taking occurs when the public entity substantially deprives a property owner of the use and enjoyment of the property. The right to just compensation under the Colorado Constitution for the taking or damaging of property is not absolute. A governmental entity, pursuant to its police powers, "controls the use of property by the owner for the public good, authorizing its regulation and destruction without compensation...." Under Colorado common law, landowners have a duty to prevent activities and conditions on their land that create an unreasonable risk of harm to others. A landowner cannot reasonably expect to put property to a use that constitutes a public nuisance, even if that is the only economically viable use for the property. A landowner has no right to pollute a stream or use property in a manner that could result in the spread of contamination.

Such uses were never part of the landowner's "bundle of rights" that are commonly characterized as property. Thus, under such circumstances, a government need not pay even for complete takeover or destruction of property if it is justified by the owner's insistence on using the property to injure other people or their property. Under principles of Colorado nuisance law, plaintiffs had no property right in permitting the continued degradation of the environment at and surrounding the Summitville site and thus creating a hazard to public health. The court held that these actions do not rise to the level of a taking (*Aztec Mining Corp., Id.* at 1032). Plaintiffs submitted that if their claims were dismissed, they would be left without an adequate remedy at law. The court responded, "Since the crux of plaintiffs' third and fourth claims for relief is in the nature and scope of the EPA's response action, plaintiffs do have a remedy in that they may challenge that action in a federal law suit" (*Aztec Mining Corp., Id.* at 1034). The Colorado Appeals Court held that the lower, trial court had not erred in dismissing plaintiffs' claims and affirmed its decision. As noted in the foregoing Colorado decision, the property owners filed a claim in 1994 in the federal court system for damages (*Aztec Mining Corp., et al., v. U.S. EPA,* Lexis 26916, U.S. Appeals [10th Cir. 1999]).

Background of federal litigation — When Galactic sought protection in December 1992 under the federal bankruptcy laws and abandoned the mine, the State of Colorado requested the EPA to initiate an emergency response action at the mine. The EPA determined that conditions at the mine "presented an immediate, imminent, and substantial threat to public health or welfare and the environment, and [met] the criteria for initiating a removal action...." The EPA requested and received from the plaintiffs a written voluntary grant of access to the mine, to undertake the necessary response actions. The plaintiff-property owners became concerned about the EPA's activities at the mine and on April 15, 1994, revoked their consent to allow the EPA voluntary access to the mine. In response, the EPA, on April 22, 1994, issued an Administrative Order for Access (Access Order), pursuant to § 104(e) of CERCLA. The Access Order stated that access to the mine "is required for the purpose of determining the appropriate response action and to effectuate response actions" at the mine and "in order to determine whether conditions at the mine ... pose a threat to human health or the environment and to conduct appropriate response activities." The Access Order commanded the plaintiffs, inter alia, to: (1) provide the EPA "full and unrestricted access" to the mine; (2) not "enter onto the access areas" without "written authorization from [the] EPA"; (3) notify the EPA in writing "at least thirty (30) days before any conveyance of an interest in all, or any portion of the mine"; and (4) ensure that any conveyance of any interest in the mine by the plaintiffs provide for continued access by the EPA to conduct, without interference, its response activities. On December 14, 1994, the plaintiffs filed suit against the EPA, its administrator, and its regional administrator. In their amended complaint, the plaintiffs alleged nine claims for relief. In four claims, the plaintiffs alleged two provisions of the Access Order violated CERCLA. In four more claims, the plaintiffs contended that the EPA actions at the mine violated the due process clause of the Fifth Amendment to the United States Constitution; in one claim, the plaintiffs charged the EPA under both CERCLA and the Fifth Amendment with improperly removing valuable metals. The latter claim was voluntarily dismissed after the EPA provided the plaintiffs with an accounting of all metals removed from the mine.

The federal court's analysis and decision — The defendants challenged the eight remaining claims by filing a motion to dismiss. The motion was referred to U.S. Magistrate Judge Bruce D. Pringle. On April 28, 1997, Magistrate Judge Pringle recommended the defendants' motion to dismiss be granted. The district court adopted the recommendation and dismissed the amended complaint. The plaintiffs appealed and on *de novo* review of the record, the court affirmed the lower court's decision (*Id.*). Thus, the property owners' claims for damages by a CERCLA action "to remove or remedy " a hazardous site were denied and to no avail.

7.3.2.4 Comments on the Summitville Mine Site

In retrospect, according to Paul C. Jones, Chairman of the Colorado Mining Association's Summitville Committee, "the mining and reclamation laws were not the principal problems" causing the Summitville disaster; "poor engineering and quality control, noncompliance, and inadequate enforcement" were the main culprits. Jones listed the key problems at Summitville as:

1. Design flaws: Related to the zero discharge design and the initial water balance which in reality required continual water discharges
2. Poor construction quality control: Poor installation of the leach pad liner; inadequate repair of liner leaks detected early in the process
3. Monitoring: Failure of the operator and the regulatory agencies to recognize and respond to the build-up of water and copper in the system
4. Regulations permit: Operator not following the permitted mining plan, including construction of the Cropsy waste dump which was not permitted, and building one large valley-located heap instead of the three smaller heaps as approved in the permit
5. Agency enforcement: Insufficient surveillance by the Mine Land Reclamation Division due to lack of state funding and staff, or for whatever reason
6. Unrealistic water quality limits: Required limits by the Colorado Department of Health in the settlement agreement with Galactic for some metals below detection levels achievable and certainly more stringent than apparent background quality in the area
7. Bonding: Scope and amount of required bonding and financial assurances not adequate to address remedial and reclamation activity required to close the site (Jones, 19xx).

7.3.3 Idarado Mine and Facility, San Miguel, Ouray, and San Juan Counties, Colorado — Case Study of a State Regulatory Remediation of a Hazardous AMD Mine and Mill Site

The 1989 case (hereafter, *Idarado*) represents a unique example of an AMD case from an inactive mining and mill site, with waste rock dumps that contained hazardous metals and mine discharging waters that were allegedly contaminating local water resources and required a remedial cleanup plan under CERCLA. It is distinguishable from other AMD and CERCLA cases in that the response and planned remedial action were initiated by a state instead of the federal government.

7.3.3.1   Legal Background for the Idarado Mine Site

In 1986, the State of Colorado filed an action, pursuant to CERCLA and as amended under SARA, for declaratory judgment against the Idarado Mining Company to fix liability of remedial action and costs on the mining company as the responsible party for several hazardous locations on their properties at San Miguel, Ouray, and San Juan Counties, Colorado. The Idarado Mining Company was a subsidiary of Newmont Mining Company and Newmont Services Ltd. The parent companies became parties to the action.

In 1987, the state issued an ROD. The U.S. Federal District Court for the District of Colorado upheld the State of Colorado's right to bring an action against a responsible party with injunctive measures to force a cleanup of the site. However, in 1990 on review, the U.S. Court of Appeals for the Tenth Circuit held that the federal district court did not have the authority under CERCLA to grant injunctive relief requiring a responsible party to carry out a remedial plan proposed by the state. The case was remanded to the district court to vacate its mandatory injunctive orders.

7.3.3.2   Brief History of the Idarado Mine Site

The Idarado mining and milling complex, known as the Telluride–Red Mountain area, is in a highly mineralized, extensively prospected and mined area, containing the base metals, gold, silver, copper, lead, and zinc. Mining for the valuable metals began around 1874 on hundreds of claims. Major mining activities occurred between 1875 and 1894, and again between 1926 and 1942. Idarado Mining Company was formed in 1939, and it began consolidating mining properties in the Red Mountain area culminating in 1953 with the purchase of the Telluride Mines. Between 1954 and 1956, Idarado began a plan of consolidation for its mines and mills by decommissioning the old Red Mountain Mill. The Telluride Valley Pandora Mill site was reconstructed and enlarged. From 1956 until Idarado closed down in 1978, all ore was run through the Treasury and Mill Level Tunnels, and then milled near Telluride at the Pandora Mill. Mines and mill remained closed until 1986 when the major part of equipment and machinery was dismantled and sold. Thereafter, only a skeleton crew remained on the property for maintenance (*State of Colorado v. Idarado*, 707 F. Supp. 1227, 1234).

7.3.3.3   Overview of the Idarado Complex

Idarado owned 10,574 acres/4279 ha, covering about 16.5 square miles located in San Miguel, Ouray, and San Juan Counties of southwestern Colorado. The mining–milling complex consisted of miles of underground workings, numerous mine adits, shafts, mine waste rock piles, and 11 mill tailing piles, located on patented and unpatented mining claims.

The complex spanned two drainage systems of the San Miguel River, and a tributary of the Uncompaghre River called Red Mountain Creek. Two of the mine portals, Mill Level and Meldrum Tunnels, discharged the major part of mine drainage into the San Miguel River. Eleven other portals discharged mine waters into Marshall and Savage Creeks, which are tributaries to the San Miguel River. Several other portals discharged mine waters into Red Mountain Creek.

The site can be divided into three distinct natural topographic areas, viz. (1) the Telluride Valley, (2) the Red Mountain Valley area, and (3) the High Country between the two. Three location maps, marked Figures 7.4 to 7.6, provide some geographic reference and relative orientation as to the physical location of the Idarado mine complex and the relationship of the various alleged hazardous sites and contaminated areas and streams discussed hereafter. Figure 7.4 shows the location of the Idarado complex within the State of Colorado.

The Telluride Valley is a canyon that contains the Mill Level and Meldrum Tunnels at its upper end at an elevation of about 9000 feet. The Pandora Mill area is next to the Mill Level portal with four filtration lagoons. Below the mill site and adjacent to the San Miguel River are six consecutive mill tailings piles that extend down the valley. The San Miguel River originates at the head of the canyon and flows over its 81-mile length through Idarado property, down the valley, and past the Town of Telluride.

The Red Mountain Valley is approximately 12 mile southwest of the town of Ouray. At an elevation of about 10,000 feet. It contains the old Red Mountain Mill

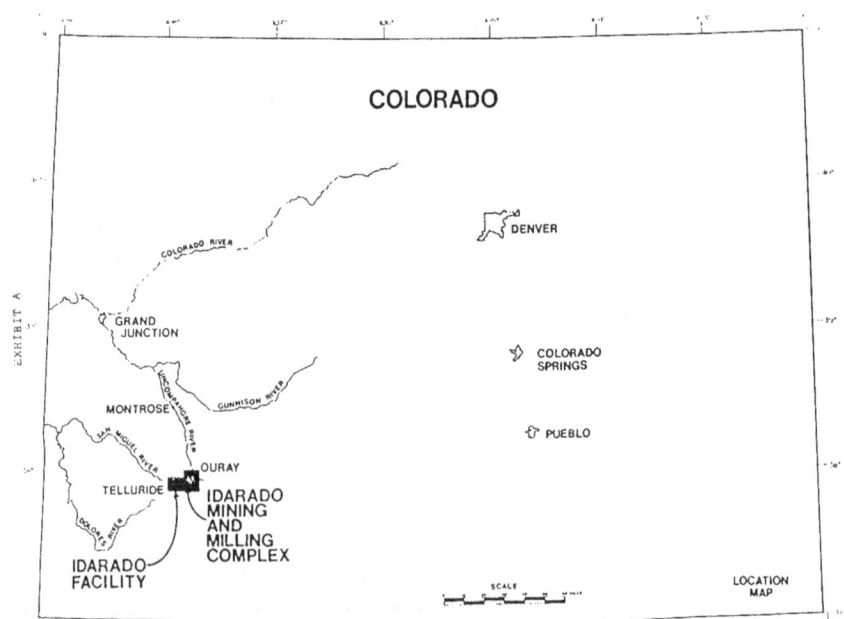

Figure 7.4   Geographic location map of the Idarado complex in Colorado.

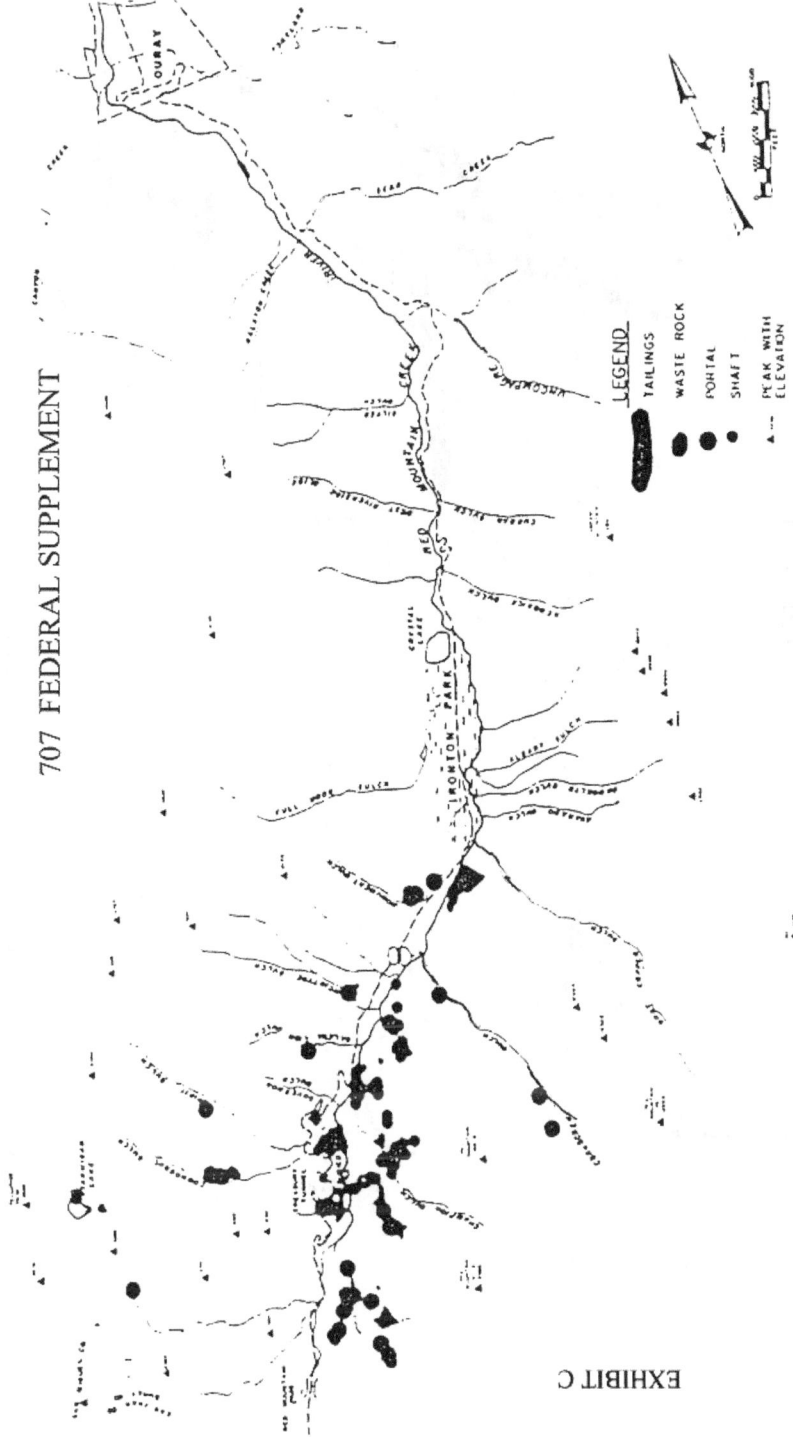

Figure 7.5 Map of Idarado's Red Mountain Creek valley polluting sources.

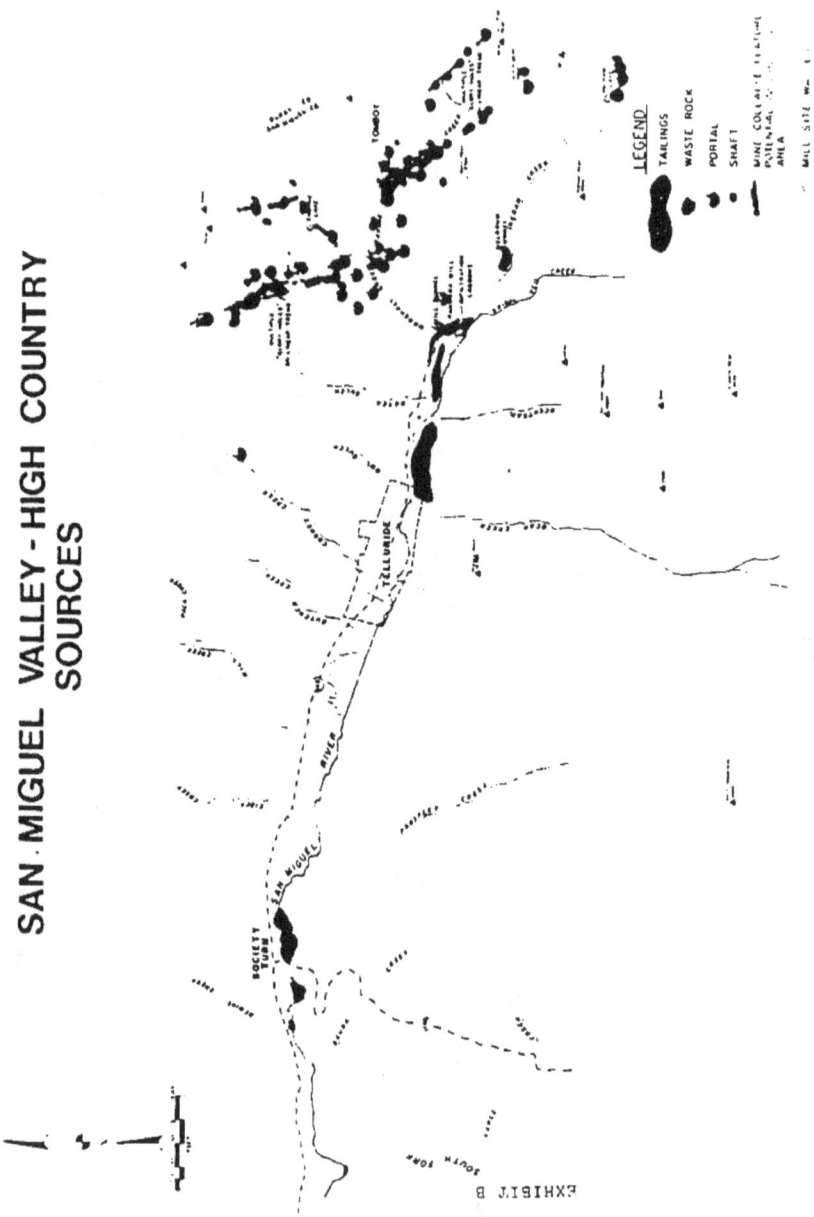

Figure 7.6 Map of Idarado's San Miguel Valley–High Country polluting sources.

area with offices and shops. Downgrade from the mill are four mill tailings piles adjacent to Red Mountain Creek. There are numerous mine adits and five tailings piles that extend down valley to Ironton Park. Ironton Park is a marshy area that adjoins the creek for about 1.5 miles, and is about 2 miles above where Red Mountain Creek empties into the Uncompaghre River above the town of Ouray (Figure 7.5).

The High Country refers to the high mountain basins between the two valleys. Although Idarado owns much of the High Country, it has never mined there. However, it contains scattered mine adits and mine waste rock piles and relics of past mining activities. Drainage from the High Country is into both valleys (Figure 7.6).

Early complaints of contamination were cited by Colorado, for example, "Numerous incidents of surface water pollution and blowing tailings have been reported from the Idarado site. In 1963, the State of Colorado advised Idarado that effluent discharging from the mill tailings piles contained quantities of lead, arsenic, cyanide, and copper above tolerable limits for drinking water. Telluride citizens voiced their concerns about blowing tailings and leaching of the metals from the site in 1966" (State of Colorado, Record of Decision for the Idarado Mining and Milling Complex, March 17, 1987, revised 03/26/87, p. 2).

The State of Colorado undertook certain remedial actions and sought to recover their incurred response costs from the responsible parties (Idarado, Newmont Mining Corporation and Newmont Services, Ltd.). In December 1983, the State of Colorado filed its first action against Idarado for natural resources damages and response costs (*Id.*, 1233–1234)

7.3.3.4 Scope of Liability under CERCLA Section 107(a)

In establishing liability under CERCLA, the state must prove (1) the defendants are owners or operators of (2) a facility (3) from which there has been a release or a threatened release of a hazardous substance that causes response costs to be incurred.

7.3.3.5 CERCLA National Contingency Plan for Remediation of a Hazardous Site

Under CERCLA, EPA is authorized to undertake pollution abatement and cleanup measures, seeking reimbursement for the incurred costs from the responsible parties. U.S. EPA has prepared a National Contingency Plan (NCP), which outlines the administrative process governing the cleanup measures by establishing procedures and standards for the specified response actions. Following the listing of a site on the NCP, a detailed environmental investigation is made. This is comprised of two parts, viz. the remedial investigation (RI) and the feasibility study (FS). The RI determines the nature and extent of the harm caused or threatened by the release of hazardous substances at the site into the environment. The FS is the evaluation of the proposed remedies (*Id.*, 1231).

CERCLA, § 121(d)(1) (42 U.S.C. § 962 (d)(1)) provides:

> Remedial actions selected under this section ... shall attain a degree of cleanup of
> hazardous substances, pollutants, and contaminants released into the environment and
> of control of further release at a minimum which assures protection of human health
> and the environment. Such remedial actions shall be relevant and appropriate under
> the circumstances presented by the release, or threatened release, of such substances,
> pollutant, or contaminant (*Id.*, 1230).

CERCLA also provides that applicable federal and state environmental and
public health requirements be identified and incorporated into the response actions
taken at the site; CERCLA § 121(d)(2)(A) provides that after notice and opportunity
for public comment, a comprehensive remedial action plan appropriate for the site
as a whole is selected by applying the data collected from the RI/FS activities. The
final remedy selected for the site is issued as the Record of Decision (ROD).
CERCLA § 121(a) requires that the governmental response be cost-effective, meaning that long- and short-term costs be weighed, inclusive of operational and maintenance costs. In addition, the NCP, 40 CFR § 300.68(i), provides that:

1. The appropriate extent of the remedy shall be determined by the ... selection of
   a cost-effective remedial alternative that effectively mitigates and minimizes
   threats to and provides adequate protection of public health and welfare of the
   environment ... This will require the selection of a remedy that attains or exceeds
   applicable or relevant and appropriate Federal public health and environmental
   requirements that have been identified for the specific site.
2. In selecting the appropriate extent of the remedy from among the alternatives ...,
   the lead agency will consider cost, technology, reliability, administrative and other
   concerns....
3. Pertinent other Federal criteria, advisories, and guidance, and State standards will
   be considered and may be used in developing alternatives, with adjustments for
   site-specific circumstances (*Id.,* 1231).

Although, generally, CERCLA authorizes U.S. EPA to undertake pollution abatement and cleanup efforts, remedial actions may be selected under § 121(d)(1), which
provides that, like U.S. EPA, **states can sue responsible parties to recover remedial
and removal costs**. To prevail, a state's response efforts must be "not inconsistent with
the NCP" (quoting *U.S. v. Northeastern Pharmaceutical and Chem. Co.*, 810 F.2d 726,
747 [8th Cir. 1986]; *State of New York v. Shore Realty Corp.*, 759 F.2d 1032 [2d Cir.
1985]; *State ex. rel. Brown v. Georgeoff*, 562 F. Supp. 1300, 1315 [N.D. Ohio 1983]).

Based on the right of a state to pursue remedial action to a hazardous site, the
State of Colorado began an investigation of the Idarado site to define the nature and
extent of the contamination. In 1986, Colorado produced its RI identifying the nature
and extent of the site problems including a framework for remedial measures, and
followed with its FS which identified a range of remedial alternatives with their
cost-effectiveness. Public comment and hearings followed issuance of both reports.
Pursuant to the NCP, Colorado issued its preliminary ROD, held and received
subsequent public comment, and then issued its final ROD in March 1987.

For a party to recover its response costs from responsible parties, there are several mechanisms provided for by CERCLA in responding to the releases of hazardous substances and the act delineates the respective powers of the federal government, the states, Indian tribes, and private parties. § 111 of CERCLA (42 U.S.C. § 9611) provides for the creation of a Hazardous Substances Superfund to finance cleanup actions at sites affected by releases or threatened releases of hazardous substances. CERCLA § 104 (42 U.S.C. 9604[a]) authorizes the federal government to take the necessary cleanup actions financed by the Superfund to respond to such releases or threatened releases. Alternatively, under § 104 (d), the federal government may enter into cooperative agreements with states, political subdivisions, or Indian tribes to conduct cleanup actions using the Superfund.

Even where no cooperative agreement exists, permitting use of Superfund money for cleanup, CERCLA § 114(a) preserves the right of a state, or other party, to proceed under applicable state law to conduct the cleanup of a site affected by hazardous substances. Thus, states that incur response costs not financed by the Superfund may bring a recovery action against the responsible parties.

In 1986, the State of Colorado filed an action pursuant to 28 U.S.C. § 2201, asserting claims under the federal CERCLA, 42 U.S.C. §§ 9601 et seq. (1980), and as amended by SARA (1986), and under the Toxic Substance Control Act (TSCA), 15 U.S.C. §§ 2601 et seq., for declaratory judgment against the Idarado Mining Company to fix liability of remedial action and costs on the mining company for several hazardous locations on their properties at San Miguel, Ouray, and San Juan Counties, Colorado. Colorado sued Idarado and its parent company:

1. To recover its response costs incurred, and to be incurred, in the cleanup of past and ongoing releases of hazardous wastes at the mining site under pertinent · provisions of CERCLA, viz., § 107(a)(1)–(4)(A) and (D)
2. For injunctive measures to implement certain remedial actions it proposes to effectuate the cleanup process under CERCLA § 121(e)(2)
3. For damages for destruction of natural resources, under CERCLA § 107(a)(4)(C) (*Note:* The Federal District Court granted Colorado a motion for a bifurcated trial of this issue.)
4. For an injunction under the TSCA compelling defendants to remedy the adverse environmental consequences caused by their alleged illegal handling of polychlorinated biphenyls (PCBs) at the site (*Id.,* 1235).

7.3.3.6 Areas of Contamination at the Idarado Facility

Colorado had identified three main sources of metals contamination at the Idarado facility, viz., mill tailings, mine drainage, and waste rock piles.

The tailings are essentially made up of gangue and waste rock from the milling process that contain a small fraction of metalliferous minerals not recovered during milling, plus some of the rock-retained chemical flotation reagents and water used in milling. Waters from precipitation, underground sources, including the mines, flow across and through the tailings picking up metals that are deposited in the area creeks and streams.

Water flowing from the mine portals not only reputedly carries hazardous metallic values from the mines but in flowing onto the surface and through the tailings piles picks up additional hazardous values. The hazardous metallic contents are reported to be cadmium, copper, lead, and zinc, and in some instances, arsenic, chromium, nickel, and silver. All are hazardous within the meaning of CERCLA, § 10(14). Seventeen mine portals at the Idarado facility are known to discharge year-round with numerous others discharging during the spring runoff (ROD, p. 7).

Seven areas of contamination at the Idarado facility requiring remediation were found to be (A) the Telluride tailings piles; (B) the Society Turn tailings; (C) miscellaneous tailings along the San Miguel River; (D) Telluride soils; (E) the Red Mountain tailings; (F) the mine portals discharges and waste rock piles; and (G) the mill site cleanup.

Following are outlines and discussions of the state's, Idarado's, and the court's proposed remediation plans for each of the seven areas, cited earlier in A through G (*Id.*, 1235).

Telluride Tailings Piles — There are six large tailings piles at this site. Piles 1 through 4 contain between 430,000 and 800,000 tons, and cover a total surface area between 14.3 and 18.1 acres. Piles 1 and 2 average 15 to 35 feet in height. Piles 3 and 4 average 35 feet in height. Merged piles 5 and 6 average 100 feet in height, contain approximately 10 million tons and cover an estimated 60 acres. Estimates of metal concentrations in Piles 5 and 6 are 30 million pounds of lead and zinc, over 5 million pounds of arsenic, and over 2000 pounds of cyanide (*State of Colorado v. Idarado Mining Co.*, 707 F. Supp. 1235, 1237 [D. Colo. 1989]).

The yearly average of mine discharge waters flowing from the Mill Level and Meldrum portals (shafts) has been measured by Colorado as 4.82 million gallons per day (ROD, p. 7). The mine discharge waters flow across and through the tailings piles. All piles show evidence of gullying and erosion as a result of surface waters and precipitation. Oxidized acidic "hot spots" were found where water is seeping out of the piles. Oxidized metals are generally more soluble and mobile in water. The flow from the piles is directly into the San Miguel River. Checks on the sediments of the San Miguel River adjacent to Piles 5 and 6 reveal that they are mainly tailings. Concentrations of zinc, cadmium, and copper were found to be relatively low upstream of the Pandora Mill, but increased downstream of the mill site. Sediment contamination was found to extend as far downstream as the confluence of the San Miguel with its South Fork. The high metal concentrations found downstream in the San Miguel have allegedly resulted in significant degradation of the water quality as a habitat for aquatic life.

"Brook trout, the resident fish, located in the San Miguel River, are reported to be fairly resistant to metals compared to other trout. Liver tissue samples taken from the fish indicated elevated metal levels with elevated cadmium and lead levels in other tissues. The metal levels were 6 to 6,000 times higher than the metals concentrations found in fish from Lake Powell (a recreational lake located across the Arizona–Utah border). But for metals concentrations in the San Miguel River, cut-throat, rainbow and brown trout would be expected to be present. Toxic effects of metals on aquatic life include death, growth reduction, diminished egg hatchability, and in the case of lead, deformities, as well as breakdowns in the metabolic system" (*Id.*, 1236).

The Telluride tailings piles are continually eroded by wind. The wind-blown dust is scattered over the downwind area, onto surface waters, and adds to the dispersal of the hazardous metals. Wind erosion is mitigated by Idarado's dust program which sprinkles water on Piles 5 and 6. However, that water is thought only to add to that percolating through piles and increasing the contaminated discharges.

In 1977, the town of Telluride drilled two water wells downgrade from Pile 6. The concentrations of chromium were elevated during the 1970s, but are now within drinking standards. The EP toxicity test identifies wastes that are capable of posing a substantial hazard to public health and the environment. Tests indicated that tailing Piles 5 and 6 are EP toxic, containing high concentrations of lead. As a result of those findings, the court found that the waters of the San Miguel River did not satisfy Colorado's Water Quality Control Act and its Safe Drinking Water Regulations. The San Miguel serves as a source of municipal water supplies for the Telluride area. Further, the court found that the releases of hazardous substances from the Telluride Mine and Mill tailings piles constituted a source of the contamination, and that a cleanup remediation plan must be implemented (*Id.*, 1237).

7.3.3.7   The State's Proposed Telluride Remediation Plan

Colorado's plan for the Telluride tailings piles was to isolate the 10 million tons in Piles 1 to 6 by:

1. Consolidating Piles 1 through 4 with 5 and 6 at the latter location
2. Constructing a multilayered cap on the pilings
3. Constructing surface water run-on diversion ditches
4. Constructing flood control measures designed to withstand the probable maximum flood event to protect the consolidated tailings impoundment from erosion and structural failure

At the sites of Piles 1 to 4, after removal of the tailings, the original soil beneath would be removed to a depth of 1 foot. Tailings would be removed by a pipeline conveyor across the San Miguel River to Piles 5 and 6. A multilayer cap would employ a low permeable layer, an erosion resistant layer, and a growth media layer.

Strangely, the state proposed using compacted tailings and slime materials from Piles 1 to 4 for the low permeability layer on the consolidated pile. It also proposed for the erosion resistant layer using coarse rock material from a mine waste rock located near the Idarado Mill site. The base of the consolidated pile was to be protected from erosion by placing riprap type, or large rock, available on the site, up to 50 feet high on the side of the consolidated pile. Colorado's estimate of cost for the Telluride remediation plan was $9.8 million (*Id.*, 1237).

7.3.3.8   Idarado's Proposed Telluride Remediation Plan

Idarado's plan was to stabilize all six tailings piles in place. The piles were to be treated with limestone to neutralize the acidic nature and hot spots. A naturally manufactured soil cover was to be made by covering the surfaces with manure and straw before seeding. The company revegetation plan had been implemented on Piles 1 through 4 with growth already started. Modification of the drainage around the piles was proposed by contouring surfaces, adding water diversion ditches, and placing riprap on the slopes of the piles. Erosion gulleys were to be backfilled and covered with erosion blankets. The overall costs for Idarado's plan were estimated to be $1.8 million (*Id.*, 1238).

7.3.3.9 The Court's Telluride Remediation Plan

In reviewing the two submitted plans, the court found that the state's remedial plan did not appear to be technically feasible or necessary. Judge Carrigan concluded "that certain components ... moving and consolidating the tailings piles, and use of a six-foot multilayer cap do not comport with CERCLA's overall scheme ... nor is the state's plan cost-effective.... the state's plan would go too far" (*Id.*, 1239).

The court saw disturbing and dangerous side effects of transporting huge amounts of tailings through the town; dispersal of tailings dust into the air, water, and environment during the removal process; and disruption of the town with heavy construction traffic for an indefinite time.

On the other hand, Idarado's plan was perceived to be insufficient and not going far enough. The "manufacturing of soil" was thought to be unsuccessful. The court then asked the parties to submit alternative plans after its constructive criticism. The court adopted the state's alternative plan subject to its submission of a cost estimate for the alternative plan. The court's adopted plan was as follows:

1. Stabilizing the six piles in place
2. Using cover construction of multilayers on each pile, using local material
3. Using cover to contain a 2-foot thick erosion-resistant (random fill) layer, amended with bentonite
4. Using cover topped by a plant growth media layer 2 feet thick
5. Sides of Piles 5 and 6 similarly covered, but without the bentonite
6. Using drainage and erosion control structures as described by the state, designed to withstand the 500-year flood event and forces of maximum credible earthquake
7. Monitoring activities for surface water, groundwater, soils, vegetation, erosion, and construction aspects as outlined by the state's plan

Society Turn Tailings — Below the town of Telluride, at the lower end of the valley, there is a bend in the San Miguel River known as Society Turn. Tailings are piled at Society Turn, adjacent to the San Miguel River, about 5 miles downstream from the Pandora Mill site. Three tailings piles, containing a total of approximately 120,000 to 160,000 tons, are located in a glacial moraine, almost entirely barren of vegetation. These piles were river-made around 1900 when tailings were dumped directly into the river upstream and carried downstream. They were trapped and impounded behind wooden timber dams built at Society Turn where the tailings settled in calmer waters behind the dam. At a later date, the dams were breached, and the San Miguel River cut a channel through the tailings leaving the present-day piles behind. Stream erosion continues to cut into the tailings material and carry the hazardous metallic components downstream. The three piles are low in height, ranging from 5 to 15 feet. The piles contain heavy metals including cadmium, copper, lead, and zinc. EP toxicity ranges from 4 to 32 times above the EP threshold for lead; the deepest pile is EP toxic for cadmium (*Id.*, 1239). Stream sediments at Society Turn are also high in metals with the highest concentration of silver for the entire course of the San Miguel River occurring there. Colorado reported, "The macroinvertebrate population at Society Turn is one-tenth of the population found upstream. This indicates that the aquatic system, overall, is in poor condition and there is probably a major disruption in the aquatic food chain for all fish species." This is borne out by the fact that brook trout are present in only one third the amount compared to the upstream population (*Id.*, 1239, 1240).

7.3.3.10   Colorado's Proposed Remediation Plan for Society Turn

The state's plan was for removal of the tailings from Society Turn to a hydro-logically and geologically suitable disposal site within a mile from the present location of the tailings. The state had proposed using a disposal site in the Mancos shale with a lined impoundment. The soils beneath and adjacent to the old tailings site would be cleaned to 1000 ppm lead or less.

However, due to voiced protests and objections on the part of the local population because the plan would have entailed substantial trucking traffic through the town of Telluride, that plan was altered to treatment in place. The state would construct a multilayered cap 6 feet deep on top of the consolidated tailings, using Mancos shale materials and covering in the same manner as for the Telluride tailings. Surface water runoff and run-on diversion ditches would be installed. The impoundment would be designed to withstand maximum credible earthquake, maximum precipi-tation, and flood waters. All construction areas would be recontoured and revege-tated. The state estimated its plan would cost $2.58 million.

Idarado did not have a proposed plan for Society Turn as they argued and disclaimed any liability for the area's cleanup, particularly since they do not own property at the site. Without admitting liability, Idarado submitted evidence that river diversion from the tailings would be possible, and then stabilization could take place covering the piles with natural glacial moraine material, buttressing the sides with riprap, and dressing the top with revegetation (*Id.*, 1242).

7.3.3.11   The Court's Plan for Society Turn Remediation

The court approved the state plan for removal and consolidation of the three Society Turn tailings piles, but with the restriction that the removal shall not involve any truck traffic traversing and disrupting the town of Telluride. In addition, the court specified that instead of the 6-foot multilayered cap, the capping shall be done in the same manner prescribed for the Telluride piles. The state shall select a removal site within 6 months.

7.3.3.12   Miscellaneous Tailings along the San Miguel River

Colorado estimated that there are between 50,000 and 75,000 tons of tailings material located along the stream banks of the San Miguel River between the towns of Telluride and Society Turn. These stream bank tailings deposits continually add to the hazardous components of the San Miguel. The state alleged that humans, livestock, and wild animals come into contact with the tailings, exposing them to the hazardous and toxic contents. The court noted that there was no direct evidence of resulting health problems.

Idarado again disclaimed any liability for the miscellaneous tailings deposits along the San Miguel, emphasizing that before Idarado's existence, the general practice for mining companies in the area was to dump tailings directly into the San Miguel River and its tributaries. The court was unpersuaded, stating that even if Idarado's contention was correct in that they only made a *de minimus* contribution of hazardous materials to the San Miguel River, their liability was not precluded. Stream bank deposits of tailings are made in periods of high water. In view of Idarado's six tailings piles above the town of Telluride having made the major contribution to contamination to the San Miguel River, part of the stream bank deposits must include wastes from Idarado's property. The court stated, "It is unnecessary for the State to 'fingerprint' waste" (*Id.*, 1243, quoting *U.S. v. Ottati & Goss, Inc.*, 630 F. Supp. 1361, 1402 (D. N.H. 1985); and as settled in *U.S. v. Conservation Chemical Co.*, 619 F. Supp. 162 (W.D. Mo. 1985), parties are responsible and "liable for cleanup even as *de minimus* polluters.")

7.3.3.13   Colorado's Proposed Stream Bank Remediation Plan

The state's plan for stream bank areas would require removal of tailings from them where the banks are not vegetated or only sparsely vegetated. The tailings would be deposited with those of the Telluride or Society Turn tailings piles. Disturbed areas would then be revegetated. The plan had not clearly identified all areas targeted for tailings removal. However, it proposed removing the tailings with track-mounted backhoes and truck haulage. Using 10-ton trucks, it would take about 5,000 truck loads to remove the 50,000 cubic yards of stream bank tailings. An additional 500 truck loads would be required to haul in cover material.

The state's plan also included aquatic habitat restoration, restocking of several trout species in the San Miguel River. All habitat improvements were designed to allow for channel stability of up to the 100-year flood level.

The state's cost estimate for the miscellaneous stream bank remediation was approximately $582,000.

7.3.3.14   Idarado's Stream Bank Remediation Proposal

Again, Idarado, disclaiming any liability for the stream bank deposits, did not submit a plan for remediation. However, Idarado argued that the state had not quantified the extent of the environmental disturbance resulting from their extensive hauling activities. Removal of trees and vegetation on private property access to private property would be required, possibly necessitating crossing of the river and construction of bridges, all of which would cause a significant impact on the environment. Hence, the state plan was not feasible or cost-effective (*Id.*, 1244).

7.3.3.15 The Court's Proposed Stream Bank Remediation Plan

The court supported the state's plan, finding it generally effective but deficient in a few aspects. The state was instructed to identify the target areas for tailings removal, where the removed tailings will be deposited, the areas to be revegetated; to set forth the methods by which the state will gain access to the affected privately owned areas; and to address the extent of the environmental disturbance resulting from the construction activities. The state would also have to submit a revised cost estimate.

7.3.3.16 Telluride Soils

The main issue, contentions, and arguments over treatment of the soils in the town of Telluride revolved about the blood lead levels of the inhabitants, and whether they were being endangered by Idarado's tailings piles. Pro and con evidence of blood lead surveys and tests was entered into the trial record and weighed by the court.

Briefly, lead is toxic to all humans, but children are at a greater risk than adults. Lead taken into the body can have various detrimental effects causing neurological defects and abnormalities. A state expert medical witness testified that the lead levels found in and around Telluride posed a threat to human life where it occurs in the soil at a level greater than 500 ppm. The potential pathways of lead exposure are primarily by ingestion and inhalation. "According to guidelines published by the Federal Centers for Disease Control, an 'elevated blood lead level' is defined as blood lead concentration of 25 micrograms per deciliter of blood (µg/dl), or greater.... and in the United States, a person's average blood level is approximately 10 µg/dl." The same expert medical witness also testified that she did not believe the current soil levels of cadmium and arsenic posed a threat to Telluride citizens.

In 1977 and 1978, the Colorado Department of Health conducted a blood level survey in Telluride with children between the ages of one and nine. An average blood lead concentration of 19 µg/dl was found. In 1986, the state sampled the soil at 100 locations using a grid pattern. Surface and depth samples were taken from areas mainly showing high soil contact, such as day care centers, playgrounds, and gardens. Of the samples taken, 30% indicated soil lead levels over 500 ppm; 20% over 1000 ppm, with the remainder under 500 ppm. Of these grid samples, 25% exceeded 3 ppm for cadmium. For comparison, the background concentration of lead in Telluride soils ranges from 45 to 110 ppm, and 0.4 ppm for cadmium (*Id.*, 1244).

Evidence showed that tailings from Idarado's mill piles had been used in Telluride for fill and grading material, bedding of underground water pipes, and sealing road surfaces. Idarado argued that several other sources contributed heavily to the soil lead content in Telluride. In 1986, at Idarado's request, a blood lead survey was conducted in the Telluride area, with samples taken from 258 persons. Participants were of various ages, both sexes, pregnant and nonpregnant females; 150 lived in the town of Telluride; 26 lived in the Pandora area near the tailings piles; and 82 lived in down valley areas near Telluride. The results showed that all children tested under 25 µg/dl, with an average of 19 µg/dl. All adults tested below 25 µg/dl. None of the pregnant women had a blood lead level over 6 µg/dl. There was no significant difference in blood levels between persons living in Telluride and those living down the valley. The court stated that the blood levels were among the lowest in the nation. Based on the evidence, the court found Idarado's blood lead testing and results credible and the area residents were not suffering adverse health effects from lead as a result of being exposed to allegedly contaminated soils (*Id.*, 1245, 1246).

7.3.3.17   The State's Proposed Remediation Plan

The state's plan was for soil cleanup in places where necessary, and for removal of soils with lead levels above 1000 ppm. Soils with lead content between 500 and 1000 ppm would be covered; treated with lime; and disked or covered, or placed under control. A health study of Telluride citizens was called for. The state's soil treatment plan was estimated to cost $1.77 million, with an additional $1.5 million for the health study.

7.3.3.18   Idarado's Proposed Remedial Plan

The company proposed a 5-year blood lead screening study, inclusive of iron deficiency, for community children under 6 years of age. Appropriate medical equipment for local physicians was included for a cost of $4000.

7.3.3.19   The Court's Remediation Plan

The court found that there was inadequate proof of causation for the state's allegations of harm to justify its plan; that the 5-year screening plan of Idarado was adequate, and the company shall bear its expense (*Id.*, 1247).

7.3.3.20   Red Mountain Tailings Piles

There are five tailings piles in the Red Mountain area of the Idarado facility. All are close to or adjacent to the Red Mountain Creek. Volume and area range from the smallest at Pile 1, approximately 13,000 tons over 0.8 acre, and 25 feet in height, to the largest, Pile 4, containing about 1.2 million tons over approximately 25 acres. Piles 1 to 3 are located near the Treasury Tunnel, whereas Pile 4 is about 2.5 miles downstream. The four piles comprise 80% of the area's tailings. The fifth pile is buried and located at a natural lake site upstream from Pile 1. All the piles are subject to runoff waters that drain the area, and notably from highway culverts draining directly onto some of the tailings piles. Erosional gullies are observed in places on the piles, particularly Pile 3, which the state has labeled the most unstable. This is evidenced by a sink hole and a 25-foot deep gulley showing drainage directly into the Red Mountain Creek. An estimated 6500 tons of tailings have washed from the deep gulley of Pile 3 into the creek.

The heavy metals composition of these tailings piles is similar to those at Telluride, but the piles are made up of 97% unmineralized waste rock. The ores mined on the Red Mountain side of the Idarado mine are reputedly more acidic and oxidized than at Telluride, causing them to be more soluble and mobile. Drainage from Pile 2 has been reported by the state as EP toxic for lead, and with high concentrations of zinc, copper, and cadmium. Thus, the Red Mountain tailings are allegedly the major contributor to Red Mountain Creek and the downstream areas at Ouray and the Uncompaghre River. However, loading analyses attempting to ascertain metals concentrations in waters of the area to establish point sources found significant nonpoint metals loading to Red Mountain Creek. Up to 50% nonpoint loading was determined in the vicinity of the buried tailings of Pile 5.

Although upstream sampling of Red Mountain Creek contained relatively low concentrations of zinc and other hazardous substances, the state alleged severe degrading of water quality with major impact downstream on the Uncompaghre River with zinc levels running about five times higher than the upstream concentration. Stream standards, based on aquatic life, and not drinking standards, were exceeded for metals in the Uncompaghre River. The state also found the Ridgeway Reservoir, 15 miles downstream from the Treasury Tunnel, to be severely impacted by high concentrations of iron and metals in the waters and sediments. The reservoir has agricultural and recreational uses. Still the state was admittedly not aware of any health risks in the Red Mountain district.

The court found that hazardous metals being released from Idarado's tailings piles into the Red Mountain Creek were endangering the aquatic environment of it and the Uncompaghre River. Remedial action was required, with Idarado liable for the cleanup costs (*Id.*, 1248–1249).

7.3.3.21 The State's Proposed Remedial Plan

The state's plan calls for consolidation of Piles 1, 2, 3 and the buried Pile 5, onto Pile 4. The consolidated pile would be covered with a three-layer cap identical to the Telluride pile. Cap material would have to be transported from the Iron Park area, but is also available from the area adjacent to Pile 4. [*Comment:* It would seem that capping material from this highly suspected hazardous area would not help resolve the release problem even if it has a low impermeability.]

Runoff and run-on diversion ditches were to be constructed to divert surface waters. Construction of remedial features were to be built to withstand the probable maximum flood event. Construction areas were to be recontoured and revegetated (*Id.*, 1249).

The estimated cost, including maintenance, was approximately $9 million.

7.3.3.22 Idarado's Proposed Remedial Plan

The company's plan calls for stabilization in place for Piles 1 through 4 utilizing construction of run-on/runoff diversion and contouring ditches; and erosional protection of the piles with side slope terraces, modification of slopes, revegetation, and riprap placed on the sides of the piles. Tailings material that eroded from Pile 2 would be scraped up and reloaded on to Pile 2 before stabilization. Future tailings migration from Pile 2 would be remedied by building a rock dike between the pile and the creek. The present decant towers on the piles would be removed and the openings plugged with bentonite. Revegetation would be by spreading an organic layer cover using manure, hay, and chemical nutrients along with seeding.

Idarado claimed its remedial measures would withstand the 500-year flood event, whereas those of the state were designed only for the 100-year event. The company also challenged the necessity of the state's plan of moving 1 million tons of material over a distance of 2.5 miles with the danger of affecting the environment; further, the company contended that the state's cost calculations were inaccurate by nearly $2 million underestimated. Idarado's cost was estimated at $930,000 (*Id.*, 1250).

7.3.3.23 The Court's Remedial Plan

The court found the state's plan more cost-effective and efficient. The major concern of the court for remediation appeared to be for aquatic life, because there was no imminent risk to human life. The court's plan was for the construction to be designed to withstand the 500-year flood event. It also ordered that the 6-foot multilayered cap was excessive and unnecessary, and that it be designed in accordance with the Telluride capping. A new and revised cost estimate was ordered by the court (*Id.*, 1250)

7.3.3.24 Mine Portal Discharges and Waste Rock Piles

Idarado's mine complex has over 100 miles of tunnels, drifts, and underground workings. Surface run-in waters enter the mine workings primarily through collapsed openings from mine subsidence (which the court has erroneously called "glory holes," unless the holes were made to be used as glory holes in the mining process); lesser amounts of surface run-in enter through geological features as joints and faults; and groundwaters also percolate through the mine workings.

There are 16 mine portals discharging water from the mine workings. Of those, 13 discharge water into the Telluride Valley drainage system of the San Miguel River; 11 portals in the High Country area discharge small amounts of water in the range of 5 to 10 gallons per minute (gpm) with a base flow of 2000 gpm. The Mill Level and Meldrum Tunnels drain the major portion of mine waters into the San Miguel River system. The Mill Level portal is located 2900 feet below the surface. Its base discharge flow is about 2000 gpm with a peak water discharge of about 13,000 gpm. Mill Level shaft waters discharge into a series of lagoons constructed by Idarado, eventually seeping into the ground and through the tailings piles. It has been estimated that Mill Level water discharge carries 91,000 pounds of zinc per year. Thus, it is considered a major source of metals loading into the San Miguel River (*Id.*, 1251).

The Meldrum Tunnel is 2000 feet below the surface and has a peak flow of 8000 gpm and a base flow of 300 gpm. The estimate for water-contained zinc discharge is about 60,000 pounds per year. The Black Bear Mine is considered the worst mine water source containing zinc concentrations of 460 milligrams per liter, as well as elevated lead levels. [*Author's Comment:* It should be noted that water samples from the San Miguel River presently meet Colorado's drinking water standards, yet failed to meet the water quality standards set forth in the state's ROD.]

On the Red Mountain Creek side of Idarado's mine, there are three tunnels (shafts) discharging mine waters. Treasury Tunnel is 1200 feet below the surface and has a peak water flow of 135 gpm and a base flow of 5 gpm. The annual zinc content contained in the mine discharge is estimated at 1500 pounds, with elevated levels of copper, lead, and cadmium (*Id.*, 1252). [*Comment:* Water samples taken from Red Mountain Creek irregularly pass and fail the state water quality standards.]

The court found that the discharge of contaminated waters from the Idarado mine workings from portals and through waste rock piles adversely affects the stream systems in both the Telluride and Red Mountain valleys to the detriment of the aquatic environment, and the judge held that Idarado was liable for the cleanup costs of those areas.

7.3.3.25 The State's Proposed Remedial Plan

The state's proposal consisted of two parts, viz., (1) diverting water away from the contamination sources; and (2) treating the contaminated waters.

7.3.3.25.1 Water Diversion

To prevent mine waters from contacting contamination sources, the state would plan to divert water around the waste rock piles, and from around the collapsed surface entries (glory holes); to plug, or close, the majority of High Country portals; and finally to divert water away from the mineralized areas within the mine.

Plugging of the Mill Level and Meldrum Tunnels (shafts) is not feasible because water pressure against the plug would be too great. [*Author's Comment*: Ironically, Idarado stated that it would not close the Stillwell Tunnel because it is a drinking water source for the town of Telluride.] To further reduce contamination, the state proposes to divert mine water underground away from ore chutes, ore passes, broken vein rock exposures, and other highly mineralized mine areas. [*Comment:* The state did not say how it proposed to divert the mine waters from those areas. Certainly cement application to the walls would not prevent rock-percolating waters.] (*Id.*, 1253).

7.3.3.25.2 Treatment of Contaminated Waters

The state's plan calls for diverting and separating the waters within the mine itself. All waters in the Idarado mine above the 1200-foot mine level would be diverted by gravity to the Treasury Tunnel on the Red Mountain side. This in-mine diversion is anticipated to reduce the flow from the Mill Level and Meldrum Tunnels on the Telluride side. Clean waters that meet the state's standards would be separated within the mine from the contaminated water within the mine, routed out the

Meldrum Tunnel, and discharged directly into the San Miguel River. Water below the 1200-foot mine level would be discharged from the Mill Level Tunnel on the Telluride side for treatment. On the Red Mountain side, all contaminated water would be discharged into Red Mountain Creek, become neutralized, and be treated in a passive reservoir. The state argues here that by diverting the contaminated mine waters into Red Mountain Creek, the stream's pH levels will be enhanced because the diverted water is more alkaline than the creek's water (*Id.*, 1253). [*Comment:* If the introduction of Idarado's mine discharge would reduce the natural acidity of Red Mountain Creek, and as the court has stated, "Red Mountain Creek is very acidic and contains high concentrations of metals," particularly before reaching the discharge of the Treasury Tunnel, then it would appear that possibly some other natural source is contributing the hazardous metals other than Idarado. The Red Mountain side of the Idarado mine complex was known to have been a naturally more acidic mineralized area. Thus, the creek would have been a naturally highly contaminated creek containing hazardous metals even without Idarado's lesser mine waters contribution that act to reduce the natural high acidity.]

The court noted Red Mountain Creek will be charged with lime or limestone to raise its pH from 3.5 to 6.5 or 7 pH. If limestone is used, the state plans to grind it with a ball mill to a fineness that will more readily mix with the creek water, using an agitator to maintain a slurry mix and prevent settling, before entering the stream via a piped slurry line (*Id.*, 1253).

The state's argument for acid neutralization is to cause precipitation of iron and aluminum in the waters, which will in turn cause sorption of the toxic metals copper, zinc, lead, and cadmium, taking them to the bottom of the treatment reservoir as precipitate. The treatment reservoir is to be located downstream a few miles and will hold water from Red Mountain Creek for 20 days while the toxic metals settle out. [*Author's Comment:* The court noted that this method of lime treatment for acid mine waters has not been successfully applied any place in the United States at this time (*Id.*, 1253).]

The state contended that this lime charging method would "...clean up twenty miles of stream, protect downstream areas from future degradation, and remove 90% of the metals from the Uncompahgre River ... and to enhance all present and future uses, ..., including irrigation, drinking water, and recreational uses" (*Id.*, 1253–1254).

For the Telluride side of Idarado's complex, a geochemical adsorption field treatment is planned. The adsorption field would utilize a series of infiltration lagoons (similar to Idarado's existing lagoon system) over a longer path of travel and employ fine glacial till particles as the adsorptive material as the precipitates settle to the bottom of the lagoons. The adsorptive field was to be constructed in a glacial debris fan approximately 100 feet above the valley floor. Treatment of the Marshall Creek drainage was proposed separately in a pond allowing for an 18-hour holding time and a capacity of 56,000 cubic yards.

The state's estimated cost for the water diversion and treatment plan for the Telluride–Red Mountain areas was approximately $18 million (*Id.*, 1254).

7.3.3.26   Idarado's Proposed Remedial Plan

Idarado proposed plugging the mine portals on its property on the Red Mountain side with the result that all mine waters would be discharged on the Telluride side. On the Telluride side, only the Black Bear portal would be plugged. The existing infiltration lagoon system on the Telluride side for the Mill Level and Meldrum Tunnels would be expanded by 25% for discharge of the mine complex. Sediments from the lagoon treatment would be disposed of on Piles 5 and 6 with subsequent cover and treatment. Their plan called for extensive mine water diversions to prevent contact between mine water and mineralized rock. Idarado opposed liability for plugging portals and diverting waters around waste rock piles that were not located on Idarado property. Idarado opposed the state's lime treatment reservoir on the Red Mountain side, arguing that it would fill with sludge in a year. They also opposed the adsorption field lower in the Telluride Valley on the glacial till debris fan as its discharge would be close to the town, and would affect the fish population and have an adverse social impact on the town, causing relocation of some of the residents. The company estimated that their proposed portal plugging and water diversion would result in a 67% reduction of zinc loading into Marshall Creek; and that their proposed internal mine work would result in a 21% reduction in zinc loading from the Mill Level Tunnel discharge, and a 53% zinc reduction from the Meldrum Tunnel discharge *(Id., 1255)*. Idarado's cost estimate was approximately $1.1 million.

7.3.3.27   The Court's Remedial Plan

The court found difficulty in deciding on its plan. The state's lime treatment plan on the Red Mountain side seemed uncertain as well as expensive, and was incomplete in several respects. The court stated, "(It) was not persuaded that plugging the mine portals on Idarado's property will either stop the harm or end their liability" *(Id., 1255)*.

The court declined to order Idarado to plug mine portals or divert water around waste rock located on property in which the company did not have a possessory interest. The court found that the state's plan should be implemented for water diversion on both sides of the High Country (i.e., the Red Mountain and Telluride sides). It also ruled for implementation of the state's water treatment plan, viz., neutralizing Red Mountain Creek waters; and construction of the Ironton Park reservoir, the geochemical adsorption field for the mine portals, and the Marshal Creek pond on the Telluride side. A revised cost estimate was ordered *(Id., 1256)*.

7.3.3.28   Mill Site Cleanup

There were two mill sites requiring cleanup, viz., the Pandora Mill on the Telluride Valley side, and the Red Mountain Mill site. Red Mountain includes warehouses and office buildings. In addition to the hazardous metals found over the complex, in 1985, polychlorinated biphenyl (PCB) transformers and PCB-contaminated soils were discovered at three surface areas at the Pandora Mill. On further investigation, a total of eight waste source areas were designated at the mill site. In general, at all eight sites, where the transformers, sludges, organic mill reagents (e.g., xanthate), and drum-stored waste oils had stained the soil, Idarado had removed most of them. The company had already excavated the stained soil areas to approximately 10 inches, placed the contaminated soil in drums, and disposed of the containerized soil, along with the transformers and trash, at off-site licensed waste disposal facilities (*Id.,* 1256–1257).

7.3.3.29 The State's Proposed Remediation Plan

PCBs and other organic materials were to be removed to a licensed waste disposal facility. Cleanup of PCBs spilled before May 5, 1987 was required to attain a cleanup of 0 to 2 ppm if practicably attainable. All tailings and soils with lead concentrations in excess of 1000 ppm were to be removed from the mill sites. Mill site areas that are disturbed by remedial activity would be revegetated. The state's estimated costs are figured at approximately $644,000. Subsequent monitoring activities would cost an additional $880,000 (*Id.,* 1257).

7.3.3.30 Idarado's Proposed Remedial Plan

During trial, Idarado had reached an agreement with U.S. EPA under the TSCA concerning the PCB spills and PCB-containing equipment at the site that allows cleanup to a level of 10 ppm. Idarado moved to reopen the evidence on the mill site cleanup in view of the new cleanup work done there.

Mill site tailings were to be covered and revegetated in place, if feasible. Through July 1987, Idarado had already spent $587,000 on cleanup activities at the mill site. The court granted Idarado a reopening for new evidence on the mill site cleanup.

7.3.3.31 Miscellaneous Remedial Matters

Four remaining remedial matters were (1) relocation of the Pandora Trailer Park and other residents; (2) revegetation of disturbed areas; (3) habitat enhancement and fish stocking; and (4) performance monitoring and compliance verification.

7.3.3.32 Remedial Plans for Relocation of the Pandora Trailer Park and Other Residents

The State's Proposed Plan — Colorado had proposed to relocate 70 to 80 residents and 18 to 24 trailers at the Pandora Trailer Park in Royer Gulch. The trailer park provides low-income housing for the town of Telluride. The trailers are located on land owned by Idarado near the tailings Piles 1 through 4.

The state contended that the material at the Royer Gulch trailer park is glacial till fan debris, is near the tailings piles, is readily available, and is a natural low-cost source of cap material. It would also eliminate the need of transporting cap material through the town. The same fan debris location at Royer Gulch would be used to construct the geochemical adsorption field as proposed by the state (see Section 7.3.3.25 on treatment of contaminated waters).

Idarado's Proposed Plan — Idarado's plans for treatment of the tailings piles and the Mill Level Tunnel discharge would not require relocating any of the trailer park residents. Further, Idarado argued that the residents could not be relocated under CERCLA § 101(24) absent a Presidential order.

The Court's Remedial Plan — The court ruled Idarado incorrect concerning removal of the residents requiring a Presidential order. Judge Carrigan held that a Presidential order is not required, and ruled that the residents must be relocated so that the state could implement the court's remedial plan. Idarado was ordered to not take any contrary action concerning the trailer property without a prior order from his court. The state was ordered to follow through with a specific plan for removal of the trailer residents and trailers (*Id.*, 1259).

7.3.3.33 Remedial Plans for Revegetation of Disturbed Areas; Habitat Enhancement, Fish Stocking and Performance Monitoring, and Compliance Verification

The state's overall remedial plan called for establishing permanent, effective, self-sustaining vegetation on tailings piles caps, tailings removal areas, construction areas, diversional water areas, borrow areas for cover material, mill sites, and areas disturbed by remedial measures.

Idarado had agreed to revegetate tailings piles and those areas disturbed during remediation, if the areas were previously vegetated. The company objected to revegetating the mill yard areas because there had not been any vegetation there for many years. The court found that under the Colorado Mined Land Reclamation Act, every operator of a mining operation is required to reclaim all affected lands (*Id.*, 1259–1260).

Habitat Enhancement and Fish Stocking — The state's proposed aquatic habitat enhancement included placing random rock and gabions in the streams to create rifles and pools, construction of diversion structures and dams, reconstruction of riparian vegetation, and planting of trees along stream banks.

For fish stocking, the state proposed to stock brook, rainbow, and brown trout in the San Miguel River. Idarado opposed fish stocking, arguing that rainbow and brown trout are not native to the San Miguel River area. Idarado contended that there is no causal connection between the release from the mining operations and the absence of rainbow and brown trout in the San Miguel River. The court concluded that habitat improvement and fish stocking were necessary to mitigate damage to the environment caused by the mining operations. Idarado's objections to attempted restoration to exact pristine conditions were without merit. CERCLA response action did not anticipate replication of exact conditions existing before the land and water were despoiled as a response action. The state's estimated cost for implementation was to be $188,000 (*Id.*, 1260).

Performance Monitoring and Compliance Verification — The court ordered the state's plan for monitoring and compliance as set forth in their ROD to be cost-effective and efficient for implementing the plan to conform with the court orders (*Id.*, 1261).

7.3.3.34   Intervention by Town of Telluride

The Town of Telluride attempted to intervene in the litigation to collect incurred pollution treatment costs in excess of $1.236 million. Telluride's attempt to intervene in the suit was denied. However, the court noted that under CERCLA, a town is a "person" and can sue on its own behalf.

7.3.3.35   The Court's Conclusions and Holdings

On February 22, 1989, Judge Carrigan of the U.S. District Court for the District of Colorado ruled that "the State of Colorado's CERCLA response was not inconsistent with the NCP, ... that applicable or relevant appropriate federal and state environmental and public health requirements have been identified and applied as set forth in the State's Record of Decision; ... the State's response efforts are cost-effective except as modified by this Order, ... effectively protects and minimizes threats to public health, welfare, and the environment; ... that the State's ROD is not arbitrary and capricious; ... that the defendants are strictly liable, jointly and severally liable, to the State of Colorado for all costs of remedial actions incurred; ... Defendants are ordered and mandatorily enjoined to conduct the remedial activities as described in the State's ROD, to the extent they are in accordance with this Order, in a cost-effective manner at the direction and under the oversight and supervision of the State of Colorado ..." (*Id.,* 1261–1262).

More specifically, the court held the following:

(A) With respect to the Telluride tailings piles, cleanup liability shall be imposed against the defendants in accordance with the court's findings and conclusions, and declaratory judgement shall be entered in favor of the State on this issue; the State is ORDERED to submit within thirty days a revised cost estimate regarding the implementation of the cleanup plan as selected by the Court.

(B) Liability shall be imposed against the defendants regarding the cleanup of the Society Turn tailings in accordance with the court's findings and conclusions, and declaratory judgement shall be entered in favor of the State on this issue; the State is ORDERED to select the tailings removal site within six months.

(C) With respect to the miscellaneous tailings, cleanup liability shall be imposed against the defendants in accordance with the court's findings and conclusions, and declaratory judgement is entered in favor of the State; the State is ORDERED to submit a more specific cleanup plan within ninety days, including a revised cost estimate.

(D) Defendants are not liable for clean up of the Town of Telluride soils, and the State's request for declaratory judgment on this issue is denied; the defendants shall, however, implement, and bear the cost of, their proposed five year monitoring plan of Telluride area residents in accordance with the court's findings and conclusions.

(E) Cleanup liability shall be imposed against the defendants with respect to the Red Mountain tailings in accordance with the court's findings and conclusions, and declaratory judgement shall be entered in the State's favor on this issue; the State is ORDERED to submit a revised plan and cost estimate within ninety days.

(F) With respect to the mine portals and waste rock piles, the defendants are liable for the cleanup of these areas in accordance with the court's findings and conclusions, and declaratory judgement is entered in the State's favor on this issue; the parties shall prepare and file within thirty days a written stipulation regarding all remedial action concerning this issue that has been agreed upon; the State is ORDERED to submit within ninety days a more specific cleanup plan and revised cost estimate, as indicated in this Order.

(G) The parties jointly, shall apply, in writing, within eleven days, for an expedited hearing date regarding the issue of the millsite cleanup; the parties are further ORDERED to meet in advance of that date to stipulate all facts as to which there is no good faith dispute and agree upon admission of exhibits.

(H) The Pandora Trailer Park shall be relocated in accordance with the court's findings and conclusions; the defendants are enjoined from taking any action regarding this property contrary to the court's findings and conclusions without an order from this court; and the State is ORDERED to file a more complete relocation plan and revised cost estimate regarding the trailer park's relocation within ninety days.

(I) Defendants shall be liable for revegetation of disturbed areas, habitat enhancement, and fish stocking as set forth in the court's findings and conclusion; declaratory judgement shall be entered in favor of the State on these issues.

(J) The State shall file within ninety days a modified performance monitoring and compliance verification plan in accordance with the court's findings and conclusions.

The renewed motion to intervene filed by the Town of Telluride is denied ...
Dated at Denver, Colorado this 22nd day of February 1989. (*Id.*, 1261–1263)

7.3.3.36 *Final Litigation and Out-of-Court Settlement*

In February 1990, Idarado filed a motion to amend the opinion and order of the Federal District Court of February 22, 1989, concerning the response costs incurred prior to December 9, 1983 (date of Colorado's filing of the suit) of $1.9 million. Defendants also notified the district court that they had appealed the injunctive provisions of the same order with the Tenth Circuit Court of Appeals.

In the district court's opinion of April 20, 1990 (735 F. Supp. 368), it said that it was unclear at this time whether there was jurisdiction under CERCLA for the court to decide the defendants' challenge against a mandatory order for the defendants to follow the court order for the response actions. The matter would have to be ruled on by the Tenth Circuit on defendants' appeal.

Concerning response costs and prejudgment interest dates, the district court held:

1. The state's motion to amend the trial court's judgment to authorize payment of prejudgment interest incurred since 1977 is denied.
2. The state may recover prejudgment interest on the response costs incurred since December 9, 1983, the date it filed its complaint in this action.
3. In all other respects, the state's motion to amend the trial court's judgment is granted.

7.3.3.37 *No Injunctive Power under CERCLA or SARA*

On its appeal by Idarado Mining Co., et al. (916 F.2d 1486, 10th Cir. 1990), the United States Court of Appeals for the Tenth Circuit decided on October 11, 1990 that the district court did not have the authority, implied or otherwise, under CERCLA to grant a mandatory injunction to a state government ordering the defendants to carry out a remedial plan proposed by the state along with certain remedial alterations made by the court. Section 106(a) of CERCLA provides for the federal government to seek "such relief as the public interest and the equities of the case may require.... Such relief may include a mandatory injunction compelling responsible parties to perform a cleanup. However, § 106(a) does not allow a state, with or without a cooperative agreement, to seek injunctive relief against responsible parties" (*Id.*, 1489). Thus, the mandatory injunctive order issued to the State of Colorado by the federal district court for enforcement of the specified hazardous response actions could not be applicable to the state government for enforcement. Therefore, the case was remanded to the district court to vacate its mandatory injunction for defendants to perform the state's cleanup plan. Idarado was, nevertheless, the responsible party for the cleanup, but not held under an injunctive order to comply in all respects with the court's ordered actions.

On the issue of the district court ordering the permanent relocation of the trailer park residents and trailers to use the trailer site as part of the remediation construction facilities, the circuit court of appeals found that this, too, was not within their jurisdiction and power under CERCLA. That issue was remanded to the district court for it to vacate its order for permanent relocation of the residents and trailers.

As of the fall of 1991, the State of Colorado and Idarado were negotiating an out-of-court settlement as to remedial measures. Although the parties had partially reached an out-of-court agreement for the response and remedial procedures and a court decree approved the settlement, the State of Colorado and Idarado Mining Company have continued the cleanup over intervening years, but not without minor disputes as recently as 1998. However, the intervening disputes have been resolved without recourse to litigation.

7.3.4 Penn Zinc and Copper Mine, Calaveras County, California

Case decisions involved:

1. *Committee to Save the Mokelumne River v. East Bay Municipal Utility District and the Regional Water Quality Control Board*; No. Civ. S-91-1372, U.S. Dist. for E.D. Cal. (1993)
2. *Committee to Save the Mokelumne River v. East Bay Municipal Utility District and the Regional Water Quality Control Board;* 13 F.3d 305, U.S. App. 9th Cir. (1993)
3. *East Bay Municipal Utility District v. U.S. Department of Commerce*, 948 F. Supp. 78, U.S. Dist. Ct. for District of Columbia (1996)

The Penn Mine site, a former zinc and copper mine in Calaveras County, California, is a unique AMD and abandoned mine case study from two standpoints, viz., (1) the site is jointly owned and operated by two California governmental entities, and (2) the entities attempted to involve the U.S. government as an "operator" responsible party liable for the remedial and response costs of the site cleanup. Also, the first litigation concerning the site's water pollution was brought under the citizens' suit provisions of the CWA.

As codefendants in these first actions, the two local government entities owning and operating the facility are the East Bay Municipal Utility District (EBMUD) and joint owner, the California Regional Water Quality Control Board (RWQCB), Central Valley Region, a California state agency.

7.3.4.1   Background History of Penn Mine

The Penn Mine had been operated for zinc and copper as far back as the 1850s. The ore was mined primarily for copper. Approximately 87% of all the ore ever produced from the mine was produced by 1919. Mining operations were intermittent thereafter until its final operations in World War II to furnish zinc for the war effort. Unsuccessful attempts for mining zinc were made as late as the Korean War period. The mine has been shut down since 1953.

Former mining operations had left spoil piles of low mineralization waste rock and mill tailings over an area of 20 acres, which became a source point for the AMD resulting from exposure of the mining wastes to oxidation and surface rainwaters seeping into the ground and flowing on the surface into nearby streams, eventually reaching and polluting the Mokelumne River.

EBMUD owns water rights on the Mokelumne River and supplies water to towns and cities east of San Francisco. In the mid-1960s, for environmental reasons, EBMUD needed to construct a storage and stream flow reservoir on the Mokelumne River downstream from the Penn Mine, to be known as the Comanche Reservoir. Authorization for the development was conditioned on EBMUD's acquisition of a small portion of the inactive mine through condemnation. Although the owner of the Penn Mines was directed to take corrective actions to mitigate the effects of the AMD, it did not. A regional environmental board then formally requested EBMUD to abate the problem by building an impoundment on the portion of land it owned at the Penn Mine site. In 1978, EBMUD, joined by the RWQCB, constructed the Penn Mine facility (the facility) in an attempt to reduce the threat of continued toxic runoff from the site. EBMUD built the impoundment, known as the Mine Run Dam and the Mine Run Reservoir, along with a series of smaller impoundments, drainage ditches, culverts, channels, etc. (*East Bay Municipal Utility District v. U.S. Department of Commerce*, p. 81).

The facility was designed to capture contaminated water flowing through the site, and to contain and evaporate the water through a ponding and recirculation system. Collected AMD from the mine site runs from the upper impoundment through a cascade of three impoundments, flowing to the lower impoundments, eventually collecting in the Mine Run Reservoir. Polluted water from Mine Run Reservoir was recirculated back to upper impoundments. EBMUD has conducted

investigations and studies for over 16 years concerning the environmental problems of the Mokelumne River, the Comanche Reservoir, the Mine Run Dam, and
the Penn Mine site. From time to time, polluted runoff had passed over the
spillway or through the dam's discharge valve into the Mokelumne River and
Comanche Reservoir in violation of the CWA. The CWA prohibited a discharge
of any pollutant into navigable waters from any point source without a permit.
Certain discharges are authorized by the EPA administrator under the NPDES.
EBMUD and RWQCB/the board were found not to have an NPDES permit for
their polluted discharges.

7.3.4.2 CWA Citizens' Suit to Save the Mokelumne River

The Committee to Save the Mokelumne River (the Committee) initiated a legal
action against the district (EBMUD) and members of the board under the citizens'
suit provisions of the CWA (33 U.S.C. § 1365). The Committee sought judgment
declaring defendants had discharged pollutants without a permit in violation of the
CWA and enjoining them from discharging further polluted waters from the Penn
Mine facility until they obtained the NPDES permit, and also seeking an ordered
remedial plan to remove and dispose of the contaminated sediments in the reservoir.
In March 1993, the Federal District Court for the Eastern District of California
granted judgment for the Committee on the issue of liability.

On appeal to the Ninth Circuit Court of Appeals, defendants raised four issues
and defenses that the district court erred on, viz.: (1) Mine Run Dam is not subject
to the discharge permit requirements of the CWA; (2) whether defendants have
discharged a pollutant within the meaning of the CWA; (3) defendants' activities in
operating the facility are regulatory and therefore do not constitute "additions of
pollutants" under the CWA; and (4) defendants, as government entities, are immune
from liability under the CWA (*Committee v. EBMUD, Id.* at 307).

The court of appeals rejected all four of defendants' arguments of error by the
district court, thus affirming its decision. Consequently, the district and the board
had to obtain an NPDES permit and conform its water quality discharges to the
specifications of the EPA into the navigable waters of the United States.

7.3.4.3 CERCLA Litigation and Decision for the Penn Mine Site

EBMUD next brought a CERCLA court action against the U.S. government
seeking recovery of certain remediation costs incurred, or future costs it may incur,
in cleaning up the hazardous waste and AMD from the Penn zinc and copper mine
located upstream of the Mokelumne River in Calaveras County.

At issue in *East Bay Municipal Utility District v. U.S. Department of Commerce*,
948 F. Supp. 78, U.S. Dist. Ct. D.C. (December 1996), was the federal government's
involvement in the activities and operations of the Penn Mine during and after World
War II. EBMUD claimed that the government's involvement derived from combined
efforts of the War Production Board (WPD), the War Manpower Commission
(WMC), the Office of Price Administration (OPA), the Reconstruction Finance
Corporation (RFC), and the Metals Reserve Company (MRC — a subsidiary of the

federal RFC agency); that these agencies issued numerous regulations and orders related to the Penn Mine's production of zinc; and that those regulations and orders allegedly controlled the type and quantity of ore to be mined, and the production of schedules, prices, marketing, and labor. In effect, EBMUD alleged that the government, beginning in 1942, exercised sufficient control over the mine to incur liability as an operator of the mine, or as an arranger of the mine's waste disposal. Under CERCLA, the federal government is deemed a person. Thus, the government is strictly liable for its actions as an owner, operator of a property; an arranger, disposing of hazardous substances at a facility; or a person accepting hazardous substances for transport or disposal of such hazardous substances.

In reviewing the factual background and evidence of the government's activities at the Penn Mine, the court noted the degree of the government's involvement in critical areas of mine operation (e.g., labor, daily operations, supervision, financing, mill quantity and quality, waste rock handling and disposal, pricing of ore, and sales and marketing).

Labor

In the summer of 1942, the WMC had declared all 12 western states a critical labor area, deeming all nonferrous metal mining as essential war production industries. No worker engaged in mining was allowed to seek other employment without a certificate of separation from the U.S. Employment Service. Area employers were prohibited from employing any former mine worker who did not have a certificate of separation. On October 8, 1942, the WPB closed all nonessential gold mines, thus freeing some 1600 miners for employment in other critical mines. Draft boards granted deferments to miners in the nonferrous mines. If a miner left his job for other employment, his deferment was revoked.

Labor regulation also involved job priority classifications, recruiting of new workers, and a nationwide job referral system to provide more miners. In 1945, the U.S. Employment Service referred 132 workers to the Penn Mine management. Of those, 60 were hired, but a shortage of workers remained and the production lagged behind schedule. After a site visit, a WMC official reported the labor shortage was affecting production.

In spite of those labor-hiring facts, the court found that the Penn management did not have to hire those referred workers. Worker referrals were merely assistance offered. The government did not hire, fire, or otherwise supervise or discipline the workforce at the Penn Mine. Thus, the government was not an employer.

Daily Operations Supervision, Mill Quantity and Quality, Waste Rock Handling and Disposal

There was no evidence that the government had any day-to-day, on-site presence at the mine, or that it was involved in operational activities. Nor was there evidence that the government sat on, or participated in, any board of directors; physically disciplined or discharged Penn employees or managers; inspected or approved the ores; dictated how the operators were to use the equipment, facilities, or the workforce; or required any particular waste disposal method or technology for the disposal or treatment of waste. Further, the company never consulted the government about the manner of waste disposal.

With regard to mine production, the Penn operation continually lagged. At the time the contract for government purchases of zinc concentrate ended, the Penn operator (Eagle Shawmut) was short 3000 tons.

Federal Financing, Pricing of Ore, Sales and Marketing

From November 1943, the RFC and the MRC had approved 14 mining projects to receive federal funds, inclusive of the Penn Mine. In 1943, Eagle Shawmut Mine, a nearby gold mining company, leased the Penn Mine. Shawmut applied for and received a loan authorized by the WPB for $120,000. At the end of the war, Shawmut paid back its loan in full to the government.

During the war period, a zinc producer could deliver zinc only by presenting an allocation certificate issued by the WPB, specifying the grades and amount that could be shipped. No buyer could accept delivery of zinc except as authorized by government regulation. Willful violations subjected the offender to a fine or imprisonment. However, the entire output of the Penn Mine's zinc concentrate, up to 10,000 tons, was under a sales purchase contract to the government (MRC).

Price controls and subsidies were also implemented by government agencies. The WPB authorized a premium price for Penn Mine zinc production and later, in 1943, authorized a subsidy in addition to the premium price that continued to 1946, but was reenacted by Congress, extended to June 1947, and administered by the RFC to encourage mineral exploration.

After the war, in February 1947, Shawmut received a $100,000 loan from the RFC for exploration at the Penn Mine. The project was terminated in 1947 and Shawmut abandoned the Penn Mine lease. The mine was reopened in 1949 by a private company, operated until 1953, and then closed again.

CERCLA Liability

Under CERCLA, a person who incurs cleanup costs is entitled to recover from anyone who qualifies as a responsible person. As stated earlier, the federal government is deemed a person under CERCLA and may be found liable as a responsible person. The federal government may not claim sovereign immunity in such a case under CERCLA.

Responsible persons under CERCLA include (1) the current owner, or operator, of the facility; (2) any person who owned or operated the facility at the time of disposal of any hazardous substances; (3) any person who arranged for disposal or treatment of hazardous substances at the facility; and (4) any person who accepts hazardous substances "for transport to disposal or treatment facilities, incineration vessels or sites" (42 U.S.C. 9601 et seq.).

When "the responsible parties are identified, the court has authority to allocate response costs among the liable parties using such equitable factors as the court determines appropriate" (42 U.S.C. 9613[F][1]).

EBMUD argued that the extent of the federal government's activities at the Penn Mine made it an operator and an arranger of disposal of waste under CERCLA. In response, the government claimed sovereign immunity. However, the court rejected that defense of the government stating, "There may be circumstances under which government regulation is so pervasive and focused that the company is, effectively, a 'subsidiary' of the government. In such a situation, the government should not be able to hide behind the shield of sovereign immunity simply because it activities may be characterized as regulatory."

The D.C. court noted that CERCLA provides, "Each department, agency, and instrumentality of the United States (including the executive, legislative, and judicial branches of the government) shall be subject to, and comply with, this chapter in the same manner and to the same extent, both procedurally and substantively, as any non-governmental entity, including liability under section 9607 of this title" (*East Bay Municipal Utility District v. U.S. Department of Commerce*, p. 88).

Further, that under SARA of 1986, "the new subsection 9620 ensured that the federal government would be liable to the full extent as private parties under § 9607 of CERCLA."

The D.C. Court noted that, "Courts have imposed operator liability under two alternative tests. First, operator liability may result from actual or substantive control over a facility, with active involvement in its activities. [See *FMC Corp. v. U.S.*, 29 F.3d p. 843 (3d Cir. 1994).] Second, operator liability may attach if the party in question has the authority to control the operations of facility or decisions involving the disposal of hazardous waste."

For this defense, the D.C. Court held that under either of the two theories, the government's conduct at the Penn Mine did not meet the standard for CERCLA operator liability. Additionally the court noted that, "In this case, ... (unlike *FMC Corp.*), the government's regulatory efforts were not directed specifically at the Penn Mine, but rather were aimed at directing the wartime economy in general.... Wartime regulations did affect the Penn Mine. However, that did not make it unique. Penn simply took advantage of many government programs available to all mines ... which provided financial incentives to meet wartime goals." (*East Bay Municipal Utility District v. U.S. Department of Commerce*, p. 89).

Consequently, EBMUD failed to show that the government exercised that degree of actual control over the Penn Mine necessary to make it an operator under CERCLA. As to EBMUD's charge of "arranger" of waste disposal at the Penn Mine, the court stated, "Although a party need not be actively involved in waste disposal to be an arranger, there must be a connection between the alleged liable party and the actual disposal." The court found no such nexus to exist in this Penn Mine case.

Citing the Second Circuit's case [*General Electric Co. v. Aamco Transmissions, Inc.*, 962 F.2d 281, 286 (2d Cir. 1992)], the court added "that the mere existence of certain economic bargaining power which would permit one party to impose certain terms and conditions on another, does not itself create an obligation under CERCLA" (*Id.*, 94). The court concluded that the government could not be held as an arranger of waste disposal at the Penn Mine.

The D.C. Court dismissed the charges against the government and granted summary judgment for the United States. Thus, EBMUD found no other responsible party to share the CERCLA response costs for cleanup of the Penn Mine site.

7.3.4.4 Comments on CERCLA's Application to Abandoned Mine and AMD Sites

CERCLA has many aspects about it that are of an *ex post facto* law nature. The party held responsible is now being punished for the acts of others as well as their own, and made alone to correct those acts, or remedy a resulting situation, that when performed was not illegal, against any statute, or wrong. Instead, the acts were accepted practices in the mining industry and by the nation and states at the time done. Now, at a later date, those same legal acts have been made illegal and punishable. *Ex post facto* laws are specifically prohibited by the United States Constitution. At the time the mining occurred, society was glad to receive the benefits of the mining operations that extracted the minerals from the earth, all without blame for the accepted process and practice. With environmental concerns of decades later, society now absolves itself and places the blame on the mining industry, when society itself was as responsible and guilty as the miners themselves for the alleged contamination by their acts of condoning the exploitation at the time, and also for accepting the benefits. A fairer way to distribute the costs of remedial response is for society to share in the expenses of cleaning up the environment, not to single out a former mine operator and punish the mining industry alone.

Percolating groundwaters flowing through toxic mineralized areas of the earth are acts of God or nature. Naturally occurring higher metal content groundwaters would result from flowing through the mineralized zones and would contaminate streams and soils of those surrounding areas with high metal values, though, granted, to a lesser contaminating degree because of the absence of oxidation of the exposed minerals by mine openings. Higher metal contents in groundwaters are to be expected in metallic provinces, whether man did or did not exploit the mineral deposits.

As an example, a high proportion of the counties in the State of Missouri show a higher than normal presence of lead in the ground and in groundwaters. This may fairly be attributed to the fact that Missouri lies within a lead-bearing, metallic geoprovince and not to the lead mines that have been opened and mined.

7.3.4.5 Legal Outlook for CERCLA as an Ex Post Facto Law

See *ex post facto* definition under Section 7.3.1.6.1, B.4. At the turn of the new century, in an article in the *National Law Journal* of January 1999, California attorneys Bruce Howard and Jennifer Harr discussed the prospects of the CERCLA statute's retroactive aspect in view of a change of direction of recent decisions by the U.S. Supreme Court. Howard and Harr noted, "Two Supreme Court cases evidence growing judicial opposition to laws that impose liability retroactively, and may signal a new willingness (of the Supreme Court) to limit the retroactive reach of CERCLA."

In its recent 1998 decision, the U.S. Supreme Court reversed the decisions of the district trial court and the affirmation of the First Circuit Appeals Court in *Eastern Enterprises v. Apfel,* 118 S. Ct. 2131 (1998). "*Eastern Enterprise* concerned the Coal Industry Retiree Health Benefit Act of 1992 (a.k.a. 'Coal Act') ... The Act established a Combined Fund, and assessed premiums against certain coal operators according to a variety of allocation formulas.

"The Commissioner of Social Security Apfel assigned to Eastern Enterprises Combined Fund premium obligations of more than 1000 retired miners who worked for the company before 1966. Eastern sued, arguing that the Coal Act violated due process and constituted a taking in violation of the Fifth Amendment."

On review by the Supreme Court, the court noted "... that the aim of the takings clause is to 'prevent the government from forcing some people alone to bear public burdens which, in all fairness and justice, should be borne by the public as a whole.' Acknowledging that a determination of 'fairness and justice' is 'essentially *ad hoc* and fact intensive,' the court delineated three significant factors: the economic impact of the regulation; its interference with reasonable investment-backed expectations; and, the character of the government action' (1998 Lexis 4213, p. 43).

Thus, the future expectations of CERCLA being declared *ex post facto* by the federal courts, at least in certain abandoned mine cases and sites, are bright. This may be particularly true where there are "orphan shares" of liability in which portions of the response costs might have been properly allocated to defunct mining companies causing the earth disturbance.

7.4 EXAMPLE OF A NON-CERCLA AMD DAMAGE COMMON LAW CLAIM SUIT

Inactive Surface Mine Owner Pays Millions for Water Pollution

by R. Lee Aston, Mining & Enviromining Law Editor,
Mineral Resources Engineering Journal (London)

Note: The following case review is notably different from acid mine drainage and water pollution cases brought under the Comprehensive Environmental Response, Compensation and Liability Act (CERCLA) because it was brought as a common law claim for damages by acts of nuisance, negligence, trespass, etc. Further, the abandoned mine site at issue has not been placed on the National Priority List of hazardous sites.

The case review is also a lesson in the maxim "a word to the wise is sufficient." Perhaps few owners of abandoned, worked-out pits, or surface mines may find themselves in a similar environmentally hazardous situation as in the following case, and their inactive mining property may not contain hazardous minerals (as pyrite). However, even inactive mining properties not containing hazardous metals may be subject to erosion that contribute sediments to nearby streams, thereby increasing their turbidity and are susceptible to environmental water pollution violations and damage claims. Another maxim applies — "an ounce of prevention is worth a pound of cure," or more likely, a few pounds of $10,000 bills! The "prevention" is in getting a hydrogeological report on whether the inactive mine property is polluting "waters of the United States" from an environmental or geological engineer before a claim is filed in court for water pollution.

In *Johansen, et al., v. Combustion Engineering Inc.*, 170 F.3d 1320 (11th Cir. 1999), sixteen neighboring landowners to an abandoned surface mine sued the former operator and present owner of the mineral property in 1991 alleging damages to their properties by acid mine drainage (AMD) contamination of their streams from acid waters flowing from the closed down mining operation. In May 1992, several other landowners joined the suit.

Since trial of the claim in 1993 in a federal district court in Georgia, the case has been litigated from a successful damages award against the abandoned surface mine owner in an amount of over $45 million to the neighboring landowners until its final decision in 1999. The claim's monetary award has been carried all the way from the Federal District Court in 1993 to the 11th Circuit Court of Appeals, to the U.S. Supreme Court, and back to the U.S. Court of Appeals, then returned to the District Court and back to the Appeals Court in 1999. In *Johansen, et al., v. Combustion Engineering Inc.*, the final award to the landowners was reduced to $4.35 million in punitive damages, plus compensatory damages of $47,000 and post-judgement interest on the awards from June 1994.

Background Case History

In the 1930s, Tiffany, a fine jewelery company, started mining the Grave's Mountain site near Lincolnton, in Lincoln County, Georgia, for rutile, a titanium oxide mineral used for polishing diamonds. About 1960, Aluminum Silicates, Inc. began mining the site by surface quarrying/benching methods for kyanite, the predominant mineral at the site. [Kyanite is an anhydrous aluminum silicate used as a refractory in fire brick, high temperature-resistant tiles (e.g., spark plugs, space craft tiles) and cements, etc.] In the mid-1960s, Aluminum Silicates sold the surface mine and property to Combustion Engineering, Inc. (CE), who mined kyanite until 1984 when it sold the property to Pasco Mining Company. Pasco continued mining for until 1986 when it defaulted on the purchase terms. The property and all environmental responsibility reverted to CE.

Graves (or Little) Mountain was essentially a large, bare rock exposure rising several hundred feet above the surrounding area. In mining, the operators mined, crushed and milled the kyanite-bearing rock to extract the kyanite. The waste rock tailings from milling was placed in tailings ponds. One of the wasted minerals was pyrite ("fool's gold"), an iron sulphide, which was the culprit in forming acid mine drainage (AMD). When water (rain or other waters) washes on pyrite that has been exposed to oxygen, the water becomes acidic. The acidic waters, thus formed, seeped into the streams that flowed through CE's Graves Mountain property lowering its quality as it ran through properties downstream.

Damage Claims and Litigation Begins

In 1991, sixteen owners of land tracts downstream from the former mine site sued CE claiming damages for trespass and nuisance by virtue of the acidic waters flowing from CE's inactive mining property through their lands. The landowners alleged that the waters from the abandoned surface mine workings causes heavy metals to be carried on to their lands, and that it also contains high sulphate levels which harmfully lowers the pH levels of the waters.

Landowner-plaintiffs presented experts in their behalf to offer a demanded plan for construction of a system of wetlands to remove the heavy metals and sulphates from the

Graves Mountain run-off waters. The pollutants would settle in the wetlands and be filtered out before continuing downstream to contaminate lands of the plaintiffs.

Plaintiffs estimated the cost of constructing the wetlands, plus 20 years of maintenance, would reach approximately $20 million. In contrast, the court noted that the estimated, aggregate fair market value of all the plaintiffs' lands, "absent contamination problems" was only $1.347 million. The greatest, estimated, diminution in value of plaintiffs' lands from contamination was no more than 50%.

The general rule for the measure of damages in actions for injuries to real property is the difference in value before and after the injury to the premises. Whether the damage is permanent must also be determined. Where there is a continuing invasion to the land, as here, several elements of damage must be distinguished: (1) those damages attributable to past invasions, and (2) those from future invasions. Further, past damages must be differentiated between those that are lasting, and those that do not leave a lasting impact, i.e., that will heal themselves. Federal courts must be guided by the state law in which they sit. Regarding the calculation of damages for future invasions, the Georgia Supreme Court has adopted § 930 of the Restatement, Second, of Torts. Under this pertinent section, the court found that the owners of land contaminated by run-off from the mine were not entitled to an award of treatment costs in excess of the diminution value of their lands, absent a showing of sufficiently unique characteristic or use of their lands (example: historical value).

In 1993, the landowners sought actual damages, costs to restore their lands and prevent future injury, exemplary damages, litigation costs and injunctive relief. The federal district court granted Combustion's motion holding that: (1) the measure of damages did not include treatment costs in excess of the diminution of value of plaintiffs' lands, and (2) evidence of treatment costs was not admissible to show unreasonableness for C.E.'s clean-up efforts.

"Originally, in 1993 the case was tried to a jury in a two-phase trial in which issues relating to punitive damages were decided separately from liability for the underlying torts (trespass and nuisance) and compensatory damages. The jury was instructed that the relevant time frame for damages purposes was the Georgia statute of limitations for trespass and nuisance during a four-year period prior to the commencement of the property owners' suit.

"In the first phase of the trial, the jury returned a total of thirteen verdicts for compensatory damages in favor of the various property owners in an aggregate amount of $47,000. The thirteen verdicts ranged from $1000 to $10,000. The jury also awarded property owners litigation costs in the amount of $227,000. In the punitive damages phase of the trial, property owners were required as a matter of Georgia law to prove by clear and convincing evidence that CE's actions 'showed willful misconduct, malice, fraud, wantonness, oppression, or that entire want of care which would raise the presumption of conscious indifference to consequences.' [O.C. G.A. § 51-12-5(b)]. To recover more than $250,000 each, property owners were required to demonstrate, again by clear and convincing evidence, that CE acted 'with the specific intent to cause harm.' [Id. at § 51-12-5.1(f)]. The jury awarded $3 million in punitive damages to each of the fifteen property owners who owned the sixteen parcels of land at issue, for a total of $45 million.

"The district court found this amount 'shocking' which, if allowed to stand would 'give the system a black eye.' The court entered an order granting CE's motion for a new trial unless property owners agreed to remit all punitive damages over $15 million. The property owners agreed to do so and the court entered separate judgments totaling $15 million in punitive damages. CE appealed to the 11th Circuit Court. During the pendency of the appeal, CE settled with three property owners, leaving an aggregate of $43,500 in compensatory damages and $12 million in punitive damages at issue in the appeal. The 11th Circuit Court approved the settlement."

CE petitioned the U.S. Supreme Court for *certiorari* (approval to hear the case) contending that the punitive damage award was still excessive. The Supreme Court deferred ruling on CE's petition pending its resolution of *BMW of North America, Inc. v. Gore*, 517 U.S. 559, 116 S. Ct. 1589 (1996). In *BMW*, the Supreme Court ruled that the U.S. Constitution does not permit "excessive" punitive damage awards. The Supreme Court granted CE's *certiorari* petition, vacated the 11th Circuit Court's judgment, and remanded the case for further consideration in light of the *BMW* decision. Thereupon, the 11th Circuit Appeals remanded the CE case to the district court. The district court reexamined the punitive award under BMW, and concluded that the Constitution would permit punitive damages in an amount no more than 100 times each plaintiff's compensatory award. Therefore, the District Court ordered the entry of judgment for each of the remaining plaintiffs in an amount equal to the jury's compensatory award plus 100 times that amount as punitive damages. This resulted in an aggregate punitive damage award of $4.35 million. The district court, however, did not afford property owners the opportunity to elect a new trial which was charged to be constitutional error and cause for appeal by the plaintiff landowners. In the final round, the CE case wound back up in the 11th Circuit Court of Appeals again. On appeal, the Court found that, "A federal court has no general authority to reduce the amount of a jury's verdict. [*Kennon v. Gilmer*, 131 U.S. 29, 9 S. Ct. 696, (1889); and that the Seventh Amendment prohibits re-examination of a jury's determination of the facts, which includes its assessment of the extent of plaintiff's injury. (Id.) A federal court may not, therefore, 'according to its own estimate of the amount of damages which the plaintiff ought to have recovered, … enter an absolute judgment for any other sum than that assessed by the jury.' Id. at 30, 9 S. Ct. 696. To do so would deprive the parties of their constitutional right to a jury." The Court also declared that, "The Seventh Amendment applies only to the federal courts.... A district court may, however, order a new trial. The authority to do this is located in the common-law principle, later codified in the Judiciary Act of 1789."

1. The Court's Appraisal of CE's Conduct

In determining whether CE's conduct was reprehensible and the punitive damages were justified or excessive, the Court appraised CE's environmental conduct. The Court noted that "the Georgia statutes express a strong interest in deterring environmental pollution. This interest would support a substantial punitive award." To determine the maximum punitive award permitted by the Constitution in this case, the Court turned to the district court's findings on the notice CE had regarding its potential environmental liability.

Reprehensibility of CE's Conduct

"The most important indicium of the reasonableness of a punitive damages award may be the degree of reprehensibility of the defendant's conduct. [*BMW*, 517 U.S. at 575, 116 S. Ct. 1589]. The relevant conduct of CE in this case involves only the four years preceding the filing of the property owners' complaint in August of 1992. The evidence was that, during that time, the mining site was not operated by CE. CE re-acquired the property in 1986 from Pasco, but never resumed mining operations. Where a trespass or nuisance is continuing in nature, an action may be maintained for all damages accruing during the four-year period preceding the filing of an action.

"During the four years at issue, the district court found that CE put into effect a land reclamation plan, which was approved by the Georgia Environmental Protection Division to resolve the problem of acidic rain water entering the streams flowing through its property. CE also cooperated with the Environmental Protection Division. Although its efforts were not entirely successful, the district court concluded that CE's 'most egregious conduct was the failure to do more to prevent the acidic water problem.'

"The district court concluded that the degree of CE's reprehensibility was not great. Given the 'developing' science of land reclamation which characterized the late 1980's, and CE's significant efforts in that regard, the district court held that there was not evidence in this case of 'that high degree of culpability that warrants a substantial punitive damages award.' Although, under Georgia law, the jury must have found 'specific intent to cause harm' to award punitive damages of this amount, the district court found that there was no direct evidence of such intent. The court concluded that the jury must have determined that CE's conduct exhibited a want of care rising to the level of conscious or deliberate indifference to the consequences of its actions. Specific intent to cause harm may properly be inferred from deliberate indifference, but the court concluded that CE's 'deliberate indifference' did not constitute severe 'reprehensibility' under [the] *BMW* (decision). Furthermore, the district court found that none of the 'aggravating factors associated with particularly reprehensible conduct" was present in this case. The district court found that: 'The evidence does not suggest that [CE] affirmatively engaged in prohibited conduct of any kind after it re-acquired the property. It did not commit illegal acts, knowing or suspecting that the acts were illegal. In fact, [CE] responded to any criticisms or penalties levied against it by the Environmental Protection Division in a positive, more aggressive manner. Hence, there is no evidence that [CE] is a recidivist that continually repeats certain misconduct. As there was ample evidence to support the district court's view of the evidence, we hold that its finding that CE's conduct was not highly reprehensible is not clearly erroneous."

2. Ratio of Actual to Punitive Damages

The Court stated that, "Punitive damages must bear a "reasonable relationship" to actual damages. [*BMW*, 517 U.S. at 580, 116 S. Ct. 1589] If the ratio of actual to punitive damages is too great, it is an indication that the defendant did not have adequate notice that its conduct might subject it to an award of this size. [Id. at 574, 116 S. Ct. 1589.] ('Elementary notions of fairness enshrined in our constitutional jurisprudence dictate that a person receive fair notice not only of the conduct that will subject him to punishment but also of the severity of the penalty that a State may impose.')

"In this case, property owners were awarded an aggregate of $47,000 in actual damages. The aggregate $15 million punitive award, therefore, was almost 320 times the amount of actual damage. The district court concluded such a ratio was constitutionally excessive because it approached the 500:1 ratio found 'breathtaking'. The district court's reduction to an aggregate of $4.35 million represents a ratio of 100:1," the limitational amount proscribed by the Constitution for punitive damages.

Final Judgement

The Appeals Court held that the Constitution permits punitive damages in this case totaling $4.35 million. Post-judgement interest on the punitive damages accrues from the date these damages were ascertained, which, on the amounts we affirm today, was on June 16, 1994. CE enjoyed no right to a set off of amounts paid by it in settlement of property owners' judgements.

CHAPTER 8

Geoscientific and Engineering Expert Witnessing and Admissible Scientific Evidence

8.1 INTRODUCTION

The field of expert witnessing, evidence and testimony in forensic geoscience and geological engineering, in fact for all engineers and experts, has been altered drastically in the American federal court systems between 1993 and 1999 as a result of three prominent cases and decisions by the U.S. Supreme Court. Testimony, opinions, and evidence by all experts are subject to much greater scrutiny by the courts than previously, particularly by the trial judge. The trial judge now has been solely empowered as the gatekeeper to qualify experts and their evidence, and to determine whether the experts' evidence is admissible to trial. It is no longer sufficient to be the well-educated scientist or engineer, even one with much and varied experience. The expert's testimony/evidence must now meet greater and more intense inspection by the judge as to the scientist's procedure of investigative methods, relevancy, reliability of results, specific knowledge and experience of the question before the bar (the case), and professional peer acceptability of the investigatory methods.

8.1.1 Definitions

An **expert witness** is fully defined as a person selected or permitted by a court of law, or parties to an action therein, who possesses specialized knowledge, education, training, skill, information, or experience concerning the art, science, profession, practice, or activity, which is the subject matter of the question before the court, to give testimony and evidence that will assist the judge or the jury in arriving at an informed decision (*Black's Law Dictionary* has several variations of the same phrase [4th ed., 1968, pp. 688–689]; emphasis added).

Expert evidence is defined by *Black's Law Dictionary* as "testimony given in relation to some scientific, technical, or professional matter by experts, i.e., persons qualified to speak authoritatively by reason of their special training, skill, or familiarity with the subject" (*Id.*, 688).

Briefly, an expert witness requires knowledge, skill, experience, training, or education to give an opinion in a trial. More briefly, the two essential requirements for an expert witness are

1. To have expertise in the questioned subject matter before the court
2. Whose testimony will aid the judge or jury in understanding the evidence and to make an informed decision

Expert witnessing in a forensic field refers to the study or application of evidentiary information and scientific knowledge to legal problems. Therefore, forensic geology or forensic geological engineering would be the field most applicable to this work.

As an aside, but relevant comment, it should be noted that in the United States, there are two sets of court systems, viz., federal and state, with each state governing and setting its own rules of procedure, rules of evidence, et al. State court systems are free to adopt and follow the federal court rules as their own, or to not adopt the federal rules and set their own. Therefore, if a state has not adopted the federal rules of evidence, its rules may be different for the qualifications of expert witnesses and the admissibility of expert testimony and evidence. It should also be noted that state court systems in the United States have not unanimously adopted the federal court system rules, and this is also true in the context of qualifications for expert witnesses and the admissibility of their evidence and testimony.

There is no doubt that medical experts have long predated all other fields of expertise in court witnessing and testimony, particularly in the need for establishing the cause of death. However, it seems reasonably appropriate to hypothesize that the modern-day use of geoscientific expert witnesses and expert geoscientific testimony and evidence may well have largely evolved primarily from litigation in the federal courts as a result of the early federal mining laws operative on federal lands. This hypothesis seems feasible because the earliest geoscientific litigation calling for expert testimony appears to have arisen primarily from the minerals field and industry operating under the early mining laws governing the public lands and domain.

In accepting this hypothesis as feasible, the rules establishing and governing qualifications for geoscientific and engineering expert witnesses became largely the interest, concern, and jurisdiction of the federal courts. This became even more true in the period after the 1960s when the federal courts had essentially ruled, with few exceptions, that almost every aspect of business in the nation was interstate commerce and, therefore, governed by the commerce clause of the U.S. Constitution and not state law. Consequently, federal law has dominated commerce and has taken the lead in establishing qualifications for all fields of expertise evidence, inclusive of geoscientific and geo-engineering experts, and the admissibility of their evidence and testimony.

8.2 A BRIEF HISTORY OF SCIENTIFIC EXPERT WITNESSING IN THE UNITED STATES

In the development, growth, and expansion period of the United States in the early 1800s, Americans migrated from the settled states east of the Mississippi River to the West Coast, where reputedly the grass was greener. Thus, vast areas of the young nation were left undeveloped between the East and the Far West coastal states and territories. This isolation of the Far West coast from the eastern United States precipitated attempts for California to become a separate republic, even entertaining secession from the United States, to become independent and later to join the Confederacy. To counteract any possibility of such an event, and also to bring about the needed settlement of the vast area in between, the federal government spawned land development acts to entice settlers to come to the in-between unsettled area and develop it.

"In the following decades of frenzied development of the Western U.S., whole-sale land grants in fee simple were made by the federal government to the public at large" (Aston, 1999, p. 34). Besides the land give-away offers by the Mining Acts and the Homesteading and Grazing Acts, "Nearly 52,812,626.5 ha (130.5 million acres) of the public domain were granted to railroad companies for construction of rail lines to encourage the westward building of them and for settlement of the acquired western lands. In some cases of railroad grants, reservations of the minerals were made by the federal government, although with frequent exceptions for 'coal, iron, and other minerals.' Thus, up until the first half of the nineteenth century, public lands had been regulated by the federal government with concern mainly for revenue and disposal for settlement. Under the common law, the property owner was the *prima facie* owner of the minerals in the land. Consequently, a large amount of mineral-bearing lands in the U.S. became privately owned. The development of federal land law by which mineral interests were retained in public lands did not start taking place until the enactment of the Mining Laws of 1866 and 1870." (*Id.*).

The evolution of geological testimony is aptly and well described by Dr. James R. Dunn of Dunn Geoscience Corporation in Albany, New York, in the following excerpt from his paper, "The Heritage of Engineering Geology" (Dunn, 1991, p. 575).

The Changing Scope of Forensic Geoscience

by Dr. James R. Dunn, Dunn Geoscience Corp.

Whereas professional geologists have always been expected to document their determinations as though they might have to defend their reports in court or in other public tribunals, actual litigation pertaining to the quality of geological works has been relatively rare. Few practicing geologists presented testimony in court cases before the 1960s. Forensic geoscience was largely limited to three areas:
  (1) relatively few appearances before legislative bodies;
  (2) litigation regarding the taking of mineral-bearing property for such public works as highways or dams (power of eminent domain), and
  (3) litigation involving engineering works, mines, or mineral resources.

The forensic work with the highest visibility and in which most geologists have participated has been litigation. Litigation and scientific forensic work in general have evolved from the mid-1800s to the present through three different periods, *i.e.*:

1. The early period: Resource oriented (began 1870s)
2. The intermediate period: Engineering oriented (began 1930s)
3. The current period: Environment oriented (began 1960s) (*Id.*)

We shall find that a subcategory, 3(a) is established in the current period that began in 1993 as a result of the *Daubert* case (see later, Section 8.3). This subcategory remains environmental oriented, but becomes *Daubert* oriented.

8.2.1   Early Period — Resources Oriented (1870–1930)

The earliest uses of expert witnesses in geoscientific litigation seem to have made U.S. appearances starting in the 1870s, shortly after the enactment of the Mining Laws of 1866 and 1870, and the General Mining Law of 1872. The apex law of lode deposits was formulated for the law of 1866, and the law of 1870 governed placer deposits. The law of 1872, more or less, superceded the former two mining laws. The apex law proved to be a prominent causation for the need of expert geological witnesses in the adjudication of mining claims. This need for geological experts in resolving apex law litigation continued over several decades, reaching, according to Dunn, a most "celebrated period about 1903 in the Butte (Montana) copper mine claims." An interesting, brief description of the Butte litigation years and the battles between the geological experts over application of the lode apex rule in the Butte area is given by Dunn (1991, p. 576).

8.2.2   Intermediate Period — Engineering Oriented (1930s–1960s)

A new era of contention calling for geoscience experts during the subsequent intermediate period enlarged the need for their expert testimony to cover engineering works involved in litigation, "primarily for determining responsibility for engineering errors, unforeseen adverse conditions, and the determination of mineral-bearing land values in eminent domain cases when land was being taken for engineering projects" (Dunn, 1991, pp. 576–577). This especially comports with the period of the 1930s when the nation was struggling to recover from the severe financial Depression by the federal government programs of fostering labor-intensive public works projects to create large-scale employment and business.

Until 1923, the U.S. courts had given no firm guidelines for determining the admissibility of scientific evidence. Admissibility of expert witnesses was generally based on their credentials, that is, their education, their renown in their area of practice and specialization, and their publications. Unless a particular expert had obtained a damaging reputation in his field of expertise at some time, untarnished credentials usually sufficed to be admitted as a qualified expert and the expert's opinions and conclusions were generally accepted by the court as an expert in the suit at bar and his testimony submitted to the jury.

However, in the 1923 federal court of appeals case of *Frye v. United States*, 293 F. 1013 (D.C. Cir. 1923), standards for determining the admissibility of scientific evidence were set up for the first time as a result of that court's decision in *Frye*.

Frye concerned the admissibility of lie detector test results. In 1923, the lie detector was considered novel scientific equipment and procedure, and its results as questionable evidence. Perhaps, it was even yesteryear's equivalent of what is now termed *junk science*. Consequently, the *Frye* court rejected the evidence because its scientific principles had neither been thoroughly proved, nor accepted as reliable in the scientific community. Thus, the *Frye* court held "that the general acceptance" of a scientific theory by peers in its relevant field was a "pre-requisite to the admissibility of expert scientific evidence" [*Frye v. United States*, 293 F. 1013, 1014 (D.C. Cir. 1923)].

The *Frye* court's decision for admissibility of scientific evidence in federal courts became the federal courts' standard for qualifying all experts, and remained the guiding criterion, or standard rule, for 51 years, until 1974.

As noted previously, the most recent and new demand for geoscientific expert testimony in this intermediate period was in the area of critiquing or in support of large-scale, engineering construction design and works (e.g., bridges, dams, railway and highway tunnels, and large roadway cuts) that had entered the national scene. This need for scientific support testimony was particularly required where there was some liability for project failure. Examples for litigated foundation design failure might require testimonial support of a geoscientific or engineering expert to substantiate a claim of failure in the design because of changed geological conditions encountered, unforeseen geological structural conditions, or insufficient base data made available to contractors for bidding purposes.

The general area of geological expert testimony and evidence was expanding and increasingly in demand as litigation also increased.

8.2.3 (Dr. James Dunn's) The Current Period — Environment Oriented (1960s–1993)

In the 1960s, as the environmental movement gained great momentum and the public clamor for government action culminated in Earth Day,

The U.S. Congress responded to the "green" movement of the sixties by passage of the National Environmental Policy Act of 1969 (NEPA), effective January 1, 1970, along with the subsequent plethora of environmental laws created for industry, including mining. NEPA gave birth and impetus to a flood of federal acts in the subsequent two decades. The more important ones that affect mining are those dealing with dust, air quality, water quality, pollution, stormwater runoff, safe drinking water, noise abatement, solid and hazardous wastes and their disposal, wetlands preservation, mined land reclamation, mine and mill site and tailings clean up, protection and preservation of fish and wildlife, endangered species, wildlife habitats, increased restrictions on mineral prospecting and mining on national forests and other public lands, mine safety and occupational health, and penalties for violations. (Aston, chap. 5, Section 4[1]).

As a result, the forensic geoscience and engineering fields realized a great surge of participation in the formulation of continuing environmental legislation. Geoscientists and engineers were in great demand by congressional, state, and lesser legislative committees and regulatory bodies for technical and scientific opinions in promulgating environmental statutes, regulations, and ordinances.

Following the ensuing environmental enactments, geoscientific experts were, and still are, greatly in demand for presenting evidence, testimony, and opinions in court trials concerning highly contentious environmental statutes and regulations. As so frequently happens with newly enacted laws, they must subsequently undergo the judicial testing process of constitutionality. During this third period of expert witnessing, which is environment oriented, the Federal Rules of Evidence were formulated and employed by the federal courts system in 1974. The new Rules of Evidence obtained in all federal litigation and in those states that had previously adopted the federal *Frye* rule. Thus, from 1974, Rule 702, as part of the of the 1974 Federal Rules of Evidence, governed the qualifications of expert witnesses and the admissibility of expert testimony throughout the federal court system. Rule 702 states, "if scientific, technical, or other specialized knowledge will assist the trier of fact to understand the evidence or to determine a fact in issue, a witness qualified as an expert by knowledge, skill, experience, training, or education, may testify thereto in the form of an opinion or otherwise."

8.2.4 The Change in Dr. Dunn's "Current" Period

In 1993, the interpretation of Federal Rule of Evidence 702 for acceptability of expert qualifications, testimony, and evidence was fundamentally changed by the *Daubert* case decision handed down by the Supreme Court of the United States [*Daubert v. Merrell Dow Pharmaceuticals, Inc.*, 509 U.S. 579 (1993)]. The *Daubert* decision did not change the wording of Rule 702, but radically revised its interpretation and set a new standard for admissibility for scientific expert testimony and evidence.

8.3 THE *"DAUBERT"* PERIOD — 1993 TO DATE (1999)

The U.S. Supreme Court opened the *Daubert* case by announcing, "In this case we are called upon to determine the standard for admitting **expert scientific** testimony in a federal trial" (emphasis added).

8.3.1 *Daubert* — Opinion Written by U.S. Supreme Court Justice Blackmun — Review of Background Facts

In *Daubert v. Merrell Dow Pharmaceuticals, Inc.*, 509 U.S. 579 (1993), "Petitioners Jason Daubert and Eric Schuller were minor children born with serious birth defects. They and their parents sued respondent in California state court, alleging that the birth defects had been caused by the mothers' ingestion of Bendectin, an anti-nausea prescription drug marketed by respondent. Respondent, Merrell Dow, removed the suits to federal court on diversity grounds.

"After extensive discovery, respondent moved for summary judgment, contending that Bendectin does not cause birth defects in humans and that petitioners would be unable to come forward with any admissible evidence that it does. In support of its motion, respondent submitted an affidavit of Steven H. Lamm, physician and epidemiologist, who is a well-credentialed expert on the risks from exposure to various chemical substances. Doctor Lamm stated that he had reviewed all the literature on Bendectin and human birth defects — more than 30 published studies involving over 130,000 patients. No study had found Bendectin to be a human teratogen (i.e., a substance capable of causing malformations in fetuses). On the basis of this review, Doctor Lamm concluded that maternal use of Bendectin during the first trimester of pregnancy has not been shown to be a risk factor for human birth defects.

"Petitioners did not (and do not) contest this characterization of the published record regarding Bendectin. Instead, they responded to respondent's motion with the testimony of eight experts of their own, each of whom also possessed impressive credentials. These experts had concluded that Bendectin can cause birth defects. Their conclusions were based upon 'in vitro' (test tube) and 'in vivo' (live) animal studies that found a link between Bendectin and malformations; pharmacological studies of the chemical structure of Bendectin that purported to show similarities between the structure of the drug and that of other substances known to cause birth defects; and the 'reanalysis' of previously published epidemiological (human statistical) studies.

"The District Court granted respondent's motion for summary judgment. The court stated that scientific evidence is admissible only if the principle upon which it is based is 'sufficiently established to have general acceptance in the field to which it belongs.' 727 F. Supp. 570, 572 (SD Cal. 1989), quoting *United States v. Kilgus*, 571 F.2d 508, 510 (CA 9 Cir. 1978). The court concluded that petitioners' evidence did not meet this standard. Given the vast body of epidemiological data concerning Bendectin, the court held, expert opinion which is not based on epidemiological evidence is not admissible to establish causation. (727 F. Supp. at 575) Thus, the animal-cell studies, live-animal studies, and chemical-structure analyses on which petitioners had relied could not raise by themselves a reasonably disputable jury issue regarding causation. (Ibid). Petitioners' epidemiological analyses, based as they were on recalculations of data in previously published studies that had found no causal link between the drug and human birth defects, were ruled to be inadmissible because they had not been published or subjected to peer review. (Ibid).

The United States Court of Appeals for the Ninth Circuit affirmed 951 F.2d 1128 (1991). Citing *Frye v. United States*, 54 App. D.C. 46, 47, 293 F. 1013, 1014 (1923), the court stated, "... that expert opinion based on a scientific technique is inadmissible unless the technique is 'generally accepted' as reliable in the relevant scientific community" (951 F.2d at 1129–1130). The court declared that expert opinion based on a methodology that diverges "significantly from the procedures accepted by recognized authorities in the field ... cannot be shown to be 'generally accepted as a reliable technique'" [*Id.* at 1130, quoting *United States v. Solomon*, 753 F.2d 1522, 1526 (Cal. 9 Cir. 1985)].

"The court emphasized that other Courts of Appeals considering the risks of Bendectin had refused to admit re-analyses of epidemiological studies that had been neither published nor subjected to peer review" (951 F.2d at 1130–1131). Those courts had found unpublished reanalyses "particularly problematic in light of the massive weight of the original published studies supporting [respondent's] position, all of which had undergone full scrutiny from the scientific community" (*Id.*, 1130). Contending that reanalysis is generally accepted by the scientific community only when it is subjected to verification and scrutiny by others in the field, the court of appeals rejected petitioners' reanalyses as "unpublished, not subjected to the normal peer review process and generated solely for use in litigation" (*Id.*, 1131). The court concluded that petitioners' evidence provided an insufficient foundation to allow admission of expert testimony that Bendectin caused their injuries and, accordingly, that petitioners could not satisfy their burden of proving causation at trial (*Daubert v. Merrell Dow*, op cit., 582–585).

The *Daubert* court unanimously voted to vacate the petition by granting judgment to respondent Merrell Dow.

8.3.2   The Meaning of *Daubert* for Expert Witnessing, Testimony, and Evidence

The *Daubert* court clearly held that the admissibility of expert opinions is governed by Rule 702 of the Federal Rules of Evidence and not by the former, long-established "general acceptance" rule of *Frye*.

Under Rule 702, in addition to qualifications for experts, the Supreme Court established two primary criteria for the trial judge to ensure that expert's admissible testimony "rests on: (1) the testimony must represent 'scientific knowledge'; and (2) the testimony must assist the trier of fact in understanding the evidence presented by an expert and in determining a fact in issue" (op cit., 590–591). The court also stated at pages 589 and 597 that the trial judge hearing the case must be assured that "any and all scientific testimony or evidence admitted is not only *relevant*, but [*also*] *reliable.*"

To assist the triers of fact (judges) in determining relevance and reliability of the proffered expert evidence, the Supreme Court proposed that the expert evidence at issue, be it theory or technique, must be subjected to four criteria:

1. Can it be, or has it been, tested? Can it be falsified?
2. Has it been subjected to peer review and publication? What are its results of peer review?
3. Has the technique a known or potential rate of error?
4. Is the technique or theory generally accepted in the subject field?

8.3.3 Post-*Daubert* "Questions and Application" Debate

A principal question that arose after *Daubert* in subsequent litigation involving expert testimony was whether the *Daubert* requirements applied to *all* expert testimony or *only* to scientific testimony. The majority of federal trial courts (the district courts) and the appellate (circuit) courts that have since had to consider the *Daubert* requirements in cases agreed that *Daubert's* gatekeeping requirements for relevancy and reliability apply to all types of expert testimony in federal courts, but the courts became divided over whether the factors outlined in Part II-C of the *Daubert* opinion apply only to scientific evidence (i.e., the four criteria questions in Section 8.3.2).

Much debate took place and many articles were written, some "527 federal cases, 256 state cases and 362 law review articles and notes" (Kazmarek, 1999) in the 3 years after the *Daubert* decision with regard to whether *Daubert* applied to *all* expert testimony or *only* to scientific testimony. One of the seemingly best and most logical arguments for *Daubert's* limitation to scientific evidence only was made in an article in *Trial* magazine (Johnson, 1997).

Daubert's limitations, however, are evident in the (Supreme Court) opinion's first sentence: "in this case we are called upon to determine the standard for admitting expert *scientific* testimony in a federal trial" (emphasis added).

The court reinforced these limits by italicizing the "*scientific*" and "*knowledge*" in Rule 702 while discussing how the rule replaced *Frye*.

After determining that Rule 702 requires expert testimony to be both relevant and reliable, the Court explained what is contemplated by the language "scientific knowledge" immediately after dropping a footnote emphasizing that *Daubert* is limited to 'scientific knowledge." "Our discussion is limited to the scientific context because that is the nature of the expertise offered here."

Also, several expert-testimony, guideline criteria developed from case law in the next few years after *Daubert* (emphasis added), for example:

1. "The expert's own 'bald assurance of validity' is not sufficient" (*Daubert v. Merrell Dow Pharmaceuticals, Inc.*, 43 F.3d at 1316 ; *Cavallo v. Star Enterprises*, 892 F.2 Supp. 756, 761 (E.D. Va. 1995).
2. "The expert's opinion must be based on the 'methods and procedures of science' rather than on 'subjective belief or unsupported speculation'" (*Id. Daubert*, 509 U.S. at 590).
3. "In short, *Daubert* commands that in Court, science must do the speaking, not merely the scientist" (*Cavallo*, 829 F. Supp. at 761).

Although one of the reportedly intended purposes of the Supreme Court hearing the *Daubert* case had been to liberalize the standard for judging expert opinion, the actual result of the decision was that experts' methodology and opinions came under far greater scrutiny, and as a consequence, far more experts' testimony and evidence have been ruled as inadmissible than had been previously.

8.4 PARTIAL CLARIFICATION OF *DAUBERT*
IN THE 1997 *JOINER* CASE

After an interim of 4 years of divided appellate decisions concerning the *Daubert* requirements for admissibility of expert evidence, the U.S. Supreme Court took another look at its *Daubert* decision in *Joiner v. General Electric Co.*, 864 F. Supp. 1310 (N.D. Ga. 1994); reversed, 78 F.3d 529 (11th Cir. 1996); reversed 118 S. Ct. 512 (1997).

In *Joiner*, the Supreme Court reviewed the issue of a district trial court's exclusion of expert testimony and evidence in a toxic tort case involving exposure to the chemicals polychlorinated biphenyls (PCBs), furans, and dioxins. Plaintiff Joiner allegedly claimed his exposure caused his diagnosed small-cell lung cancer. Both Joiner's medical experts qualified as experts based on their education, years of experience, physical examinations of the claimant Joiner, reliance upon specific studies relating to the carcinogenic effects of PCBs, and familiarity with literature in the field. Their methodology had included traditional medical diagnostic technique (i.e., differential diagnosis), and relied on numerous scientific studies and authorities, epidemiological surveys and studies, including animal studies. Joiner's experts testified that PCBs and their derivatives furans and dioxins can promote cancer and opined that Joiner's exposure to them was likely responsible for his cancer (*Joiner, Id.* at 1314).

The district court trial judge, in exercising her gatekeeping duty, first concluded that "there was no evidence that Joiner was exposed to furans and dioxins" and any reference to them was irrelevant to the case and "must be excluded" (*Id.*, 1315). The judge next concluded that the proposed expert opinions relating to the animal studies were unreliable. The judge rejected studies that tested high levels of PCBs on mice could cause cancer in humans (*Id.*, 1324). The extension/extrapolation from mice to men had fundamental deficiencies that rendered the opinion inadmissible to conclude that "low levels of PCBs could cause cancer in human beings." The district court excluded the medical experts' opinions and granted the defendant manufacturers of transformers and dielectric fluids summary judgment (i.e., no cause).

On appeal, the Eleventh Circuit Court of Appeals reversed the trial court's exclusion of the experts' evidence as error and an invalid exclusion under the *Dauber* tests, stating that the trial court "had erroneously reviewed the evidence and assessed the conclusions of the plaintiff's experts rather than limiting its analysis to the experts' methodology."

8.4.1 Experts, Beware of Extrapolation of Your Data

In reviewing *Joiner* and the Eleventh Circuit's reversal decision, the U.S. Supreme Court reversed and specifically approved the district court searching evaluation behind the experts' opinions [*Joiner,* 118 S. Ct. 512 (1997)]. The Supreme Court held that it was "within the District Court's discretion to review the allegedly supporting literature (of the experts) and conclude that the studies upon which the experts relied were not sufficient, whether individually or in combination, to support their conclusions that Joiner's exposure to PCBs contributed to his cancer ... nothing in *Daubert* or the Federal Rules of Evidence requires a district court to admit opinion evidence which is connected to existing data only by *ipse dixit* of the expert (lit. "he said it himself"; a bare assertion resting on the authority of an individual alone). **A court may conclude that there is simply too great an analytical gap between the data and the opinion proffered."** (*Id.,* at 519; emphasis added).

8.4.2   *Joiner* Court Refines the *Daubert* Requirements for Expert Evidence

The Supreme Court recognized that in *Daubert* it had distinguished between the experts' methodology and their conclusions (opinions) by stating that the trial court must evaluate the expert's methodology rather than his conclusions (*Id.*) In *Joiner*, the court refined its position by recognizing that "conclusions and methodology are not entirely distinct from one another" (*Id.*).

8.5 *KUMHO TIRE CO. LTD. V. CARMICHAEL* — 1999 — FURTHER REFINEMENT OF *DAUBERT'S* REQUIREMENTS FOR EXPERT OPINIONS

As noted in Section 8.3.3, "A principal question that arose after *Daubert* in subsequent litigation involving expert testimony was whether the *Daubert* requirements applied to *all* expert testimony or *only* to scientific testimony." The *Joiner* decision did not address that question because it was not an issue. As indicated in Section 8.3.3, the federal courts had become divided in their opinions and decisions that certain technical, skill and experience types of expertise did not require the scrutiny of meeting the *Daubert* requirements for admissibility of expert evidence (e.g., publication for peer review and acceptance).

8.5.1   Background Facts of *Kumho Tire Company, Ltd. et al. v. Carmichael and Carmichael et al. v. Samyang Tire, Inc.*, 131 F.3d 1433 (1997)

In 1993, after a tire on a minivan blew out and the minivan overturned, one passenger died (Patrick Carmichael) and the others were injured. The survivors and the decedent's representative, claiming that the failed tire had been defective, brought a diversity suit in the United States District Court for the Southern District of Alabama against the tire maker and distributor. The plaintiffs rested their case, in significant part, on the depositions of a mechanical engineer, an expert in tire failure analysis, who intended to testify that, in his expert opinion, a defect in the tire's manufacture or design caused the blowout. The expert's opinion was based on: (1) a visual and tactile inspection of the tire; and (2) the theory that in the absence of at least two of four specific physical symptoms indicating tire abuse, the tire failure of the sort that occurred in the case at hand was caused by a defect. The district court granted Kumho's motion to exclude the expert's testimony as well as a motion for summary judgment against the plaintiff.

Carmichael appealed the decision to the Eleventh Circuit. In *Carmichael v. Kumho Tire* at the first appeal level, the Eleventh Circuit Court held that there was a difference between a scientific expert and an expert who relied on skill- or experience-based observations as the basis for an opinion; thus, it reversed the decision of the lower, trial court. [See *Carmichael v. Samyang Tire, Inc.*, 131 F.3d 1433 (1997). It "reviewed ... *de novo* (anew) the district court's legal decision to apply *Daubert* (131 F.3d at 1435). It noted that "the Supreme Court in *Daubert* explicitly limited its holding to cover only the 'scientific context,'" adding that "a *Daubert* analysis" applies only where an expert relies "on the application of scientific principles," rather than "on skill- or experience-based observation" (*Id.,* 1435–1436).]

The court concluded that Carlson's testimony, which it viewed as relying on experience, "falls outside the scope of *Daubert,*" that "the district court erred as a matter of law by applying *Daubert* in this case," and that the case must be remanded for further (non-*Daubert*-type) consideration under Rule 702 (*Id.,* 1436).

On remand by the Eleventh Circuit to the district court, the district court (1) agreed with the defendants that the district court ought to act as a *Daubert*-type reliability gatekeeper, even though the testimony at issue could be considered technical instead of scientific; (2) examined the expert's methodology in light of the reliability-related factors that *Daubert* had mentioned; and (3) concluded that all those factors argued against the reliability of the expert's methods (923 F. Supp. 1514). On reconsideration, the district court, although acknowledging that the *Daubert* factors ought to be applied flexibly and were simply illustrative, affirmed the earlier rulings on the ground that there were insufficient indications of the reliability of the expert's methodology of tire failure analysis.

The United States Court of Appeals for the Eleventh Circuit, in reversing after remand, expressed the view that the district court had erred as a matter of law in applying the *Daubert* factors to the tire expert's testimony, because (1) *Daubert* was limited to the scientific context, and (2) the testimony in question relied on experience instead of the application of scientific principles [131 F.3d, 1433 (1997)].

Kumho Tire Ltd. petitioned the Supreme Court for *certiorari* (review), asking it to determine whether a trial court may consider *Daubert's* specific factors when determining the "admissibility of an engineering expert's testimony." On petition for *certiorari* the Supreme Court stated, "We granted *certiorari* in light of uncertainty among the lower courts about whether, or how, *Daubert* applies to expert testimony that might be characterized as based not upon 'scientific' knowledge, but rather upon 'technical' or 'other specialized' knowledge."

The Supreme Court in *Kumho Tire Co., Ltd. v. Carmichael,* 119 S. Ct. 1167 (1999), reversed the Eleventh Circuit Court of Appeals and answered that important question that the circuit courts had been awaiting, by holding that "**all experts**, whether scientists, engineers, technical and specialized, experience-based analysts for tire failure, automotive designers, chemical sniffers, *et al.*, **must *all* pass *Daubert* scrutiny**" (emphasis added).

According to the *Kumho Tire* court, the *Daubert* test factors for evaluating expert evidence "may or may not be pertinent to a reliability determination, depending on the nature of the issue at hand, the area of expertise, and the subject of the expert's testimony" (*Kumho Tire* at 1170). In an opinion by Breyer, J., joined by Rehnquist, Ch. J., and O'Connor, Scalia, Kennedy, Souter, Thomas, and Ginsburg, JJ., and joined — as to points (1) and (2) that follow — by Stevens, J., it was held that "(1) a federal trial judge's gatekeeping obligation under the FRE (Federal Rules of Evidence), to insure that an expert witness's testimony rests on a reliable foundation and is relevant to the task at hand, applies not only to testimony based on scientific knowledge, but rather to all expert testimony, that is, testimony based on technical and other specialized knowledge; (2) in determining the admissibility of an expert's testimony — including the testimony of an engineering expert — a federal trial judge may properly consider one or more of the specific *Daubert* factors, where doing so will help determine that testimony's reliability; and (3) in the case at hand, the District Court's decision not to admit the expert testimony in question was within the District Court's discretion."

8.6 EXAMPLES OF EXCLUSION AND ADMISSIBILITY OF EXPERT OPINIONS IN ENVIRONMENTAL COST RECOVERY SUIT CHALLENGES

In *Thomas v. FAG Bearing Corp.*, 846 F. Supp 1382 (W.D. Mo. 1994), a groundwater contamination–waste migration case, the defendant alleged that hazardous substances had migrated through groundwater and contributed to the contamination of the cleanup site. The court found **the expert's testimony was inadmissible** and based on assumptions that contamination from the PRP's (potentially responsible party) facility had entered the groundwater and traveled via an uncharted aquifer to the cleanup site was "**concocted on impermissible bootstrapping of speculation upon conjecture**" (emphasis added) (*Thomas* at 1394).

In *Kalamazoo River Study Group v. Rockwell International Corp.*, 171 F.3d 1065 (6th Cir. 1999), **the federal district court rejected a scientific expert's opinion** on the cause of PCB contamination in the Kalamazoo River. In reviewing the basis for the trial court's exclusion of an expert's evidence and opinion, the Sixth Circuit Appeal court stated:

> The district court did not "choose sides" to resolve a battle of the experts. Instead, the court focused on the factual underpinning of Dr. Brown's conclusions. Citing (*Daubert*), the court stated that it "may, indeed, must look beyond the conclusions (of the experts) to determine whether the expert testimony rests on a reliable foundation."

> When the court did so, it found that Dr. Brown's conclusion was based on "speculation, conjecture, and possibility" and that "the inadequate factual basis makes Dr. Brown's affidavit scientifically unreliable" (*Id.* at 1072).

B.F. Goodrich v. Betkoski, 99 F.3d 505 (2d Cir. 1996), was a cost recovery suit against nonsettling municipality defendant PRPs to recover response costs for the cleanup of two Connecticut landfills designated as Superfund sites. The two landfills contained hazardous wastes from unknown area generators, including the defendant municipalities in the suit.

The plaintiff's expert witness was an extensive-publishing agronomist on hazardous wastes disposal. This expert had prepared lists of classes of defendants who routinely deposited waste during the pertinent time frame at the landfills in question. This expert used his lists of waste depositors along with his reviews of the reports on the landfills in trying to make possible correlations of the specific hazardous wastes found in the landfills with suspect generators.

Relying on *Daubert*, the **trial court found the plaintiff's expert's evidence inadmissible**. However, on appeal, the Second Circuit Court disagreed. In reviewing the expert's procedure, the court pointed out that the expert had explained the type of hazardous substances commonly found in normal tire products and "that even though the expert could not swear with absolute certainty that the tire products disposed of by a specific defendant/appellee were similar to normal tire products, such absolute certainty is not required to oppose successfully a summary judgement motion …" (*Id.* at 525).

In *Woll & Co. v. Fifth & Mitchell Street Corp.,* No. 96-5973, 1999 WL 75059 (E.D. Pa. February 4, 1999), a similar circumstance occurred as in *Betkoski*. Defendant Fifth & Mitchell Street Corp. (hereafter F&M), was the former owner and lessor of a contaminated industrial site. "In order to tie the hazardous substances to operations during the defendant's ownership, the plaintiff's expert testified that [defendant's tenant] conducted precisely the sort of operations that produce the kind of contamination found on the subject property, and because disposal of hazardous substances in a manner causing that kind of contamination was commonplace in the tenant's industry at the time, the tenants more likely than not contributed to the release of hazardous substances during the time the [defendant] owned the property…" (*Id.* at 6).

The expert based his conclusion on evidence that the same hazardous substance, from the same tenant's operation, had leaked into the site during the ownership of another party.

The court stated that "absent another logical explanation, it is not unreasonable to infer disposal of a substance by a party from its use of significant quantities of the substance while operating a leased facility on a property and evidence of a subsequent release of that substance at the property" (*Id.* 6).

The court also noted that the defendant, in answering the expert's affidavit, had not presented any denials to the allegation, or counterevidence that its tenant had not leaked hazardous substances during the defendant's ownership. Finally, the court observed that "the CERCLA (Comprehensive Environmental Response, Compensation and Liability Act) plaintiff was not required to prove its case with scientific certainty and to require more would allow many PRPs to evade CERCLA's broad remedial goals."

The approach taken by the courts in *Betkoski* and *Woll* bears a completely opposite holding of the district court case for a landfill expert's testimony in *Dana Corp. v. American Standard, Inc.,* 866 F. Supp.1481 (N.D. Ind. 1994). In *Dana*, as in *Betkoski* and *Woll*, the court considered the admissibility of an expert's affidavit offered in another CERCLA cost recovery action.

The defendants challenged the sufficiency of the expert opinion which had determined that each of the defendants had disposed of hazardous substances at the target site. But the *Dana* expert was unable to show any support for his conclusions that the defendants had disposed of hazardous waste at the landfill rather than simply non-hazardous waste. In attempting to establish whether a particular defendant's waste contained hazardous substances, the expert was only able to opine as to the hazardous constituents of a generic class or type of waste, like flashlight batteries or paint, but did not confine his opinion to the particular brands or products actually produced by the defendants. Likewise, the expert could not state that all classes or types of a particular product contained a hazardous constituent (*Dana*, at 1501). (Marionneaux, 1999, pp. 01–014).

The *Dana* court, relying on Daubert, refused to consider the plaintiff's expert's affidavit and held it "**unreliable**" (emphasis added). The court stated that:

Expert opinions premised on speculation and conjecture are insufficient to create a genuine issue of material fact to survive summary judgement (a dismissal of the suit in favor of one of the parties). When basic foundational conditions themselves are conjecturally premised, it then behooves a court to remove the answer from one of admissible opinion to one of excludable speculation (*Dana*, p. 1499) (emphasis added).

The *Dana* decision and reasoning for excluding the "conjectural" expert's opinion are in marked contrast to the Second Circuit's reasoning for reversal of the *Betkoski* district court's holding of the expert's opinion as inadmissible for similar findings, and also for the *Woll* district court's decision to accept the expert's somewhat similar conjectural testimony/affidavit.

The *Dana* court noted that, "despite CERCLA's broad remedial purposes, a plaintiff must nevertheless prove that a particular defendant's hazardous waste was disposed of at a site, and that hazardous substance similar to those of were present at the time" (*Id.* at 1481). "Because the expert's affidavit was unreliable and, therefore inadmissible, the plaintiffs failed to carry this burden and summary judgement for the defendants was proper" [Marionneaux, op cit., pp. 01–015].

Expert witness fees: In *Davis County Solid Waste Management & Energy Recovery Special Service District v. EPA*, 169 F.3d 755 (D.C. Cir. 1999), the Special Service District and others challenged EPA standards governing combustion of municipal solid waste (MSW). The D.C. Circuit vacated the standards in part and remanded the case to the Environmental Protection Agency (EPA). The EPA acknowledged that the district was entitled to recovery of litigation costs, but the EPA and district could not agree on amounts. The district petitioned the D.C. Court for an award of attorney fees and costs under the act. The court held that "fees charged by an 'expert witness,' as opposed to a 'technical consultant,' were recoverable to the extent that they reflected the expert's time necessary for preparation of technical affidavits submitted with motions to the court, and (2) the local rates of the District's Utah attorneys, rather than the rates of D.C. attorneys, governed the experts' fee awards" (Mansfield, 1999, p. 199).

8.7 AN EXPERT GEOSCIENTIFIC WITNESS CASE — AN EXAMPLE OF INADMISSIBLE GEOLOGICAL EXPERT TESTIMONY

McLENDON v. GEORGIA KAOLIN CO., C.A. 85-338-2-MAC (WDO),
C.A. 85-338-2-MAC (WDO), UNITED STATES DISTRICT COURT FOR
THE MIDDLE DISTRICT OF GEORGIA, MACON DIVISION, 841 F.
Supp. 415 (1994); January 14, 1994, Decided; January 14, 1994, Filed

O.L.McLENDON, *et al*, Plaintiffs, v. GEORGIA KAOLIN CO., INC., Defendant.
GRANT SMITH, Plaintiff v. GEORGIA KAOLIN CO., INC., Defendant.
(Cases combined)

C.A. 85-338-2-MAC (WDO), C.A. 85-338-2-MAC (WDO)
UNITED STATES DISTRICT COURT FOR THE MIDDLE DISTRICT OF GEORGIA,
MACON DIVISION

841 F. Supp. 415 ; 1994 U.S. Dist. January 14, 1994, Filed;
January 14, 1994, Decided

JUDGE: OWENS, W.D., JR.
OPINION BY: WILBUR D. OWENS, JR.;

Before the court is defendant's motion *in limine* (at the beginning), to exclude the testimony of plaintiffs' expert, M. Eugene Hartley ("Hartley") (a geologist), concerning the quantity, quality, commercial uses, and value of the kaolin deposit on the property in question ("Smith property"). Pursuant to Federal Rule of Evidence 104(a), the court issues the following order on the admissibility of Hartley's expert testimony (emphasis added for engineers and geoscientists).

CASE FACTS

Plaintiffs' claims involve similar facts but due to distinct legal issues will be tried in two separate trials. The suit developed after plaintiffs, heirs of Edward D. Smith, conveyed their interests in a tract of land in Wilkinson County, Georgia, to defendant, Georgia Kaolin Company, Inc., between 1969 and 1971. Plaintiffs claim that defendant fraudulently concealed the true value of the property while under a duty to disclose material information.* In addition, plaintiff Grant Smith has an equitable claim to set aside the judgment of the ordinary which approved the sale of his interest in the land. In both trials, counsel for plaintiffs seek to present the testimony of Hartley as a scientific expert on the value of the kaolin deposit. Defendant challenges Hartley's qualifications as an expert and questions the bases of Hartley's opinion.

The court held a lengthy hearing on December 7, 1993, where counsel for both parties were provided an opportunity to *voir dire* Mr. Hartley as to his qualifications and opinions. Hartley is expected to testify concerning damages; specifically, he will offer an opinion of the quantity of the kaolin deposit on the Smith property as well as its quality, commercial uses, and value in the late 1960's and early 1970's.

* The 325 acres of land in central Georgia contain a large kaolin deposit.

DISCUSSION

The court has broad discretion in determining the admissibility of expert testimony. Salem v. United States Lines Co., 370 U.S. 31, 35, 82 S. Ct. 1119, 1122, 8 L.Ed.2d 313 (1962); Evans v. Mathis Funeral Home, Inc., 996 F.2d 266, 268 (11th Cir. 1993); Polston v. Boomershine Pontiac-GMC Truck, Inc., 952 F.2d 1304, 1309 (11th Cir. 1992). Defendant challenges Hartley's testimony on two grounds, (1) that Hartley is not qualified as an expert under Federal Rule of Evidence 702, and (2) that his opinion is based solely on the opinion of other experts rather than on facts normally relied upon by experts in the particular field.

A. **Hartley's Qualifications as an Expert**
Federal Rule of Evidence 702 governs the admissibility of expert testimony (emphasis added):

If scientific, technical, or other specialized knowledge will assist the trier of fact to understand the evidence or to determine a fact in issue, a witness qualified as an expert by knowledge, skill, experience, training, or education, may testify thereto in the form of an opinion or otherwise.

Rule 702.

Rule 702 sets forth a two tier test for admissibility of expert testimony. First, the expert must show that he is qualified by "knowledge, skill, experience, training, or education." Second, the testimony must "assist the trier of fact to understand the evidence or determine a fact in issue." The expert opinion is not required to be generally accepted in the field, but must be relevant and meet a standard of "evidentiary reliability." Daubert v. Merrell Dow Pharmaceuticals, Inc.,U.S., 113 S. Ct. 2786, 2795, 125 L.Ed.2d 469 (1993).*

The expert testimony must be based on "scientific ... knowledge." Id. at, 113 S. Ct. at 2795. "The adjective 'scientific' implies a grounding in the methods and procedures of science. Similarly, the word 'knowledge' connotes more than subjective belief or unsupported speculation." Id. The court should focus "solely on [the expert's] principles and methodology, not on the conclusions" generated. Id. at, 113 S. Ct. at 2797.

Mr. Hartley testified that he is an economic geologist and an expert in the field of industrial minerals, of which kaolin is a part. He received his Bachelors and Masters degrees from the University of Georgia where he took the standard survey courses in geology, but did not attend a course on clay mineralogy or otherwise study the mineral kaolin in any depth. (Transcript of December 7, 1993, hearing, pp. 14–15. See also affidavit of Dr. Hurst, May 6, 1993.) Hence, Hartley's training and education alone do not qualify him to testify as an expert on kaolin (emphasis added).

Still, Hartley contends that his knowledge, skill, and work experience, in combination with his schooling, qualify him to testify as an expert on kaolin (emphasis added). Mr. Hartley has minimal work experience with kaolin evaluation. Hartley testified that he assessed kaolin reserves for RTZ, a mining corporation. However, cross-examination

---

* In Daubert, the Supreme Court clarified the standard for admitting expert testimony. The Court held that the Federal Rules of Evidence provided the test for determining the admissibility of scientific expert testimony. Moreover, the rules occupy the field, thus superseding the "general acceptance" test set forth in Frye v. United States, 54 App. D.C. 46, 293 F. 1013 (D.C. Cir. 1923). Daubert, U.S. at, 113 S. Ct. at 2793. The "general acceptance" of the proffered scientific theory or technique can influence the court's analysis, but is not necessarily determinative. Id. U.S., 113 S. Ct. at 2797.

disclosed that the RTZ project did not include an evaluation of core samples but, instead, involved a review of leases and maps. Hartley has some additional experience in valuing kaolin deposits for Dorfner, a mining company. Despite this slight experience with kaolin, Hartley lacks sufficient knowledge of kaolin, the processes used to make the mineral commercially usable, and its market value to testify as an expert.

Hartley admitted that the kaolin deposit on the Smith property is commercially unusable at its present color, and that it would have to be brightened in some manner before it would be commercially acceptable.* (Hearing Transcript, p. 66.) Kaolin can be brightened through bleaching, whitening, or blending. However, the expert showed that he was rather unknowledgeable about these brightening processes. (Id. at 52–67.) The expert could not adequately explain the criteria used in evaluating kaolin. (Id. at 42–53.) Importantly, Hartley was not aware of when various kaolin cleaning processes, such as ozonation and certain bleaching agents, were developed and used commercially.** (Id. at 53–54.) Because the transaction occurred in the late 60's and early 70's, the expert must be knowledgeable as to the commercial practices and market value of kaolin at that time in order to assist the trier of fact. Absent knowledge of how and if the clay can feasibly be brightened to an acceptable level, one can not accurately estimate its quality and commercial uses.

At best, Hartley is capable of estimating the quantity of kaolin on the Smith property. Defendant's expert, Dr. Hurst, testified that Hartley's estimate of quantity included sand, dirt, and bauxite. While a difference of opinion between experts is expected and does not support the disqualification of an expert, serious questions concerning Hartley's ability to assess the commercial quality of kaolin also bring his estimate of quantity into question. Hartley has not demonstrated that he knows the outer limits of usable kaolin. Hence, the court is concerned that Hartley's estimate of quantity is inflated and includes elements which do not amount to commercially usable kaolin. Furthermore, an estimate of the amount of kaolin without an assessment of its quality or value would not assist the trier of fact in valuing the kaolin deposit.

Hartley's estimate of the 1969 value of the kaolin is flawed for several reasons. First, having found Hartley's estimate of quality unreliable, the valuation which is based upon the assessment of quality is equally suspect. Second, Hartley estimated the 1969 value of the kaolin at $4.2 million by using a royalty of three percent of the finished product. He testified that a royalty of three percent of the finished product is used to value other industrial minerals. However, he had no knowledge or evidence of royalties used in kaolin leases either current or in 1969. In addition, questions as to Hartley's estimate of quantity certainly would affect his valuation.

The second requirement of Rule 702 is that the expert opinion assist the trier of fact to determine a fact in issue (the value of kaolin). In his testimony, Hartley concluded that the deposit was valuable because defendant is holding the property, drilling on it, and defending the lawsuit. (Id. pp. 59–60.) The conclusion is that of a lay person and requires no expert reasoning or knowledge (emphasis added).

---

* Kaolin is used in coating paper and must be light in color to be used commercially. In addition to brightness, the factors of viscosity, particle size, and grit content affect the quality, commercial uses, and value of kaolin.

** In evaluating the Smith deposit, Hartley relies primarily on a memo dated March 20, 1969, from John Smith, to E. J. Grassmann. Smith writes that some of the samples would not respond to normal brightening processes, but did brighten in the research lab using "sodium hypochlorite at three gallons per ton." See Plaintiffs' Exhibit JS-17, attached to defendant's motion in limine. Hartley stated that he was unaware whether sodium hypochlorite could be used commercially in 1969 or even now.

Plaintiffs and their expert contend that one does not need to be a kaolin specialist to accurately value a kaolin deposit. However, a certain degree of specialization is necessary. The court is less concerned about whether Hartley specializes in the kaolin area and more concerned with his knowledge of the subject matter. The court finds Hartley not sufficiently knowledgeable to accurately assess the quantity, quality or value of kaolin to be of assistance to the trier of fact (emphasis added).

The Court, in Daubert, recognized that courts have additional safeguards available should expert testimony be admissible under Rule 702. For example, cross-examination, presentation of contrary evidence, and jury charges on the burden of proof and causation should suffice to attack questionable but admissible expert testimony. However, these devices are only sufficient safeguards where the scientific testimony meets the standards of Rule 702. See Daubert, U.S. at, 113 S. Ct. at 2798. Hartley is not sufficiently qualified in the field of kaolin exploration and valuation to offer testimony concerning the quantity, quality, commercial uses, and value of the Smith property kaolin deposit in the late 60's and early 70's or in today's market. Therefore, the court finds Hartley not qualified under Rule 702. Cross-examination would expose Hartley's lack of expertise, but at the risk of prejudicing and misleading the jury. Rule 403 (emphasis added).

B. The Basis of Expert Hartley's Testimony

Hartley's testimony must also be excluded because it is based upon unreliable hearsay, the opinions of others, rather than on data reasonably relied upon by experts in the particular field. Testimony can be admitted only if the facts or data upon which the opinion is based are "of a type reasonably relied upon by experts in the particular field in forming opinions or inferences upon the subject." Rule 703. See also Daubert, U.S. at, 113 S. Ct. at 2798. Rule 703 is an exception to the exclusion of hearsay evidence because of the expert's professional knowledge and training. United States v. Williams, 447 F.2d 1285, 1290 (5th Cir. 1971). Nevertheless, a "court cannot unjustifiably defer to any expert's assertion that a particular proffered opinion has a basis in fact." Smith v. Ortho Pharmaceutical Corp., 770 F. Supp. 1561, 1573 (N.D. Ga. 1991). The court must make the ultimate determination of whether a reliable basis underlies the expert opinion. Id.

In Hartley's affidavit of December 3, 1993, he states that he has based some of his opinions, namely his opinion of the quality of the kaolin clay, on "admissions" by "a knowledgeable employee of defendant." (Hartley affidavit, December 3, 1993, para. 3. See also Hearing Transcript, p. 33.) When questioned at the hearing, Hartley testified that he was referring primarily to a March 20, 1969, memo from John M. Smith to Grassmann discussing the processes used to brighten the kaolin extracted from the Smith property. (Plaintiff's Exhibit, JS-17, attached to defendant's motion in limine.) In the memo, Smith explains that some of the samples would not respond to normal brightening processes, but did brighten in the research lab using "sodium hypochlorite at three gallons per ton." (Id.) Hartley stated that he was unaware whether the use of sodium hypochlorite to brighten kaolin was commercially practicable in 1969 or even now. Despite this lack of knowledge, Hartley relies on Smith's opinion of the clay quality. Of concern is the fact that Hartley did not consider that some of the kaolin was brightened to acceptable levels in a research lab; he is unaware whether the use of sodium hypochlorite was experimental or not. (Hearing Transcript, pp. 57–58.) Yet, he assumes that because the kaolin was brightened in a research lab that the method is commercially feasible.

In addition, Hartley relied upon a memo from A.G. Bowman to a Georgia Kaolin executive, dated February 21, 1967, which states that Bowman has "no doubt in (his) mind that (the

Smith) property contains a large tonnage of clay similar to the clay on the adjacent Califf property…." (Hearing Transcript, p. 31.) At the time, Georgia Kaolin had an interest in the neighboring Califf property which also contained a kaolin deposit. Hartley relied upon Bowman's assessment without knowing Bowman's occupation, training, or background. (Id. p. 41.) Moreover, Hartley is unfamiliar with the data upon which Bowman derived his opinion of the kaolin's quality. (Id.)

The above two memoranda are not the type normally relied upon by scientific experts. Hartley's opinion that these documents constitute admissions as to quality on the part of defendant is a matter for the court to decide. Hartley's opinion on the quality of the Smith kaolin must be excluded because it is not based upon data of a type reasonably relied upon by experts in the field of kaolin evaluation (emphasis added).

CONCLUSION

Plaintiffs' expert, M. Eugene Hartley, has not shown that he is qualified by "knowledge, skill, experience, training, or education" to offer an opinion as to the quantity, quality, commercial uses, or value of the Smith kaolin deposit. Furthermore, under Rule 703, Hartley's opinion on the quality of the Smith kaolin must be excluded because it is not based upon data reasonably relied upon by experts in the field of kaolin evaluation. Accordingly, the court must exclude Hartley from testifying as to his opinion of the quantity, quality, commercial uses, or value of the Smith kaolin deposit. Defendant's motion in limine is GRANTED (emphasis added).

SO ORDERED, this 14th day of January, 1994.
WILBUR D. OWENS, JR., UNITED STATES DISTRICT JUDGE

8.8 CONCLUSIONS

The trilogy of cases, *Daubert* (1993), *Joiner* (1997), and *Kumho Tire* (1999), have clearly had a profound effect in revising the interpretation of Federal Rules of Evidence, § 702, governing the admissibility of all expert opinions, testimony, and evidence, whether scientific, engineering/technical, or skill and experience-based (emphasis added).

Under the sole, gatekeeping power of a federal trial judge to determine admissibility of expert evidence in pretrial hearings, admission and acceptability are far more difficult to obtain than formerly before 1993 and the *Daubert, Joiner*, and *Kumho Tire* decisions.

Expert testimony submitted by qualified experts for a trial no longer automatically goes to open court for the jury to hear, or for the opposing counsel to cross-examine as in previous decades. The qualified expert's evidence is now screened and scrutinized in pretrial hearings by the gatekeeper/trial judge for its reliability and relevancy; the methodology employed in forming the expert's opinion; acceptance by peer community and reviews; whether the evidence has been subjected to the *Daubert* four criteria (see Section 8.3.2); and finally, whether the evidence will assist the trier of fact in arriving at an informed decision, all before it is heard by the jury (emphasis added).

It has taken nearly 100 years (1870s to 1970s) for geoscientific expert witnessing to develop to its present status of importance and growth in a wide variety of geotechnical and geoscientific involvement in litigation. The future involvement of geoprofessionals and engineers in environmental litigation appears to be one of continued expansion until the national environment and health have been restored to a likely unobtainable, pristine condition.

8.9 A FEW POINTERS FOR BEING A SUCCESSFUL GEOSCIENTIFIC EXPERT WITNESS IN COURT

8.9.1 Procedures in Forensic Geoscience

The following are recommendations for engineers to observe when serving as expert witnesses.

Quality assurance of samples, and handling, that is, its chain of custody: "A very high order of measurement precision and structure is absolutely essential because geologists may have to defend every research step, procedure, and conclusion to a cross-examining attorney. Written procedures (protocols) are often mandatory. Detailed procedures/protocols for drill site selection, any type of sample collecting and handling, from retrieval through analysis to sample disposal, and finally, interpretation are generally required.

- "Geologists may have to answer such questions as, 'who collected the samples?'" (whether they be core, cuttings, contaminated dirt, water, or other).
- 'Were they ever out of that person's sight?'
- 'What was the condition of the samples on arrival at the laboratory?'
- 'Can you prove there was no tampering with the samples?'
- 'Are the chain of custody forms complete and accurate?'

Questions of this type must be anticipated for every stage of the investigation.

"Geological intuition, speculation, extrapolation of data, and basic research ... should rarely be a visible part of the final product or opinion" (Dunn, 1991, p. 581).

"Innovative research techniques, while not ruled out, must be so thoroughly rationalized in scientific terms that their use is discouraged. Preferably, procedures should follow governmental recommendations by such agencies and organizations as the EPA, the American Society of State Highway Officials, the American Society for Testing and Materials (ASTM), the American Institute of Professional Geologists (AIPG), or the Association of Engineering Geologists (AEG)" (*Id.*).

8.9.2 The Expert's Witness Stand and Testimonial Behavior

Pointers for trial appearances are

1. Listen to and follow your attorney's instruction.
2. Remain COOL — maintain natural composure on the witness stand in court and during depositions.
3. Be sure of your facts, putting them forth without hesitation or faltering.
4. Speak convincingly and authoritatively.
5. Do not let the cross-examining attorney rattle you in your testimony, or rouse your anger.

6. Do not retreat or compromise in your statements or testimonial position.
7. Do not volunteer extra information that was not asked for in the question. When answering a question under oath, answer only sufficient information to adequately respond to the questions asked by either attorney (i.e., either yours or the opposing attorney).
8. If asked by an examining attorney, if it is possible that there might be an exception to your hypothesis or opinion, do not answer "anything is possible," or the like, for, then, you may have blown your, thus far, air-tight conclusion and positive opinion. The possibility of your admission to an "exception" gives the opposing counsel a crevice to open in your opinion and break it down to invalidation of it. Don't let him/her create a crevice or weak spot in your testimony (emphasis added).

The best answer to 8 would be to reply that you would be speculating in your answer. The court frowns on speculation and conjecture from experts. Refusing to speculate for the opposing attorney only strengthens your opinion and position as an expert witness (emphasis added).

8.10 FALSE STATEMENTS

In *U.S. v. Fern,* 117 F.3d 1298 (11th Cir. 1997), the owner of an asbestos testing and consulting firm was indicted and convicted for, inter alia, knowingly making false statements on Notification of Demolition and Renovation forms in violation of § 113(c)(2)(A)of the Clean Air Act (CAA).

Defendant Fern argued that the indictment charging him with falsification in violation of CAA § 113(c)(2) was insufficient because it did not expressly state that willfulness and materiality are elements of the violation and did not specify the subsection of § 113(c)(2) with which he was charged. The court held that (1) willfulness is not an element of a § 113(c)(2)(A) violation, and (2) the indictment, by referencing § 113(c)(2) and using the phrase *false statement* was sufficient because defendant could not read the pertinent section and conclude that he was charged with a violation of anything else. The indictment sufficiently identified the statements in the notice that were false. "The Court declined to determine whether the U.S. Supreme Court's holding in *United States v. Gaudin,* 515 U.S. 506 (1995), that it is reversible error if a trial court does not submit to the jury the question of materiality of the false statement, applies to § 113(c)(2) cases. Instead, the Eleventh Circuit held that, whether the *Gaudin* case applies or not, *Fern* was not prejudiced by any *Gaudin* error because the evidence showed that no reasonable juror could have concluded that the false statements were not material" (Mansfield, 1997, p. 170).

8.11 AN EXAMPLE CASE EMPLOYING NUMEROUS EXPERT WITNESSES

The following case, though litigated in New South Wales, Australia, serves as an excellent example of environmental litigation between two parties with each side employing numerous geoscientific expert witnesses. The fact that the case is from an Australian jurisdiction (still an English common law nation), does not detract from its value in expert witnessing, or for its excellent environmental analysis by the Australian Court in the new-mine permitting process involved. It might well have taken place in a U.S. court. In its decision, the Land and Environment Court issued a conditional mine permit subject to many conditions for the protection of the environment. The case is highly instructive in environmental protection as well as in expert witnessing.

NEW SOUTH WALES SILICA SAND OPERATION HIVEN CONDITIONAL MINING PERMIT
A study of a court's analysis of surface mining's environmental impact

(From ASTON'S MINING LAW CASE REVIEWS™, *Mineral Resources Engineering Journal*, Vol.10, No.1, Jan.–March 2001)

In June 1999, New South Wales Glass and Ceramic Silica Sand Users Association, Inc., (hereafter, "Applicant") applied to the Port Stephens Council (hereafter, "Council") for development consent to mine a high-purity, white silica sand from the northern dunes at Tanilba on the Tilligerry Peninsula. The Council refused to grant development consent on 22 December 1999 for environmentally impacting reasons. Applicant appealed the Council's decision to the New South Wales Land and Environment Court.

In *New South Wales Glass and Ceramic Silica Sand Users Association Ltd v. Port Stephens Council*, [2000] NSWLEC 149 (15 August 2000), the appeal was lengthy, involving voluminous reports of scientific invesitigations by 22 geo-scientists, engineers and other professionals; 10 for the Council and 12 for the Applicant.

"White, silica sand has been extracted from parts of the northern dune for many years by ACI Industrial Minerals Division ("ACI"). Currently ACI is extracting white silica sand from the western part of the northern dune pursuant to development consent which was granted on appeal by this Court in 1990 (*ACI Operations Ltd v Port Stephens Shire Council and Ors* (Bignold J, NSWLEC, 14 December 1990, unreported)." The Court referred to the land in that development consent as "the 1990 consent area." The present proposal related to the central and eastern part of the northern dune. (*New South Wales Glass and Ceramic Silica Sand Users Association Ltd v. Port Stephens Council*, [2000] NSWLEC 149 (15 August 2000), § 7)

The proposed development involved the extraction of sand by stripping the topsoil and removing the underlying white silica sand. No more than 3 hectares (7.41 acres) were to be disturbed at any one time; one hectare being cleared in anticipation of extraction; one hectare being mined, and one hectare being rehabilitated. The sand was to be transported by trucks to ACI's sand treatment plant at Salt Ash, about 11 km. from the northern dune. The subject land of the development application ("the site") is principally Crown land, but is in part owned by the Hunter Water Corporation ("HWC"). The site falls within the Tomago Sandbeds Water Catchment area, and about 1500 m. from the water treatment works of the HWC. (Id. §§ 8, 9, 10)

Port Stephen Council's (hereafter, "Council") case was that development consent should be refused principally because of the uncertainty of the impact of the proposed development on the groundwater system. Altogether, the issues raised were: (1) ground-water quality and availability; (2) impact upon the vulnerable species, the wallum froglet; (3) archaeology — aboriginal relics; (4) the planning context; (5) rehabilitation;

(6) economic benefit; and (7) the inadequacy of the environmental impact statement ("the EIS"). (Id. § 14)

The Court's Analysis

The Court stated that "The impact of the proposed sand extraction upon the quality and availability of groundwater is the critical issue which underpins this case.... an adverse impact upon the groundwater system may lead to adverse impacts upon the wallum froglet species, may prevent proper rehabilitation of the site, and may have an impact upon the extent of the resource." (Id. § 16)

(1) Groundwater quality and availability

"The site comprises a coastal sand dune which provides a fresh groundwater system in the form of an aquifer. One of the characteristics of the groundwater system is that fluctuations in the water table will occur, particularly as a consequence of rainfall. There is a risk therefore that the water table may rise above the base of the extraction area. This could cause the formation of a non-permanent or semi-permanent body of water, known as a mirror lake, which is precisely what has occurred in the 1990 consent area. The formation of a mirror lake is likely to have serious environmental consequences. It may have a contaminating effect upon the groundwater, by exposing the water table to the atmosphere, and it may, as appears to have occurred in the 1990 consent area, damage or impede rehabilitation of the post extraction landform. It may also impact upon the wallum froglet, in that it would become a host for a species known as *Gambusia holbrookii* or mosquito fish ("*gambusia*"), which is likely to be a predator of the wallum froglet.

"A method of avoiding any such consequence is to provide a buffer between the base of the sand extraction and the water table, that is, to require extraction not to proceed below a certain level. The Council claims, however, that there is insufficient data and information available for, first, assessing satisfactorily the groundwater characteristics of the site; secondly, for predicting with accuracy either the peak or average water table levels; and, thirdly, for establishing the extent of a buffer which will avoid the potential problem. This claim is based upon the opinion of Mr Beck (Council's hydogeologist), who considered that 12 months site specific monitoring of the groundwater system was essential in order to provide sufficient information and data.... The applicant's response is that there is sufficient data and information upon which to assess the groundwater characteristics of the site and to set an appropriate limit upon the depth of extraction so as to provide a buffer. The applicant relied upon the expert evidence of Mr. Jewell (applicant's hydrogeologist), who assessed the groundwater system by extrapolating from data and information relating to the Tomago sandbed system regionally and who made recommendations about the design of the mining surface so that it will not pass beneath the water table." (Id. §§ 17, 18, 19)

Judge Pearlman, for the Court, stated, "I have come to the conclusion that there is no justification for refusing development consent by reason of the groundwater issue. That follows from my analysis of the respective positions taken by Mr. Beck and Mr. Jewell. Mr. Beck did not go so far as to say that it was impossible to assess the groundwater system and set a sufficient buffer. Rather, he was critical of extrapolating regional data to derive site specific characteristics, and he recommended, as I have indicated, that the assessment and consequent limit of mining be based on a site specific monitoring pro- gramme carried out over 12 months. Mr Jewell said that site specific monitoring over a period of 12 months was desirable, but he said, in giving oral evidence, that site specific monitoring would merely have the effect of refining the details of the assessment which he had carried out, and that it would not have a significant effect upon the recommendations which he had made. I accept Mr. Jewell's opinion." (Id. § 20)

The Court continued, "The issue, then, is the nature of the conditions of consent which should be imposed. Both parties furnished draft conditions which would involve a programme of site specific monitoring of the groundwater for a period of 12 months, and which will set out or refine limits on the depth of extraction. There are three crucial differences between their respective draft conditions.

"First, the council seeks a deferred commencement condition under s 80(3) of the Environmental Planning and Assessment Act 1979 ("the EP&A Act"). That would mean that the development consent would not become operative until the council was satisfied that the groundwater monitoring programme had been carried out and that the consequent mapping and setting of extraction limits were appropriate. The applicant, on the other hand, does not wish development consent to be deferred, but instead seeks a condition which would require a 12 month groundwater monitoring programme to be completed before extraction operations commence. I accept Mr. Jewell's analysis of the groundwater conditions on the site and I accept that his assessment is based on information sufficiently certain to assess the likely impact of the proposed development. I also accept that a 12 month monitoring programme is desirable for the purpose of refining the maximum level of extraction. Therefore, I do not consider that the groundwater characteristics of the site are so uncertain as to require a deferred commencement condition, and I propose to adopt the applicant's draft condition so far as this aspect is concerned.

"The second crucial difference is the intensity of the 12-month monitoring programme. Mr. Beck recommended that monitoring be carried out on a weekly, and in certain circumstances daily, basis; Mr. Jewell recommended monthly periods. It can, however, be inferred from Mr. Jewell's evidence that the daily or weekly monitoring which Mr. Beck espouses is not necessary to establish the levels of the water table, since those levels are already established to a reliable degree by the results of monitoring over five years in both dry and wet conditions, being results upon which Mr Jewell relied. The purpose of the proposed 12 month monitoring programme is only to refine the details of the groundwater characteristics already established, and to confirm or adjust the extraction depths to be followed in the sand mining operation. Accordingly, I accept the applicant's draft condition in preference to the draft condition of the council.

"The third crucial difference relates to the maximum depth of extraction. The council proposes that the 'limit of mining' must be whichever is the higher of RL 9.75 AHD or 1.25 m of the 'peak mean annual water table' meaning 'the highest of the mean (average) annual Water Table positions determined by the Groundwater Assessment'. The applicant, on the other hand, proposes that the maximum extraction depths comply with the following requirements:

(a) No extraction is to occur below RL 9.0 m on any part of the Extraction Area.

(b) Extraction at any given point should not occur beneath the maximum predicted elevation of the water table at that point.

(c) The maximum extraction depth at any given point should have a minimum clearance of 1 metre above the mean position of the water table at that point." (Id. §§ 21–24)

Summing up the conditions to be imposed for groundwater quality and availability, the Court stated, "I propose to adopt the applicant's draft condition, and my reasons for so doing are as follows. The level of RL 9.75 m suggested by the council has two components. One is a level of RL 9.50 m recommended by Mr Beck as being 1 m above the average peak mean water table level of RL 8.5 m which he derived by correlating the water table data presented by Mr Jewell and the average annual rainfall figures which were set out in Dr Clements' report. The second component is a level of 0.25 m recommended by Mr Beck to account for the predicted rise in the water table following removal of vegetation. I accept Mr Jewell's opinion that both these components are unnecessarily conservative, and they are, I think, influenced by Mr Beck's thesis that the information and data currently available about groundwater characteristics is too uncertain to be reliable. Mr Jewell calculated a mean or average water table level of RL 8.0 m, and, taking into account the necessity for a 1 m buffer, derived a base level of RL 9.0 m. That is also the base level required by HWC and the Court in relation to the 1990 consent area. As to the 0.25 m component, Mr Jewell accepted a previous study indicating that the water table will rise to a maximum of 0.25 m after the removal of vegetation, but he said that it was taken into account in the 1 m buffer upon the basis that the 0.25 m rise would be localised, that it would reduce to zero moving away from the mining path, and that revegetation would decrease it further.

"There is, of course, a question of ensuring that the limits upon the depth of extraction are complied with, especially since the creation of the mirror lake on the 1990 consent area appears to have occurred because extraction was deeper than the limits which were set in the conditions of development consent which related to that area. Mr Marshall (applicant's geologist) gave evidence of laser technology and procedures available to survey the lateral extent of extraction and the depth of extraction. Conditions of consent requiring the use of that technology and those procedures are incorporated into the conditions of consent which I propose to adopt." (Id. §§ 26–27)

(2) Impact upon the wallum froglet habitat

"The wallum froglet (*Crinea tinnula*) is listed as a vulnerable species under sch 2 of the Threatened Species Conservation Act 1995 ('the TSC Act') and is the only threatened or vulnerable species considered likely to occur on the site. I have concluded that the proposed development is not likely to have a significant impact upon the habitat or population of the wallum froglet, and development consent should not be refused on this ground." (Id. § 28)

Both Dr. Mahoney (Council's biologist) and Dr. Smith (applicant's ecologist) found populations of the wallum froglet in areas adjacent to but outside of the site. The Court noted that "A more critical aspect concerning the wallum froglet, however, was the question of whether the proposed development would impact upon the groundwater conditions. Dr. Mahoney pointed out that alteration of the groundwater conditions was likely to have an indirect adverse impact upon populations of the wallum froglet for two reasons. First, if sand extraction results in a lowering of the water table and drier conditions, then there would be an impact upon the habitat of the wallum froglet, because the species is reliant upon wet habitats. Secondly, the formation of mirror lakes presents a habitat for the gambusia which is a voracious predator and likely to threaten the local populations of wallum froglet. This potential risk from an alteration in groundwater conditions was also recognised by Dr. Robertson (Council's ecologist). (Id. §§ 30, 32)

"Dr. Smith considered that the potential risk from the gambusia was overstated by Dr. Mahoney. He said that the wallum froglet breeds in shallow temporary pools, whilst the gambusia 'appears to favour permanent streams and deep open pools'. He also said that the drying out of ephemeral wetlands after the removal of the adjacent dune 'is unlikely to affect the wallum froglet'. I think that Dr. Smith is correct, given that I have accepted the evidence of Mr Jewell that there is not likely to be a significant alteration of the groundwater conditions, and given that, if the conditions controlling the depth of sand extraction are observed by the applicant, there are not likely to be mirror lakes or a consequent preying by the *gambusia*." (Id. 33, 36)

(3) Archeology

According to Mr. Baker, an archeologist for the Council, there had been at least 11 archeological surveys in and around the general region of the site. The surveys of 1989 and 1993 indicated one area of some interest which a part was located within the mine site.

The Court stated, "The council seeks, and the applicant is prepared to be bound by, a condition which requires subsurface archaeological assessment to be carried out on the site, in accordance with a methodology which Dr. Craib (applicant's archeologist) furnished. The only issue between the parties is whether that condition should be a deferred commencement condition, ... or a condition requiring the assessment to be completed before any extraction operations commence upon the site.... I am not persuaded that the condition should be a deferred commencement condition because, for two reasons, I do not think it is necessary for the council to be further satisfied about the archaeological issue.

"First, some doubt was cast by the evidence upon Mr. Baker's assertion that the results of a further assessment of the site are likely to be significant. Dr. Craib was of the opinion that what is likely to be found is no more than 'common components of the archaeological landscape in this region' which Dean-Jones had reported upon. Furthermore, while Associate Professor Short agreed with Mr Baker that the northern dune is a Pleistocene transgressive dune, he was of the opinion that the dune had not been 'reworked', that is, affected by marine and aeolian (wind blown) processes after aboriginal occupation, which made it less likely, as he said, in giving oral evidence, that material could be buried.

"Secondly, so long as the results of the assessment are furnished to the National Parks and Wildlife Service ('the NPWS') (as the draft conditions contemplate) and so long as a protocol for the discovery of relics is developed in consultation with the Worimi Local Aboriginal Land Council and approved of by the NPWS before the commencement of extraction (as the draft conditions also contemplate) then damage or destruction of any aboriginal relic is likely to be avoided." (Id. §§ 38–42)

(4) The Planning Context

"There are three matters which require consideration in the planning context. The first relates to the permissibility of the proposed development, and requires consideration of the Port Stephens Local Environmental Plan 1987 ('the 1987 LEP') and the draft Port Stephens Local Environmental Plan 1999 ('the draft LEP'). The second relates to whether or not the proposed development is consistent with the planning objectives of:
 (a) the 30 Year Plan for Port Stephens ('the 30 year plan');
 (b) the Hunter Regional Environmental Plan 1989 ('the Hunter REP'); and
 (c) the NSW Coastal Policy 1997 ('the Coastal Policy').
The third relates to the proposed inclusion of the site within the Tomago Sandbeds Nature Reserve. (Id. § 43)
 Permissability:
Under the 1987 LEP, the Court found the majority of the site was zoned Environment Protection 7(c), and a small portion of the site was zoned Rural 1(a). "The proposed development is prohibited in both zones, but it becomes permissible with consent under cl 37 of the 1987 LEP. Clause 37 applies to specified land, which includes the site, and it relevantly provides that the specified land may be developed for the purposes of silica

extraction with the consent of the council. However, a new, re-drafted LEP has been drawn and cl 37 is not intended to appear as part. Although "The draft LEP has been publicly exhibited.... the Department of Mineral Resources has a number of times opposed the deletion of cl 37, stating, in a letter to the council dated 12 January 2000, that it opposed the draft LEP 'because of the adverse impact of the proposed Zone No. 7(a) — Environment Protection on extractive resources and mineral deposits of regional and state significance on the Tilligerry Peninsula'. Mr. Whitehouse (applicant's geologist) gave evidence that he was present at a meeting in March 2000 between the council, the Department of Urban Affairs and Planning and others at which the Department of Mineral Resources recommended the retention of cl 37 and, at the council's request, the issue of the retention of cl 37 was deferred. Mr. Mitchell (a consulting town planner) conceded in cross examination that the draft LEP had not been submitted to the Minister and that it was still in the process of negotiation."

Summarising, the Court added, "Having regard to that evidence, I am not satisfied that the making of the draft LEP in its present form is either certain or imminent. There is no basis for giving the (new) draft LEP determinative weight, and accordingly the proposed development remains permissible with consent under the 1987 LEP." (Id. §§ 48, 49)

Consistency with Planning Objectives:

"The planning instruments which I have set out above have a number of interlocking objectives relevant to the proposed development. I am satisfied that the proposed development does not fail to take those objectives into account, nor does it fail to comply with the planning principles espoused by them. (Id. § 50)

"...As to tourism, Ms. Artist (the council's corporate strategic planner) gave some evidence outlining local tourism opportunities on the Tilligerry Peninsula. However, as revealed in cross-examination, she was unable to point to any particular tourism feature of the area which includes the northern dune, and I am satisfied that sand extraction from the site will not impact upon local tourism opportunities. (Id. § 54)

The Tomago Sandbeds Nature Reserve:

"The council claims that the development consent should be refused because, if it was to be granted, the site will not be included in what is proposed as the Tomago Sandbeds Nature Reserve. Alternatively, it claims that the prospective inclusion of the site as part of the area proposed for the Tomago Sandbeds Nature Reserve should be taken into account as signifying the importance of the natural conservation values of the site.

The Court noted that the history of the proposed Tomago Sandbeds Nature Reserve, has, thus far, failed to make any serious progress between the agencies considering its formation. Consequently, the Court commented, "This history raises serious doubt as to whether the area which includes the site will ever be the subject of a proclaimed nature reserve. At the very least, it indicates that the Tomago proposal is far from implementation. On that basis, therefore, I place little weight upon the Tomago proposal in the assessment of the development application." (Id. §§ 56–58)

(5) Rehabilitation

The Court found that there was no issue in this case concerning threatened species of flora. "Dr. Clements (Council's plant ecologist) noted that two species of both State conservation significance and national significance had been recorded on or near the site, that is, *Eucalyptus parramattensis* subsp. *decadens* was recorded approximately 200 m south of the site and 'potentially' occurs on the site, and *Acacia baueri* subsp. *baueri* was found on the eastern end of the northern dune in 1986 by Ms. du Preez (applicant's botanist). As to the former, *Eucalyptus parramattensis*, Dr. Pickard (applicant's ecologist) was unable to locate it within 10 km of the site, but he applied the eight point test of significant effect set out in s 5A of the EP&A Act, and concluded that the proposed development 'is unlikely to have any significant impact on this species.' As to the latter, *Acacia baueri*, the evidence is that it has not been listed as threatened or vulnerable under the TSC Act. In any event, Ms. du Preez stated that the species had naturally regenerated on the 1990 consent area (although it was subsequently destroyed by the inundation of the mirror lake)." (Id. § 59)

"The critical issue is, however, the rehabilitation of the site. Ms. du Preez and Dr. Smith (applicant's ecologist) prepared a rehabilitation plan, and ... They agreed upon 10 separate points which were substantially incorporated into the draft conditions about rehabilitation which the council and the applicant have respectively proposed."

Judge Pearlman continued, "The first concerns the likelihood of rehabilitation being successfully carried out ... the council is concerned that a failure to establish and then comply with conditions controlling potential groundwater impact will result in the creation of a mirror lake similar to that which has occurred on the 1990 consent area, and that in turn will impede or prevent rehabilitation. For the reasons I have earlier set out, I consider that this result is unlikely to occur. However, the council is also concerned that the applicant will not properly carry out or supervise conditions of consent concerning rehabilitation, because it is not in dispute that rehabilitation of the 1990 consent area, quite apart from the mirror lake, has not been wholly successful. In response to this concern, the applicant has offered to put up a bond in the amount of $100,000, and that I think is sufficient to allay the council's concerns.

"The second matter concerns the extent to which rehabilitation can restore the pre-extraction vegetation. The EIS stated that the aim of the rehabilitation programme which was proposed was 'to re-establish the natural vegetation of the site.' Ms. du Preez said that 100 per cent re-establishment of native vegetation was not achievable, because the landform will change post-extraction. Both Ms. du Preez and Dr. Clements considered that it was reasonable to aim to re-establish 95 per cent of the structural components of vegetation, meaning the species which form the upper, middle and lower storeys of vegetation. In Ms. du Preez's opinion an achievable standard for the re-establishment of characteristic species (meaning species which characterise the vegetation on the site) was 82 per cent. Mr. Kildea (Council's barrister) submitted that conditions which aimed at less than 100 per cent recovery were setting a lower standard than that which was set in the EIS. However, as Ms. du Preez pointed out in oral evidence, the setting of numerical standards is unrealistic in itself, and I agree that it is more appropriate simply to require the re-establishment of pre-extraction vegetation. The condition of consent concerning the objectives of the rehabilitation programme is based on such a proposition, taking into account that regeneration may occur in different proportions. The parties agree on this aspect of the condition. (Id. §§ 59–62)

"I turn now to the draft conditions about rehabilitation which each of the parties has proposed. I propose generally to adopt the conditions which the applicant proposes and my reasons for so doing and the crucial differences between the respective drafts are as follows:

(1) The council proposes that rehabilitation should be carried out in accordance with a rehabilitation programme which is prepared after the groundwater assessment (the subject of a proposed deferred commencement condition) has been completed. The applicant proposes compliance with the rehabilitation programme already prepared by Ms. du Preez. In view of my findings about the groundwater issue, I can see no reason to require the preparation of a further rehabilitation programme. I am fortified in that conclusion by the plaudits about the competence and professionalism of Ms. du Preez which were bestowed

upon her by both Dr. Clements and Dr. Pickard. (applicant's ecologist) (I must say that I have never before heard such statements of praise about the work of another expert given by experts in the witness box.)

(2) The council proposes that the appointment of both the plant ecologist or botanist who is to supervise the rehabilitation and the rehabilitation officer who is to carry it out should be approved by the council. The applicant opposes such a requirement. I do not consider it to be necessary, so long as it is made clear that both those persons are to be appropriately qualified for the task;

(3) Both the council and the applicant propose monthly site visits by the plant ecologist or botanist in the first two years of sand extraction, but the council proposes more frequent visits depending upon climatic conditions and seasonal variations. The applicant opposes more frequent visits. I can see no justification for more frequent visits, since monthly visits were acceptable to both Dr. Clements and Ms. du Preez;

(4) Both the council and the applicant agree on the requirements for quadrats and transects for the recording of species in the early years of rehabilitation, but the council seeks a similar provision for the period after plants have become established, and the applicant seeks a provision in relation to sampling pre-extraction. I consider that the condition should encompass all three periods, and I propose to amend the applicant's draft accordingly;

(5) The parties are in disagreement about the condition requiring the provision of a bond. I propose to adopt the applicant's version, because I agree with the applicant that:

(a) the condition should require the bond to be maintained for the duration of development;

(b) the form of the bond should not be prescribed by the condition, but should be as acceptable to DLAWC;

(c) a standard of reasonableness as to the application of the bond by the DLAWC is appropriate; and

(d) DLAWC should not be able to utilise the bond to itself rehabilitate in circumstances where the applicant is satisfactorily complying with the rehabilitation conditions.

(6) Generally, the condition sought by the council is more prescriptive as to time than that sought by the applicant. For example, the council seeks the furnishing of a report by the plant ecologist or botanist within 14 days of each anniversary of commencement of extraction operations rather than 28 days as the applicant proposes. I can see no justification for a more prescriptive approach;

(7) Finally, the council proposes that members of the local community may attend an annual site inspection of the rehabilitation with both DLAWC and the council. The applicant opposes such a condition. Once again, I can see no justification for it. Reports as to the progress of rehabilitation and compliance with the programme are to be furnished to the council which is the representative of the community's interest, and this is not a case where representatives of the local community have sought such a condition." (Id. § 63)

(6) Economic Effect

"The council claimed that it was in the public interest to refuse to grant development consent because the economic benefit to be derived from the silica sand as a resource for the glass manufacturing industry did not outweigh the conservation values of the site. The council contends that the proposed development will destroy the conservation values of the site because it will adversely impact upon those four elements. First, the northern dune will be in effect removed by the sand extraction; secondly, it will be rendered useless as a tourist attraction; thirdly, there will be a reduction of, and impact upon flora and fauna; and, fourthly, it will not be included in the Tomago Sandbeds Nature Reserve.

"In addition, the council claims that there are alternative sand resources which are economically viable, and that the extent of the silica sand resource upon the site is less than predicted. The council claims that, when those matters are considered in conjunction with the destruction of the conservation values of the site, it must follow that it is in the public interest to preserve the site from sand extraction and that must outweigh the significance of the resource.

"However, the evidence does not support the council's claims for such high conservation values for the site, nor is it free from doubt that alternative resources are economically available, nor is the extent of the resource significantly less than predicted." (Id. §§ 64, 66–68)

Alternative Silica Sand Sources for Glass-making:
"The question of alternative sources and their economic viability is a complex one. White silica sand is a preferred component of the making of colourless glass because it has a low iron content. Sand with a higher iron content is used for the manufacture of amber glass and other coloured glass. The significance of the sand on the Tanilba northern dune in the applicant's estimation is that it constitutes a source of white silica sand reasonably close to its main customers, which are the ACI plant at Penrith and the Pilkington (Australia) Ltd plant at Ingleburn.

"There is, however, no doubt that, through a complex beneficiation process, the iron content of friable sandstone can be reduced so that it is likely to meet the industry specification. The process is more complex than the simple process currently utilised at ACI's Salt Ash plant. The thesis which the council endeavoured to establish was that a simple process a long distance from the glass manufacturing plants is less economically viable than a more complex process nearer to them, and that there are sources of friable sandstone much nearer to those plants than is the site." (Id §§ 70–71)

"Evidence as to the council's position was given by Mr. Stitt (Council's mining and geological consultant) and by Dr. Sorrentino (Council's economist and chemical engineer). Mr. Stitt took some samples from four plants which produce construction sand from friable sandstone resources. They were located at Maroota, Newnes Plateau, Somersby and Mittagong, each of which is closer in distance than Tanilba to the main customers of the applicant. He put each of those samples through some laboratory testing and concluded that the sand from Maroota and Newnes Plateau would meet the industry iron content specification of 0.03 per cent. Then, based on various assumptions as to costs and other relevant data, he ran models to reach a preliminary financial analysis which indicated that a plant co-sited with the sand operations at either Maroota or Newnes Plateau could be profitable and could allow for cost reductions to be made for sand supplied to glass manufacturers.

"Dr. Sorrentino carried out a similar exercise to conclude that silica sand obtained from friable sandstone 'will cost users no more than what they are currently paying' and, on the assumptions which he made, would be more likely to deliver the product to the applicant's customers at a price between 4 per cent and 17 per cent lower than that currently charged by the applicant.

"The applicant took issue with the financial modelling carried out by Mr. Stitt and Dr. Sorrentino. Mr. Alste (applicant's scientist) stated that their assumptions as to the prices exworks and delivered from Salt Ash were much higher than is in fact the case. The actual prices were the subject of documents which were tendered under a confidentiality order, and were accordingly not available to the council's experts, but, applying the actual costs to the graphic representation of 'levellised cost' which Dr. Sorrentino produced, Mr. Alste stated that there would be a negative and not a positive return. Mr. Alste also pointed out that the council's witnesses assumed that the total tonnage per year produced by the ACI though its Salt Ash plant is 250,000 tonnes when in fact it is 200,000 tonnes, and that they failed to take into account the loss of other production from Salt Ash if it is to be relocated. Mr. Alste said that the variability of the quality of the friable sand deposits and the high risk of investing in a full scale plant would require a pilot trial, and it would take about five years to prove the reliability of any such alternative source, to secure planning approvals and conduct the trial.

"Mr. Wait also had reservations about the evidence of Mr. Stitt and Dr. Sorrentino. He concluded that the models which they adopted were not inappropriate, but that there were errors in their application, and 'some generalisations about inputs which require verification and some reservations about assumptions employed.'

"There were other matters which also cast doubt upon the evidence of Mr. Stitt and Dr. Sorrentino. First, the current maximum iron content specification is 0.02 per cent not 0.03 per cent, and accordingly the results of Mr Stitt's laboratory analysis would indicate that sand from Newnes Plateau and Maroota would fail to meet a specification of 0.02 per cent. Secondly, Mr. Higginbotham (applicant's analytical chemist) noted that the maximum level of iron is currently being exceeded by a significant amount of the Penrith plant. He conceded in cross examination that the sand itself was only a small part of that problem since it was also a consequence of high levels of iron in other components of the processing such as external cullet (broken recycled glass) and pickup (cross contamination from other batches mixed in the plant). He also conceded that it was a technical problem capable of a technical solution. But Mr. Higginbotham said that a technical solution comes at a cost, and, for example, changes to the batch plant at Penrith to eradicate or reduce the problem could cost in the vicinity of $2 to $3 million dollars. Thirdly, although Mr. Whitehouse (applicant's geologist), giving evidence from the perspective of the Department of Mineral Resources, thought that friable sandstone might be a suitable alternative resource in the long term, he said that, on the information presently available to the Department, he could not say to what extent that might occur or whether or not it might be viable at all.

"There is, therefore, considerable doubt as to the economic viability of friable sandstone deposits as an alternative resource for white silica sand, and I am unable to conclude that the white silica sand at the site is anything but a significant resource which is appropriate for exploitation." (Id. §§ 72–77)

The Northern Dune Reserves of White Silica Sand:

"The extent of the white silica sand at the site is also a factor in its significance. The estimate in the EIS was that the size of the resource is about 1,650,000 tonnes, and the life of the resource, at a rate of extraction of about 100,000 tonnes per annum, is approximately 16 years. Mr. Beck's initial calculations were that the total volume was likely to be between 560,000 tonnes and 640,000 tonnes, and accordingly, at that rate of extraction, the life of the resource less than six years. That estimate was based upon the adoption of all his recommendations regarding the groundwater conditions and the depth of extraction. Accordingly, if Mr. Beck's calculations were to be adopted, the short term life of the resource would reduce its significance. But Mr. Marshall (applicant's geologist) gave some estimates too. He thought that the size of the resource was about 1,500,000 tonnes based upon the assessment and recommendations of Mr. Jewell and, at the same extraction rate, the life of the resource was about 15 years. Mr. Beck did not substantially disagree with Mr. Marshall's calculations if Mr. Jewell's assessment and recommendations were adopted. Since I have accepted Mr. Jewell's evidence, as I have earlier explained, it follows that the extent of the resource may be accepted as being 1,500,000 tonnes based on Mr. Marshall's calculations. That does not differ substantially from the estimate of 1,650,000 tonnes predicted in the EIS, and does not lead to a conclusion that the life of the resource is so short that it must be regarded as insignificant.

"My ultimate conclusion, therefore, is that the conservation values of the site do not outweigh the significance of the resource. (Id. §§ 78–79)

"I should add for completeness that the council tendered a report from Mr. Abelson (Council's economist), who concluded that the proposed development would not increase employment in the sand mining industry in New South Wales, but, even if it did, it would not necessarily confer social and economic benefits on the people of New South Wales. Mr. Abelson was not called for cross examination, and his evidence was not the subject of any emphasis in the proceedings. Accordingly, I do not place significant weight upon it, and it does not alter my conclusion about the significance of the resource." (Id. § 80)

(7) Adequacy of the Environmental Impact Statement (EIS)

The proposed development is an extractive industry and falls within the list of designated development in sch 3 of the Environmental Planning and Assessment Regulations 1994 ("the Regulation"). As such, the development application was required to be accompanied by an environmental impact statement (EIS) pursuant to s 78A(8)(a) of the EP&A Act. Clauses 54 and 55 of the Regulation prescribe the contents of an environmental impact statement and, in particular, cl 54A requires an environmental impact statement to include the matters referred to in any specific guidelines which apply to the development. The specific guidelines which apply are called "Extractive Industries — Dredging and Other Extraction in Riparian and Coastal Areas" ("the EIS guidelines"). (Id. § 81)

The Court noted that "the crux of the Council's claim is that the EIS fails in a number of respects to comply with those principles. Mr. Kildea contended that it contained a number of omissions and inaccuracies which had the result that the EIS was inadequate to alert the council and the public to the inherent problems of the proposed development and that the EIS was insufficiently specific, comprehensive or objective." (Id. § 82)

The Court's opinion of the inadequacy of the EIS proceeded to reiterate its formerly discussed arguments and evaluation points on groundwater, archeology, the wallum froglet, economic impact, requirements of authorities, changes in the proposal, and inaccuracies.

In summation of the adequacy of the EIS, the Court stated, "In summary, I have come to the conclusion that the EIS is far from perfect, but a standard of absolute perfection is not required. It has shortcomings in matters of detail and emphasis, and contains some inaccuracies, but, as pointed out by Hutley JA in *Prineas v Forestry Commission of New South Wales and Ors* (1984) 53 LGRA 160 at 163, '... it is almost impossible to conceive an E.I.S. which literally complies with everything which the regulations require'. In my opinion, the EIS substantially complied with the relevant statutory requirements, and did contain sufficiently comprehensive information about the proposed development to alert the Council and the public to its inherent problems." (Id. § 84)

The Court's Imposed Conditions of Consent

Concerning Conditions of Consent to implement the approval of the mining permission, the Court stated: "Part 13 r 16(e) of the Land and Environment Court Rules 1996 requires the council to file and serve any conditions which it would seek to impose if the Court were to grant consent, and the applicant may file conditions in response. Sets of draft conditions were respectively filed in compliance with that rule. Directions were given at the close of the hearing before judgement was reserved for the filing of submissions on those respective sets of draft conditions. That direction resulted in the filing of submissions and alternative sets of draft conditions. I have considered all the documents which were filed but I have focussed attention on the council's second amended draft conditions filed on 5 July 2000 ('the council's draft'), and the applicant's updated draft conditions dated 7 July 2000 ('the applicant's draft'). (Id. § 85)

"I have decided to substantially adopt the applicant's draft rather than the council's draft. The reason for so doing is that I do not propose to impose deferred commencement conditions for the reasons which I have set out when dealing with the groundwater, archaeology and rehabilitation issues. The consequence is that the form of the council's draft, based on deferred commencement conditions, is not generally appropriate and it does not readily lend itself to redrafting so as to excise the deferred commencement conditions.

"However, it is important to note that, of the council's 58 draft conditions, the applicant's draft incorporates 35 of them, although not in precisely identical terms.

"The principal difference is that the council's draft is based upon the council taking a close supervisory role, or, to adopt Mr. Kildea's submission, the council giving its approval to each crucial stage of the development before inappropriate action is taken by the applicant which later cannot or may not be able to be remedied by the applicant. The council's stance is based upon the council's assertion that environmental damage has occurred on the 1990 consent area and its occurrence has not been explained by the applicant. In particular, the council points to the creation of the mirror lake. There is no doubt that a mirror lake was formed — it is there for all to see, and the applicant does not deny that it occurred because sand was excavated below the maximum permitted extraction depth. There is also no doubt that it is desirable that a mirror lake should not be formed because of the consequences I have earlier set out. But I agree with Mr. Robertson's submission that the Court has not examined in detail the environmental consequences of the mirror lake or any other alleged environmental damage occurring in the 1990 consent area. Those matters are irrelevant to this appeal. This appeal is concerned with the proposed development, and I am satisfied that the applicant's conditions do impose obligations upon the applicant which, if they are complied with, will serve to avoid or mitigate environmental harm. Moreover, the applicant will be required to put up a bond to secure the rehabilitation programme, and, of course, there are remedies available to the council or any person if the applicant should breach the conditions of consent.

"I have, therefore, adopted the applicant's conditions, but I have made some alterations which are as follows:

(1) I have inserted a definition of 'these proceedings,' since that expression appears in a number of conditions, and ought to be clearly defined;

(2) As a general matter at the commencement of conditions, I have inserted a reference to the development being carried out in accordance with the EIS as amended by the conditions or by any assessment, surveys or plans prepared in accordance with the conditions. Although Mr. Robertson asserted that the applicant's draft and the annexures to the conditions in effect replace the EIS, I consider that the EIS sets out the basic parameters of the proposed development and ought to be incorporated into the development consent by way of condition;

(3) I have amended condition 1 to refer to a cultural heritage management plan, and I have made a consequent amendment to condition 22. The applicant did not object to this course;

(4) It bears repeating that the groundwater issue is the most crucial issue in these proceedings, and any impact upon groundwater quality or quantity is likely to result in other environmental impacts. Conditions 12–19 set out a groundwater monitoring and management regime with which I am satisfied, as I have earlier explained. However, I have made two additions. First, I have inserted a general condition requiring all practicable measures to be taken to ensure that there is no adverse impact upon groundwater. This follows generally the council's draft condition 19. Secondly, I have inserted a requirement to furnish the annual report concerning the results of the groundwater monitoring programme to the council as well as to DLAWC. This accords with the applicant's submissions;

(5) I have made reference to the suitability of the qualifications of the hydrogeologist and the rehabilitation officer in conditions 13 and 28 respectively;

(6) I have altered condition 32 in relation to sampling and recording of vegetation in the way I have earlier outlined.

"As I have said, the applicant's draft incorporates 35 of the 58 conditions in the council's draft. I have dealt with the remaining 23 conditions in the council's draft in the following way:

(a) Six conditions appearing in the council's draft have been incorporated substantially into the conditions. They (and the place where they now appear) are 6 (22), 11 (general — preamble), 12 [47(a)], 19 (groundwater — preamble), 55 [46(b)], and 56 [43(b)];

(b) Conditions 9, 26 and 27 have been omitted entirely because they relate to the council's stance on the close supervisory role, which I have rejected as explained above;

(c) Conditions 1, 2 and 3 are deferred commencement conditions which I have rejected for the reasons earlier given. Conditions 16, 17 and 18 relate to groundwater monitoring and conditions 21, 22, 23, 24 and 25 relate to rehabilitation, but, for the reasons earlier given, I prefer the applicant's draft in relation to these issues;

(d) The applicant objected to condition 5 as merely restating the law and thus as being unnecessary; and conditions 29 and 30 which provide for buffer zones on land between the extraction area and Oyster Cove Road and the Tanilba Bay residential area respectively, on the basis that they cover land which is not in the control of the applicant. I agree with the applicant's objection and I have omitted these conditions.

"Finally, the conditions which I propose to impose refer to a number of annexures. For example, condition 5 refers to the extraction area plan as being 'annexure 1.' It is, of course, impractical, if not impossible, to attach each of those annexures to this judgment. I propose, therefore, to prepare a full set of conditions, including the annexures, which I will sign for identification and which will remain with the file. An exact replica will be given to each of the parties. For the purpose of this judgement, the conditions are annexed and they refer to the annexures, without them being attached." (Id. §§ 86–91)

The Court's Decision

"In accordance with the foregoing, my final orders are as follows:
(1) The appeal is upheld.
(2) Development consent for the extraction of white silica sand from the northern dune at Tanilba on the Tilligerry Peninsula is granted subject to conditions. The conditions are those which appear in the document entitled 'Conditions of Consent' signed for identification and placed with the Court's file.
I make no order as to costs."

New South Wales Glass And Ceramic Silica Sand Users Association Ltd v Port Stephens Council
Re. Property: Part of Northern Dune, Tanilba
Conditions of Consent

DEFINITIONS

Extraction Operations:

*includes the extraction of sand, the transport of extracted sand on and from the site, the reshaping of land from which extraction has occurred and the prior clearing of vegetation to enable those activities to be carried out;

*but does not include groundwater monitoring, sub-surface archaeological sampling or rehabilitation works.

These proceedings — means the class 1 application No 10013 of 2000 between New South Wales Glass And Ceramic Silica Sand Users Association Ltd (ACN 002 890 226) as applicant and Port Stephens Council as respondent

NOTE: Conditions 16, 17, 20, 22 and 36A below require certain activities to be carried out before the Extraction Operations may occur.

GENERAL

The development shall be in accordance with the proposal contained in the Environmental Impact Statement dated May 1999 (which is annexure 2 to volume 2 of Exhibit A in these proceedings) modified by these conditions and any assessment, surveys and plans prepared in accordance with these conditions.

ENVIRONMENTAL MANAGEMENT PLAN

1. An Environmental Management Plan must be prepared and must incorporate the following:

(a) Procedures to be followed by the manager of the Extraction Operations and those persons undertaking the extraction. Those procedures must, as appropriate, incorporate the requirements of this consent concerning groundwater monitoring, extraction depth and area monitoring, aboriginal archaeology, and rehabilitation.

(b) The Erosion and Sediment Control Plan referred to in condition 59 below.

(c) The Cultural Heritage Management Plan referred to in conditions 22 below.

(d) Such other provisions as are necessary to facilitate compliance with the conditions of this consent.

2. A copy of the Environmental Management Plan must be provided to the Port Stephens Council ("the council") and a further copy must be kept at the Salt Ash Plant for the duration of the Extraction Operations.

3. The Environmental Management Plan must be reviewed, and if necessary updated, annually during the Extraction Operations. The Rehabilitation Supervisor and the Hydro-geologist must be involved in that review process.

4. Before a person is permitted to undertake extraction activities on the site he or she must receive training in relation to the procedures set out in the EMP and referred to above in condition 1(a).

EXTRACTION DEPTH AND AREA

5. Extraction Operations may only occur in the extraction area (the "Extraction Area") marked on the Extraction Area Plan that is part of Exhibit 16 in these proceedings and Annexure 1 to these conditions.

6. A 20 metre buffer zone, as measured by the applicant and approved in writing by the council, shall be established around areas "7" and "10" as identified in Figures 1 and 2 of the report prepared by Michael Mahony — Assessment of the Status of the Wallum Froglet (*Crinia tinnula*) in the Northern Dune Area Tanilba Bay that forms part of Exhibit F in these proceedings.

Figures 1 and 2 are Annexure 2 to these conditions.

7. The limits of that part of the Extraction Area, in which a part of the Extraction Operations are about to commence, must prior to that part commencing be measured and marked on the site by a qualified geologist or surveyor using a global positioning system in accordance with the procedure set out in the report of Andrew Marshall that is Exhibit 13 in these proceedings and Annexure 3 to these conditions.

8. Extraction of sand, at any given point on the Extraction Area, must not occur below the maximum depth indicated for that point on the Post Extraction Landform Plan that is Exhibit 26 in these proceedings and Annexure 4 to these conditions. The Post Extraction Landform Plan may be amended from time to time by the Hydrogeologist in accordance with condition 18 below, in which case the amended plan prescribes the maximum extraction depth.

9. Before extraction occurs on any part of the land within the Extraction Area, the height (metres AHD) of that part of that land must be measured using a laser level in accordance with the procedure described in the report of Andrew Marshall that is Exhibit 13 in these proceedings and Annexure 3 to these conditions. Markers must then be placed on and around that part of that land to the extent necessary so as to ensure an operator of extractive machinery can clearly identify the height or heights of that part of that land.

10. During the Extraction Operations the height of the land from which sand is being extracted is to be monitored by taking regular measurements using the laser level in accordance with the procedure referred to in the previous condition. Further markers are to be placed to the extent necessary so as to ensure an operator of extractive machinery can clearly identify the depth to which extraction is occurring.

11. Measurements using a laser level in accordance with the above conditions of consent may only be undertaken by a surveyor or geologist with training in surveying techniques, or a person who has been trained in the use of that level by a surveyor or such a geologist.

GROUND WATER — General

All practicable measures must be taken to ensure that the Extraction Operations do not adversely impact on the supply or quality of ground and/or surface water on the Extraction Area on adjoining land.

12. A groundwater monitoring programme (the "GMP") is to be undertaken by the applicant for the duration of the Extraction Operations.

13. The GMP is to be designed and supervised at all times by a suitably qualified hydrogeologist (the "Hydrogeologist"). The same person need not perform the role of Hydrogeologist for the entire duration of the Extraction Operations.

14. The GMP should generally incorporate the provisions of the Groundwater Monitoring Plan (including the requirement to provide the Department of Land and Water Conservation ("DLAWC") with an annual report concerning the results of the GMP) that is Annexure 5 to these conditions and is contained in Appendix D to the report of Chris Jewell that is Exhibit 5 in these proceedings. A copy of the annual report concerning the results of the GMP must be provided to the council.

15. The GMP must include monitoring bores within the Extraction Area.

16. Groundwater levels in all monitoring bores being utilised as part of the GMP are to be measured at least monthly for one year prior to the Extraction Operations commencing, and for the following four years. Measurement frequency after the expiry of that period is to be in accordance with the directions of the Hydrogeologist following consultation between the Hydrogeologist and DLAWC.

17. Prior to the commencement of the Extraction Operations, and whenever he or she deems it necessary, the Hydrogeologist must review the maximum extraction depths in the Post Extraction Landform Plan that is Exhibit 26 in these proceedings and Annexure 4 to these conditions. The purpose of such a review is to ensure the maximum extraction depths depicted on the plan comply with the following requirements:

(a) No extraction is to occur below RL 9.0 m on any part of the Extraction Area.

(b) Extraction at any given point should not occur beneath the maximum predicted elevation of the water table at that point.

(c) The maximum extraction depth at any given point should have a minimum clearance of 1 metre above the mean position of the water table at that point.

18. If the Hydrogeologist forms the view that the maximum extraction depths shown on the Post Extraction Landform Plan do not comply with the above requirements, the Hydrogeologist must:

(a) amend that plan to achieve such compliance;

(b) provide the manager of the Extraction Operations with a copy of the amended plan and direct him or her to comply with it in relation to all future extraction; and

(c) provide DLAWC and the council with a copy of the amended plan.

In amending the Post Extraction Landform Plan, the Hydrogeologist may not permit extraction at any given point to a greater depth than that shown on that plan in the form which is Exhibit 26 in these proceedings and Annexure 4 to these conditions.

19. DLAWC and the Hunter Water Corporation shall be informed immediately of any actions or occurrences which may adversely affect the quality of groundwater flowing from the site to the extent that they would render the groundwater unsuitable for collection for the purpose of use as drinking water.

ABORIGINAL ARCHAEOLOGY

20. Sub-surface archaeological testing is to be carried out prior to the commencement of the Extraction Operations in accordance with the proposal of Bonhomme Craib

and Associates that is Exhibit 8 in these proceedings and which is Annexure 6 to these conditions.

21. That proposal may be modified, as required by the New South Wales National Parks and Wildlife Service ("NPWS"), to obtain approval for any permit or permits required under the National Parks and Wildlife Act 1974 in relation to that testing.

22. Prior to the commencement of the Extraction Operations, the results of that testing are to be presented in a report to the NPWS prepared in accordance with any relevant NPWS guidelines existing at the time the report is produced. The council and the Worimi Local Aboriginal Land Council are to be provided with a copy of that report.

Based on that report, a Cultural Heritage Management Plan must be produced in consultation with the NPWS and the Worimi Local Aboriginal Land Council.

Meaning: in the following two conditions the terms "relic" and "Aboriginal place," and "consent to destroy" have the meaning ascribed to those terms as used in section 90 of the National Parks and Wildlife Act 1974 ("Section 90").

23. Upon a relic or Aboriginal place being discovered during the excavation operations, other than a relic or place covered by a consent under Section 90, the manager of the Extraction Operations shall immediately notify the NPWS of the discovery and observe any instructions given under the National Parks and Wildlife Act in connection with the preservation of the relic or Aboriginal place.

24. A protocol for the discovery of relics during the Extraction Operations shall be developed in consultation with the Worimi Local Aboriginal Land Council. The protocol shall be submitted to and approved by the NPWS prior to the commencement of the Extraction Operations.

REHABILITATION

25. Areas of the site disturbed by the Extraction Operations are to be rehabilitated in accordance with the following further conditions of consent, and Smith and du Preez's Rehabilitation Proposal that formed part of Exhibit 14 in these proceedings and which is Annexure 7 to these conditions of consent. To the extent of any inconsistency between the Rehabilitation Proposal and the following further conditions, the further conditions prevail.

26. The objectives of the rehabilitation are to re-establish:

(a) the original vegetation community types, as defined in the vegetation mapping of the site carried out by Nerina du Preez in 1986, which may regenerate at different proportions to the original vegetation due to the lowered post-extraction landform;

(b) the structural components of the vegetation, greater than 2 metres in height, comparable with the pre-extraction vegetation at similar elevations (metres AHD); and

(c) similar species composition to pre-extraction at similar elevations.

27. All rehabilitation is to be directed and supervised by a qualified plant ecologist/botanist with experience in rehabilitation (the "Rehabilitation Supervisor"). The same person need not perform the role of Rehabilitation Supervisor for the entire duration of the Extraction Operations.

28. A suitably qualified person (the "Rehabilitation Officer") is to be employed to implement the rehabilitation on a day to day basis under the direction of the Rehabilitation

Supervisor. The same person need not perform the role of Rehabilitation Officer for the entire duration of the Extraction Operations.

29. The rehabilitation directed by the Rehabilitation Supervisor should, wherever practicable, utilise current Best Practice in rehabilitation. Where trials of alternate rehabilitation techniques demonstrate those techniques are likely to be more successful than existing rehabilitation techniques, the alternate techniques are to be utilised

30. During the first two years of the Extraction Operations the Rehabilitation Supervisor is to conduct at least 12 site visits per year, for the purpose of quality checking and continuous open communication and consultation with the Rehabilitation Officer and staff managing the Extraction Operations.

31. Only locally collected seed should be used in the rehabilitation.

32. Monitoring of rehabilitation by the Rehabilitation Supervisor to assess compliance with the above stated objectives should be at least 6 monthly for the first two years of the Extraction Operations and then at least annually. Monitoring in both pre- and post-extraction areas should be conducted along fixed transects at 10 metre intervals. In addition:

(a) for the early years of rehabilitation post-extraction, sampling should be in 1 metre × 1 metre quadrats and also include the recording of tree species (number of individuals and heights) in 10 metre × 10 metre contiguous quadrats;

(b) after plants have become established, sampling should be in 5 metre × 5 metre quadrats and also include recording of tree species (number of individuals and heights) in 10 metre × 10 metre contiguous quadrats;

(c) pre-extraction sampling should be in 5 metre × 5 metre quadrats at 20 metre intervals.

33. Once every year, from the commencement of the Extraction Operations, the Rehabilitation Supervisor is to conduct an inspection of all rehabilitation areas with the Rehabilitation Officer. That inspection may constitute one of the site visits required under condition 30. The Rehabilitation Supervisor must invite an appropriate representative from each of DLAWC and the council to attend the inspection, and must give reasonable notice to those persons of the relevant time and date.

34. The Rehabilitation Supervisor must produce a report on the rehabilitation within 28 days of each anniversary of the commencement of the Extraction Operations. That report must:

(a) include the results of monitoring;

(b) make recommendations concerning future rehabilitation; and

(c) state whether the rehabilitation is being carried out in accordance with the instructions of the Rehabilitation Supervisor.

A copy of that report must be provided to both DLAWC and the council.

35. If the Rehabilitation Supervisor considers at any time that the rehabilitation is not being carried out in accordance with this consent, he or she may issue a written direction to the manager of the Extraction Operations to do or not do one or more things to ensure the rehabilitation is carried out in accordance with this consent.

36. If the Rehabilitation Supervisor considers that direction has not been complied with in a reasonable period of time, he or she may report the non-compliance to council and/or DLAWC.

36A. Prior to the commencement of the Extraction Operations, the legal entity undertaking those operations on behalf of the applicant must lodge a $100,000 bond with DLAWC in a form acceptable to DLAWC. The bond must be maintained in that amount until the expiration of 5 years from the completion of all Extraction Operations on the site, unless DLAWC releases the bond on an earlier date. If the rehabilitation of the site fails, at any point in time, to meet the requirements of this consent, DLAWC may do whatever is reasonably necessary to meet those requirements, and may draw on the bond to recover its costs of doing so.

TRAFFIC

37. The intersection of Oyster Cove Road and the proposed sand extraction road shall be clearly identified with a side road junction warning sign and advance truck warning signs.

38. Advanced truck warning signs must be installed at the intersection of Oyster Cove Road and Lemon Tree Passage Road and on Oakvale Road adjacent to the sand processing plant.

39. The proposed sand extraction road adjacent to Oyster Cove Road shall be sealed a minimum length of 40 metres.

40. A weighbridge shall be provided at the Salt Ash Plant. Records of the tonnage of sand transported from the site to that Plant shall be kept and submitted to council 14 days after the last calendar date of each month.

41. A monetary contribution is to be paid to council, pursuant to Section 94 of the Environmental Planning and Assessment Act 1979, towards the maintenance and upgrading of the existing road system. The amount of the contribution is to be 4 cents per tonne of raw material carried per kilometre travelled on the haul route identified on Figure 5.5 of the Environmental Impact Statement dated May 1999. The amount of the contribution shall be varied annually with movements in the Consumer Price Index as published by the Australian Bureau of Statistics and as notified in writing by the council. The contribution is to be paid monthly, due 14 days after the last calendar date of each month.

OPERATIONS

42. All practicable measures must be taken to prevent and minimise harm to the environment as a result of the construction, operation and, where relevant, decommissioning of the development.

43. (a) The Extraction Operations shall comply with all safety requirements of the Mines Inspection Act 1901 (as amended).

(b) The Applicant is responsible for full costs associated with any alteration to or relocation or enlargement of public utilities whether caused directly or indirectly by Extraction Operations.

44. Final landform batters are not to exceed 1:4 (vertical:horizontal).

45. The Applicant shall identify and appropriately protect all hollow-bearing trees in white silica sand depths of 2 metres or less. A hollow is deemed to be a cavity that is at least 10 centimetres in diameter and 30 centimetres deep.

46. (a) Fuel, oil and grease are not to be stored on the site.

(b) Before any mining machinery (other than trucks which transport sand) enters the Extraction Area it must be cleaned to remove all soil and plant material and sprayed with fungicide to limit the introduction and spread of soil pathogens on the Extraction Area. Any truck which transports sand which had soil or plant material on it from any place other than the Extraction Area must be similarly cleaned and sprayed before entering the Extraction Area.

47. (a) The council must be given not less than 14 days written notice of the commencement of Extraction Operations.

(b) The Extraction Operations may only be conducted between the hours of 7 am and 7 pm, Monday to Friday excluding public holidays.

48. Work may be conducted outside approved hours where the delivery of material is required outside the specified hours by police or other authorities for safety reasons. Notification in writing of the times and circumstances of such mining must be provided to the council and the Environment Protection Authority.

49. The approved hours may be varied with the written consent of the council and the Environment Protection Authority only where they are satisfied that the amenity of the residents in the locality will not be adversely affected.

HUNTER WATER CORPORATION INFRASTRUCTURE

50. The position of the existing 200 mm diameter watermain shall be located and marked prior to commencement of Extraction Operations. Safeguards shall be implemented to ensure the integrity of the watermain is not compromised during sand extraction. Access to the watermain and associated surface fittings shall be maintained at all times. The Hunter Water Corporation shall be consulted regarding this matter and supplied with all relevant details.

51. The position of all spear points located within the application area shall be marked and safeguards implemented to ensure the continued operation of and access to those points. The Hunter Water Corporation shall be consulted regarding this matter and supplied with all relevant details. One or more of those spear points may be removed or relocated with the permission of the Hunter Water Corporation.

52. A gate shall be installed on all sand extraction roads that join Oyster Cove Road. The gates shall be locked at all times other than during operating hours. A key shall be made available to Hunter Water Corporation.

AIR — Dust

53. Activities occurring at the site must be carried out in a manner that will minimise emissions of dust.

54. Temporary topsoil and/or material stockpiles and handling areas must be maintained at all times in a condition which minimises wind-blown or traffic-generated dust.

55. Vehicles transporting material less than 7 mm in size must have their tailgates securely fixed and must be covered at all times after loading and before unloading to prevent wind-blown emissions or spillage. If the nature of the material being transported is such that wind-blown emissions will not occur during transportation, then the vehicle need not be covered.

NOISE — Noise Level Criteria

56. Levels of noise emissions from:

(a) the vegetation clearing and soil stripping phase of the Extraction Operations must not exceed a noise emission criteria of 45dB(A) LA10 (15 minute) at any residence at Tanilba Bay and 41dB(A) LA10 (15 minute) at any residence at Oyster Cove during the day (7 am to 4 pm); and

(b) mining must not exceed an LA10 (15 minute) noise emission criteria of 41dB(A) LA10 (15 minute) at the nearest residence at Tanilba Bay and 36dB(A) at the nearest residence at Oyster Cove during the day (7 am to 4 pm).

The criteria for noise emission is normal meteorological conditions (winds up to 3 m/s), except under conditions of temperature inversion. Noise impacts that are or may be enhanced by winds exceeding 3 m/s or temperature inversion must be addressed by:

*Documenting noise complaints received to identify locations receiving any higher level of impacts.

*Where levels of noise complaints indicate a higher level of impact then actions to quantify and ameliorate any enhanced impacts under temperature inversions conditions should be developed and implemented.

Definition

LA10 (15 minute) is the sound pressure level that is exceeded for 10 per cent of the time when measured over a 15 minute period.

The LA10 (15 minute) noise level must be measured or computed at any point at the residential boundary over a period of 15 minutes using "FAST" response on the sound level meter. 5dB(A) must be added to the measured level if the noise is substantially tonal or impulsive in character.

57. Now not used.

WATER — Pollution of Waters

58. Except as may be expressly provided in the licence, the licensee must comply with section 120 of the Protection of the Environment Operations Act 1997 prohibiting the pollution of waters.

Stormwater and Sediment Control

59. An Erosion and Sediment Control Plan ("ESCP") must be prepared and implemented which describes the measures that will be employed to minimise soil erosion and the discharge of sediment and other pollutants to lands and/or waters during construction activities. The ESCP must be modified during the Extraction Operations as required, to minimise soil erosion and discharge of sediment and other pollutants to lands and/or waters during extraction activities. The ESCP should be consistent with the requirements of such plans outlined in Managing Urban Stormwater: Soils and Construction (available from the Department of Housing).

60. Any water discharged to the waters of Big Swan Bay must meet the provisions of the Clean Waters Regulations 1972 for Class P: Protected Waters in accordance with Schedule 2 — Restricted Substances and must:

(i) have a biochemical oxygen demand of not more than 20 milligrams per litre; and

(ii) have a non-filterable reside of not more than 30 milligrams per litre; and

(iii) be visually free of greases, oils, solids and unnatural discolouration and free of settleable matter; and

(iv) not be of pH value of less than 6.5 or more than 8.5 and must not induce a variation of the pH value of the receiving waters of more than 0.5.

8.12 UPDATE ON RECENT EXPERT WITNESSING CASES

The following are recent expert witnessing cases:

1. *Moore v. Ashland Chemical, Inc.*, 151 F.3d 269 (5th Cir. 1998 — en banc), concerned a toxic tort action in which the court considered whether the district court below abused its discretion in excluding the opinion of a physician/expert witness on the causal relationship between plaintiff's exposure to industrial chemicals and his pulmonary illness. The appeals court found no abuse of discretion and upheld the findings of the district court.
2. *Curtis v. M&S Petroleum, Inc.*, 174 F.3d 661 (5th Cir. 1999), concerned circumstantial evidence by an expert showing a temporal connection between exposure to toxics and illness. The court held that such expert evidence should be given little weight for causation. Dismissal of the toxic claims by the lower court was upheld by the appeals court.
3. *Castello v. Chevron USA*, 97 F. Supp. 2d 780 (S.D. Tex. 2000), concerned the unsupported opinions of experts concerning exposure to benzene and benzene derivatives causing fatal blood disease in an alleged wrongful death claim. The judge barred admission of the testimony of the experts because it was not supported by scientific literature.
   For another benzene exposure and causation illness claim involving expert testimony, see *Chambers v. Exxon Corp.*, 81 F. Supp. 2d 661 (M.D. La. 2000).
4. *Diehl v. Danuloff*, 618 N.W. 2d 83 (Mich. App. 2000), concerned the potential liability of a court-appointed expert for alleged negligence in performing his duties as an expert. Danuloff, a licensed psychologist, was ordered by a Michigan trial court judge to perform a family evaluation in a divorce proceedings. He submitted his report to the court with recommendations, which the court adopted. A party to the suit then alleged negligence by the expert. The trial court held that the court-appointed expert was entitled to absolute judicial immunity under the state statute. The party appealed. |
   On appeal, the appeals court affirmed the trial court's decision but qualified the expert's immunity status stating that "the court depends on the court-appointed expert to exercise discretionary judgement to render an evaluation and make a recommendation." The court held that such judgment is "a hallmark of a position functionally equivalent to that of a judge." It noted too, that the court-appointed

expert acts essentially as a neutral fact finder and fact-finding is an integral part of the judicial process, a function normally associated with judges and juries, both of whom are granted immunity from suit. Thus, the expert was entitled only to "quasi-judicial immunity."

5. *Dombrowski v. Gould Electronic, Inc.*, 85 F. Supp. 2d 456 (M.D. Pa. 2000), concerned a learning disability claim based on experts' testimonial evidence for a worker at a battery lead-recycling company. In spite of misgivings about the expert testimony, the court made an award to the plaintiff.

Iron Mountain Mine Site, Further Detailed Description

Credit for the following description of the Iron Mountain Site is given to D. K. Nordstrom, author, and to the editor of *Mining Environmental Handbook*, Jerrold J. Marcus, and the editor, F. R. Banta, for Chapter 18, Environmental Case Studies from the Hard Rock Industry, and to Imperial College Press, United Kingdom, the publisher.

---

(1) THE GEOLOGY AND HYDROLOGY OF THE IRON MOUNTAIN SITE

The topography of Iron Mountain is steep and rugged. Iron Mountain rises about 3000 feet above the Sacramento Valley (approximately 3585 feet above sea level). The summers are hot and dry, the winters cool and rainy with occasional snow. Average annual rainfall at the top of Iron Mountain is estimated to be about the same as that measured at Shasta Dam: 60 inches over a 47-year period (range on annual average for 1944–90 is 28–130 inches; Alpers and others, 1992). Slickrock Creek and Boulder Creek drain the south and north sides cff Iron Mountain, respectively. These two tributaries of Spring Creek carry the acid mine drainage and eroding waste and tailings piles from the mountain and the Spring Creek drainage transports them to the Sacramento River.

Before reaching the Sacramento River, Spring Creek is retained by the Spring Creek Debris Dam built in 1963 as part of the Central Valley Project. The releases from this dam are metered at an amount that should be sufficiently diluted by Shasta Dam releases so that no fish kills should occur. On several occasions since 1963, however, the capacity of Spring Creek Reservoir has been exceeded and uncontrolled releases over the spillway have caused fish kills. During February/March 1992, an additional 100,000 acre-feet of water were released from Shasta Dam to provide the necessary dilution of a Spring Creek Reservoir spill and to prevent fish kills below Keswick Dam on the Sacramento River. This event occurred when deliveries of water to farmers from the Central Valley Project were at an all-time low due to 6 years of consecutive drought, so the cost to the U.S. Bureau of Reclamation in terms of lost revenue was substantial. Natural landslides as well as erosion of waste piles and tailings piles has also occurred on the 4,400 acres of mining property.

The mineral deposits are primarily massive sulfides composed of large single masses of up to 95% pyrite with variable amounts of chalcopyrite and sphalerite to average about 1% copper and about 2% zinc. Some disseminated sulfides occur along the south side of Iron Mountain. Trace quantities of several other metals and metalloids occur in the mineral deposits including gold, silver, lead, cadmium, arsenic, antimony, vanadium, cobalt, and thallium (based on relative concentrations of these constituents in the acid effluent). The deposits are of the Kuroko type having been formed along an island arc in a marine environment (Albers and Bain, 1985). The country rock is the Balaklala rhyolite, a spilitized Devonian rhyolite that has undergone regional metamorphism during episodes of tectonic collisions of oceanic crust with continental crust. The nature of the altered igneous bedrock gives rise to a predominance of fracture-flow hydrology at Iron Mountain. The Copley greenstone, an altered basalt, underlies the rhyolite and is approximately contemporaneous. Part of the region, shown in Figure 6, to the south of Iron Mountain is the Mule Mountain stock, a trondhjemite-albite granite, considered to be cogenetic with the Balaklala rhyolite (Albers and others, 1981).

The mineral composition of the rhyolite is albite, sericite, quartz, kaolinite, epidote, chlorite, and minor calcite. Studies by Kinkel and others (1956), by Reed (1984), and those in the special issue of Economic Geology (1985, vol. 80, no. 8) have documented the chemical composition of both the ore minerals and the non-ore minerals. These studies also provide information on relative abundances of minerals and isotopic compositions.

Weathering of massive sulfides near the surface has given rise to a large gossan outcrop at the top of Iron Mountain, enriched in gold and silver. The extent of the exposure suggests that significant quantities of sulfides were oxidized during weathering and eroded before humans discovered the site. Some of the gossan material is found several hundred feet below the surface (Kinkel, and others, 1956). Secondary enrichment in the upper zones of the massive sulfides and just below the gossan resulted in high concentrations of copper (5–10%) and silver (about 1 oz./ton). These observations suggest that large quantities of metals have been mobilized over geologic time.

Three main massive sulfide mineral bodies, known as the Brick Flat, the Richmond, and the Hornet occur at Iron Mountain. These are thought to have been a single massive sulfide body about a half-mile long (well over a half-mile if the offset Old Mine mineral deposit is included) over 200 feet wide and over 200 feet high but offset by two normal faults (see Figure 7). All three of these bodies have been mined and the consequences of mining include altered groundwater conditions and highly contaminated surface waters originating from portal effluent waters (Banta, F. R. and Nordstrom, D. K., 1997, pp. 683–684).

(2) ENVIRONMENTAL CONTAMINATING CONDITIONS AT IRON MOUNTAIN

Acidic mine waters contain three necessary ingredients: pyrite, oxygen, and water. The amount and rate of acid production can depend on factors such as the concentration of pyrite, the temperature, the oxygen transport rate, the water flow rate, flow path, and the microbial growth rate conditions.

Conditions at Iron Mountain are nearly optimal for the maximum production of acid mine waters from pyrite oxidation. The concentration of pyrite is nearly 100% in single large masses excavated by tunnels, man ways, and stopes that allow rapid transport of gaseous oxygen advection. The massive sulfides are at or above the water table so that moisture and oxygen are always present. The airflow is probably aided by thermally driven convective cells due to high heat generated from pyrite oxidation. About 1,500 kilojoules of heat are released per mole (or about 120 grams) of pyrite. The average discharge from the Richmond portal indicates that about 2,400 of pyrite oxidize every hour, producing about 1 kilowatt of power, or almost 9,000 kilowatts per year. In the early days of mining the Iron Mountain massive sulphides, fires were frequent and before proper ventilation was installed, temperatures of 430° F (221°C) were recorded at the ore's surface (Wright, 1906).

The presence of the mine workings draws down the water table and pulls water toward the sulfide deposits at Iron Mountain where the pyrite oxidation reaction occurs. The resulting acid mine waters drain by gravity flow out the major portals of the Richmond mine, the Hornet mine (Lawson tunnel), the Old mine, and the No. 8 mine. Sulfide oxidation in the Richmond mine workings has led to the most acidic effluent (pH = 0.02–1.5) and the highest concentrations of metals and sulfate for any surface water in the area; sulfate has been measured as high as 118 grams per liter (Alpers and others, 1992). In addition, the Richmond portal discharge has had the highest recorded flow rates (as high as 800 gallons per minute) of any mine portal on Iron Mountain. Note that the major cations are iron, aluminum, and zinc, and that most trace metals are present at very high concentrations. There is very little capacity of the bedrock to neutralize these highly acidic waters. Other important discharges of metals are the seep from the vicinity of the Old Mine and No. 8 mine portals, the "Big Seep" discharge in Slickrock Creek, and discharges from the Brick Flat open pit. These sources, as well as downstream sites on Spring Creek, have been monitored and the relative contribution of each site to the total pollution load for copper, zinc, and cadmium has been established. Just under 2,000 pounds of these three metals are leaving the site per day, about 300 tons per year. In terms of pyrite weathering, it has been estimated that 2,500 tons of pyrite are oxidizing every year from the Richmond mine workings alone (Nordstrom and Alpers, written communication, 1990).

During the second Remedial Investigation phase (1986–1992) of EPA's Superfund activities, the Richmond tunnel and part of the Richmond mine workings were made accessible to underground surveys. On September 11, 1990, water and mineral samples were collected during one of these surveys that resulted in the discovery of extremely acidic seeps with pH values as low as –3.4 and a total dissolved solids concentration of about 935 grams per liter. These acid iron-sulfate waters were precipitating or effloriscing soluble iron-sulfate salts, often coating tunnel walls and muck piles with a colorful array. These waters are the most acidic ever reported anywhere in the world. The only other recorded pH values of natural waters comparable in magnitude are acid crater lakes found in Japan, New Zealand, Alaska, and Costa Rica (e.g., pH ≤ 0, Rowe, 1991). The development of such extreme acidic conditions is due to optimal conditions for pyrite oxidation combined with considerable evaporation from the heat during oxidation and several years of draught conditions in California. The formation of extensive efflorescent salts means that acid solutions are being temporarily stored in a solid form until climatic conditions become wetter. Wet climatic conditions will cause dissolution of the salts and some flushing of the stored acidity out of the mine workings. Rapid increases of copper concentrations up to a factor of 2 or 3 have been reported as a result of heavy rainstorm events early in the wet season (Alpers and others, 1992).

Waters with high concentrations of metals, especially copper, zinc, and cadmium, have drained from the mine portals and leached tailings and waste piles, entering Boulder and Slickrock Creeks and joining the Spring Creek drainage. The Spring Creek Reservoir (built in 1963) receives the discharges from Boulder and Slickrock Creeks with some dilution and iron oxidation. Waters stored in Spring Creek Reservoir typically have pH values in the range of 2.5–3.5, but they are not always well-mixed and often show chemical gradients with depth. Detailed temporal depth measurements to investigate the seasonal patterns have not been done. From the reservoir, the waters are released in controlled amounts so that the dilution with water released upstream from Shasta Dam prevents fish kills (Lewis, 1963). Over twenty fish kill events have occurred since 1963 with at least 47,000 trout killed during one week in 1967 (Nordstrom and others, 1977). The fish kills have occurred because the reservoir capacity has been overwhelmed by high-rainfall events and a large load of metals discharged. These high metal flows, even during low-flow conditions, have led to adverse aquatic conditions in Keswick Reservoir and the Sacramento River. Water-quality objectives for the Sacramento River basin, based on laboratory and on-site toxicity studies of chinook salmon, have been adopted and approved by the Regional Water Quality Control Board (RWQCB), the California State Water Resources Control Board and the EPA to protect against both chronic and acute toxicity to aquatic life. Using these criteria, both acute and chronic toxicity, studies on chinook salmon, steelhead, and rainbow trout in the Sacramento River system have shown both actual and potential harm to these species from the acid mine drainage originating at Iron Mountain (EPA, 1992). (Banta, F. R. and Nordstrom, D. K., 1997, pp. 684–685).

(3) SITE REMEDIATION STEPS PRIOR TO LITIGATED CERCLA PROCEEDINGS

From 1939 to the present, various studies of the environmental impact of Iron Mountain have been conducted by the California Department of Fish and Game, the U.S. Fish and Wildlife Service, the RWQCB, the U.S. Geological Survey, the U.S. Bureau of Reclamation, and the EPA. Since 1983 studies have been conducted as part of the Superfund investigations authorized by the Comprehensive Environmental Recovery Compensation and Liability Act (CERCLA). These studies have documented discharges of metal from Iron Mountain, the occurrence of fish kills, the results of toxicity tests on anadromous fish in the Sacramento River, the lack of benthic and aquatic organisms in parts of the drainage system, the siltation problems of the drainage, the geology, hydrology, and geochemistry of the area, and the effects of water management engineering practices on the drainage system.

In 1950, Keswick Reservoir was completed to provide further flood control and hydro-electric power below Shasta Dam. Much sediment deposition occurred in the Keswick Reservoir and the Spring Creek Debris Dam was constructed to reduce these high sedimentation rates as well as to provide some regulation for the acid mine drainage entering the Sacramento River (Prokopovich, 1965). Continued fish kills have kept the RWQCB actively pursuing remediation of the site. Following a thesis study at the site (Nordstrom, 1977), a cleanup and abatement order was issued to the mine owner, Stauffer Chemical Company. On December 17, 1976, the property was purchased by Iron Mountain Mines, Inc., the present owners. From 1977 to 1989 six orders were issued to reduce toxic metal discharges that were in violation of state law. The orders to cease and desist as well as for emergency treatment measures have been through both the Shasta County and the State of California courts. Stauffer Chemical Company became part of Rhône-Poulenc who then became liable for the site under CERCLA.

Iron Mountain was officially listed on the EPA's National Prioity List for Superfund in 1983 and the first remedial investigation/feasibinility study (RI/FS) began. The remedial investigation report (1985b) identified the five major point sources of pollution discharges through a comprehensive surface water sampling survey. The greatest discharge source was identified as the Richmond portal effluent. EPA (1985b) also documented the occurrence of increased concentrations of copper, zinc and cadmium from portal effluents following heavy rainstorm events and related to this phenomenon to rapid flow of surface water into the mine workings through areas of subsidence. The feasibility study (EPA

1985b) considered more than a dozen alternative treatment possibilities and estimated the costs and anticipated benefits from each individual alternative as well as several possible combinations. The alternative options are listed below in simplified form:

A. No action.
B. Diversion of surface flows: divert upper Spring Creek to Flat Creek, upper Slickrock Creek around Big Seep, and South Fork Spring Creek to Rock Creek.
C. Lime/limestone neutralization: treat major point sources with conventional neutralization treatment plant.
D. Capping: implement partial or complete capping of the mountain to prevent infiltration to the underground mine workings by laying down an impermeable liner.
E. Enlargement of the Spring Creek Debris Dam.
F. Intercept groundwaters through a system of drainage tunnels and drillholes surrounding the ore body.
G. Mine plugging.
H. On-site leaching and mineral extraction technologies (proposed by owners).
I. Combined alternatives.

A Record of Decision was issued by EPA (1986) that initiated five main recommendations:

A. Partial capping of cracked and caved ground above the Richmond ore body.
B. Construction of surface water diversions for upper Spring Creek, Slickrock Creek, and South Fork Spring Creek.
C. Initiate hydrogeological studies and produce a ground water model for the site. This step would include rehabilitation of the Richmond mine for subsurface investigations. The subsurface investigations were motivated by the decision to test and demonstrate the feasibility of filling mine workings with low-density cellular concrete.
D. Install perimeter controls as necessary to avoid direct contact with contaminants.
E. Evaluate other source controls as appropriate based on the hydrogeologic investigations.

At that time, mine plugging was not considered a serious option because of questions relating to the physical integrity of the mountain to contain the mine water. When Rhône-Poulenc acquired Stauffers assets through a complicated and proprietary arrangement with ICI Americas, ICI Americas began working on the remediation possibilities and mine plugging was seriously reconsidered. The initial alternative from private industry was to plug the Richmond workings and to allow them to fill up and flood the ore body to prevent further oxidation. Subsequent investigations by the U.S. Geological Survey and by CH2M Hill for the EPA demonstrated that a mine pool of about 20 million cubic feet would be created and would have a composition very similar to the present Richmond effluent composition. This acid mine pool would be sitting on top of the current water table and would travel to Boulder Creek through the bedrock in something less than 100 years. This concern led to the development of a highly refined plugging scenario in which lime would be added before plugging, a lime slurry and various additives would be injected to chemically neutralize and immobilize the acid waters and their dissolved metals. Considerable debate has ensued as to the effectiveness and costs of such a procedure. Indeed, the most difficult task has been to assign defendable risks and to develop methods that would evaluate the effectiveness of any of the proposed alternative treatments and their various combinations.

The EPA has evaluated the modified mine plugging alternative as part of the second RI/FS completed in 1992 (EPA, 1992). The EPA has also considered air sealing but has favored a complete capping treatment as the most cost effective solution in conjunction with the surface water diversions that have already been initiated. Emergency treatment procedures which collect Richmond portal effluent during periods of high flow and neutralize it in a temporary lime neutralization plant near the portal have been instituted. The capacity of this plant was increased from 60 to 140 gallons per minute in December 1992. This plant will be removed once a more permanent solution has been found. Meanwhile, the EPA and responsible parties are currently in legal contention over the appropriate treatment to be used and the consequent costs. Some question of the federal government's share of the liability has also arisen because of the network of dams built by the U.S. Bureau of Reclamation in the drainages receiving the acid mine waters (Banta, F. R. and Nordstrom, D. K., 1997, pp. 685–687).

Decision must be made as to whether include or exclude the following.

Concluding remarks:

Control and remediation of the mine waste contamination at Iron Mountain, including prevention of some of the most acidic mine waters in the world, has proven to be an extraordinarily difficult and complex task. The physical and chemical nature of the site with all of its heterogeneities, complexities, and unknown aspects, the difficulty in assessing the effectiveness of any alternative or combined treatment, and the difficulty of assessing the relative risks and costs of alternative treatments and their contingencies all contribute to the formidable challenge of remediation. Hence, there is no clear solution to the problem and opposing parties will have inevitable differences of opinion on how to proceed and how much it should cost. Under such circumstances, it would seem prudent to proceed with remediation in stages with the most risk-free and least expensive treatments (especially in terms of operating costs) while continuing to monitor the site so that evaluations can be revised and improved.

Modern methods of mining can rehabilitate an area during production and after mining has ended with a considerable reduction of overall environmental remediation costs. The experience gained in studying Iron Mountain certainly underscores this fact. Estimated costs of cleanup at Iron Mountain start at about $25 million and exceed $100 million (1985 dollars) for the most effective combination of treatments.

The story of this site is not yet over after 100 years of mining activity and 54 years of investigation and regulation. However, more progress has been made at this site than at almost any other mining Superfund site in the United States. The recommended remedial measures may be very site-specific, but the general strategy on how the site was investigated and the difficulties uncovered during the Superfund investigations should provide insight and examples that will be useful for other mine sites.

A Very Simplified Environmental Guide for Starting a Mining Operation, or for a Takeover of a Mining Operation

As a word of caution, to draw a simplified outline for permitting a new mining project is a nearly impossible task. Each new site must be treated on an ad hoc, site-specific basis because of numerous variables from site to site. There is no "one size fits all" guideline formula. A new mining project permitting program may range from simple, small, short time, and straightforward to intensely complex, involving many agencies and authorities, and taking years to accomplish. The only common factor for expediting all permitting programs and having maximum chances for success of the project to be approved is to employ experienced permitting personnel. In a very general way, the following, over-simplified outline for an environmental permitting program may be helpful:

I. Recruiting a permitting project staff involves several aspects.

The following key personnel for a proposed new mining project are either essential or extremely helpful in expediting the permitting program. They should consist of technical experts (engineers, geologists, hydrologists/groundwater experts), governmental and public relations experts, and legal experts, all with mine project permitting experience.

A. An experienced, mine permitting project manager should be selected.

B. A mining geologist should be selected to evaluate the deposit and report on any environmentally offending minerals/metals present that must receive primary mitigating consideration for the project.

C. An experienced environmental-permitting specialist should be selected to serve as assistant project manager (an environmental coordinator-permitting specialist) who will determine the governmental authorities and agencies to deal with for local zoning, land use, mining, air, and water discharge permitting. This person will also be responsible for determining the investigations necessary for possible miscellaneous environmental permits required (e.g., local or area endangered species, wildlife habitats and preserves, and archeological and historical preservation programs).

D. The permitting manager will assist the overall project manager by recruiting the following essential permitting program personnel:
   1. Engineering specialists with expertise in environmental planning for minimizing and mitigating the special permitting processes for clean air, clean water, NPDES, et al. are needed.
   2. Legal counsel, preferably one experienced in mine permitting, but at least, one with good political contacts with local governmental officials and agencies is required.
   3. Depending on the climate and temperature of the local community as to feelings about possible opposition to mining projects, a public relations and communications specialist may be necessary to make the permitting process flow more smoothly.

II. It is important to keep a low profile from the beginning and throughout the entire process, emphasizing only through the news media, and at all times that the mining company is one that is very environmentally conscious during it operations and through restoration and site reclamation.

References

Aston, R. Legal Ed. (1992), Concerned citizens fail to stop quarry, permit, legal briefs, *Pit Quarry*, January 1992, 20.

Aston, R. L. (1999), *Surface Mining Law and Reclamation by Landfilling*, Imperial College Press, London.

Aston, R. L. (2001a), Inactive surface mine owner pays millions for water pollution, Aston's Mining Law Case Reviews™, *Miner. Resour. Eng. J.*, 10(2) June 2001.

Aston, R. L. (2001b), New South Wales silica sand operation given conditional mining permit, Aston's Mining Law Case Reviews™, *Miner. Resour. Eng. J.*, 10(1), March 2001.

Banta, F. R. and Nordstrom, D. K. (1997), Environmental case studies from the hardrock industry, in *Mining Environmental Handbook*, Marcus, J. J. Ed., Imperial College Press, London, chap. 18, section 18.2.

Black, H. C. (1968), *Black's Law Dictionary*, 4th ed., West Publishing, St. Paul, MN.

Dunn, J. R. (1991), in *The Heritage of Engineering Geology*, (The First Hundred Years: Geological Society of America, Centennial Special Vol. 3, reproduced in Forensic Geoscience for Engineering Works; Litigation, Hearings and Testimoney, Kiersch, G. A., Ed.), 575.

Eng. Min. J., (1996), August,

EPA (1985b),

EPA (1992),

Grad, F. P. (1978), *Environmental Law*, Vol. 2, Matthew Bender, New York.

Johnson, L. R., Six, S. N., and Hamilton, P. A. (1997), Deciphering Daubert, *Trial*, November 1997, 71.

Jones, P. C. (19XX), What Went Wrong at Summitville?, Colorado Mining Assn.,

Kazmarek, E. A. (1999), Kilpatrick Stockton LLP, Atlanta, GA, in Use of Scientific Evidence in Georgia Courts: Comparison with Federal Practice, Environmental Law, ICLE Seminar, Atlanta, GA.

Mansfield, M. E., Managing Ed. (1993, 1995, 1996, 1997, 1999), XII The Year in Review, Natural Resources, Energy, and Environmental Law Section, American Bar Association, and National Energy-Environmental Law-Policy Institute, Univeristy of Tulsa College of Law, ABA, Chicago.

Marionneaux, R. W. (1999), Admissibility of Expert Testimony, Powell, Goldstein, Frazer & Murphy, LLP, Environmental Law Summer Seminar, ICLE Institute, Atlanta, GA, August 6, 1999.

(1993), *Min. J.*

Nordstrom, D. K. (1997),

Prokovich, (1965),

Record of Decision for the Idarado Mining and Milling Complex (1987), State of Colorado, March 17, revised March 26, 1987, p. 2.

Salvato, J. A. (1992), *Enviromental Engineering and Sanitation*, 4th ed., John Wiley & Sons, New York.

Schoenbaum, T. J. and Rosenburg, R. H. (1991), *Environmental Policy Law*, 2nd ed., Foundation Press, Westbury, NY.

Tarlock, A. D. (1994), *Law of Water Rights and Resources*, Clark Boardman Callaghan, New York.

Index

A

Aboriginal archaeology, 281
Acacia baueri, 271
Acid deposition control, 86, 98
Acid mine drainage (AMD), 173, *see also* Mine
   sites, water pollution by abandoned
 contamination, 238
 statutory treatment of water pollution by, 175
 -tainted water, 193
Acid mine waters, formation of, 179
Acid neutralization, 224
Acid rain, 98
Acid Rain Program, 98
Action enforcing, 45
Activated sludge, 112
Adjective rules, 19
Administration Procedure Act (APA), 126
Administrative law, 6
 basic rule in, 20
 judge (ALJ), 19, 144
 statutory law vs., 17
Administrative Procedure Act (APA), 18, 22
AEC, *see* Atomic Energy Commission
AEG, *see* Association of Engineering Geologists
Affidavits, contradicted, 29
Agency(ies)
 decision, deference of, 161
 fact-finding procedures, 31
 federal, decision-making process of, 44
 rule(s)
  interpretive, 18
  legislative, 18
  making, 18
 violations, of procedural rights granted to
  plaintiffs, 26
AIPG, *see* American Institute of Professional
  Geologists
Air
 contamination, United States, 64

quality, 54
 control regions, 68
 criteria, 88
 deterioration of, 3
Air pollutant(s)
 emissions
  man-made point sources of, 68
  total, 69
 EPA NAASQ primary, 66
 hazardous, 85, 97, 98
 man-made sources of, 63
 primary, 63
 standards for hazardous, 73
Air pollution, *see also* Clean Air Act, air pollution
  and
 emissions, EPA checking of charges of, 72
 nuisance claims related to, 11
Air Quality Act (AQA), 65
Alaska Statehood Act, 26
ALJ, *see* Administrative law judge
Ambient air standard, 65
AMC, *see* American Mining Congress
AMD, *see* Acid mine drainage
American court system, 14
American enthusiasm, 1
American Institute of Professional Geologists
  (AIPG), 263
American Mining Congress (AMC), 138
American Premier Underwriters, Inc. (APU), 168
American Revolution, 1
American Society of State Highway Officials, 263
American Society for Testing and Materials
  (ASTM), 263
APA, *see* Administrative Procedure Act
Appellant-respondent, 34
Appellee-petitioner, 34
APU, *see* American Premier Underwriters, Inc.
AQA, *see* Air Quality Act
Aquatic life
 protection of, 126

299